Thank- you for all your continuing support + motivation, from all of us at the " Pandey Lab".

CELL APOPTOSIS RESEARCH TRENDS

CELL APOPTOSIS RESEARCH TRENDS

CHARLES V. ZHANG
EDITOR

Nova Biomedical Books
New York

Library of Congress Cataloging-in-Publication Data

Trends in cell apoptosis research / Herold C. Figgins, editor.
   p. ; cm.
 Includes bibliographical references and index.
 ISBN 1-60021-424-X
 1. Apoptosis. 2. Cancer cells. 3. Antineoplastic agents. I. Figgins, Herold C.
 [DNLM: 1. Neoplasms--drug therapy. 2. Apoptosis--physiology. 3. Apoptosis Regulatory Proteins--therapeutic use. QZ 267 T7943 2006]
 QH671.T74   2006
 571.9'36--dc22   2006027600

Published by Nova Science Publishers, Inc. ✦ *New York*

Contents

Preface

Apoptosis is the regulated form of cell death. It is a complex process defined by a set of characteristic morphological and biochemical features that involves the active participation of affected cells in a self-destruction cascade. This programmed cell death plays a critical role in physiological functions such as cell deletion during embryonic development, balancing cell number in continuously renewing tissues and immune system development. Additionally, a dysregulation of apoptosis is underlying in numerous pathological situations such as Parkinson, Alzheimer's disease and cancer. A number of studies have pointed out an association between consumption of fruits and vegetables, and certain beverages such as tea and wine, which are rich in polyphenols, with reduced risk of chronic diseases, including cancer. Apoptosis is also the regulatory mechanism involved in the removal of unnecessary cells during development and in tissue homeostasis in a wide range of organisms from insects to mammals. This book presents new and exciting research in this related field.

Chapter I - Apoptosis is a genetically conserved cell suicide program that is essential for proper development and the maintenance of tissue homeostasis. This cell death program removes unwanted cells, such as those harboring harmful mutations. Therefore, the genetic and functional integrity of molecules crucial for apoptosis is potentially significant for oncogenic processes. "Total cell kill" is essential in anti-cancer treatment, and apoptosis is the most extensively characterized mechanism for cancer cell killing. Recent studies reveal that apoptosis induced by anti-cancer agents is mostly the result of interactions between anti-apoptotic members of the Bcl-2 protein family and BH3-only proteins, a pro-apoptotic subgroup of the Bcl-2 family. In this context, BH3-only proteins act as initiators of cell death signals initiated by anti-cancer drugs by binding to anti-apoptotic Bcl-2 proteins, eventually leading to the activation of the mitochondria-dependent apoptosis pathway via the stimulation of Bax and Bak, disruption of mitochondrial outer membrane permeability and activation of the caspase cascade which causes dismantling of the cell. At least eight BH3-only proteins have been identified to date, most of which are involved in cell responses to anti-cancer agents. Since the expression profiles of BH3-only proteins are dependent on the type of cell, tissue, and nature of cellular stress, their degree of participation in cellular responses to various cancer treatments may also vary widely, according to the type of cancer and therapeutic approach. Here, we review current knowledge on the regulation of the Bcl-2 family, particularly the role of BH3-only proteins in apoptosis induced by anti-cancer agents, including recently developed molecular-based therapeutics, such as imatinib mesylate used to treat chronic myelogenous leukemia, and proteasome inhibitors as general anticancer drugs. Additionally, specific BH3-only proteins are likely to collaborate in specific cellular settings. Recent data suggest that the loss of function of crucial BH3-only proteins leads to resistance to certain anti-cancer drugs, confirming their significance in cancer treatment. We describe recent progress in translating these findings into novel therapeutic strategies, involving the direct manipulation of Bcl-2 proteins. These fundamental approaches directed towards understanding the diverse roles of BH3-only proteins in oncogenesis and cancer treatment should facilitate the development of more rational therapeutic approaches against cancer.

Chapter II - Cancer is an insidious disease, in which virtually every aspect of cellular control can be subverted to allow the uncontrolled, invasive cellular growth that defeats multicellular cooperation and kills an organism. In testimony to the essential role for proper execution of cell death in tumor suppression, apoptosis is widely recognized as an essential tumor-suppressor system. Indeed, defects in apoptosis are considered a hallmark of cancer, and are known to render the tumor resistant to immunosurveillance and therapy. Prion infections represent a fascinating biological phenomenon which has elicited at the interface between neuroscience and immunology. Although Prion protein (PrPc) is well known for its implication in transmissible spongiform encephalopathy, recent data indicated that PrPc may participate in program cell death regulation. PrPc would be correlated to the acquisition of a resistance phenotype by tumor cells to cytotoxic effectors or antitumor drugs. This review revisits the molecular mechanisms of tumor resistance to apoptosis and the implication of the PrPc in this phenomenon.

Chapter III - Pancratistatin is a natural compound extracted from the Hawaiian spider lily (*Hymenocalis littoralis*) known to have anti-cancer capabilities. Previous studies with Pancratistatin have demonstrated that this compound rapidly and efficiently induces apoptosis (programmed cell death) in various types of cancer cell lines, including breast, colon, prostate, neuroblastoma, melanoma and leukemia. Most importantly, when Pancratistatin was tested for toxicity on peripheral white blood cells from healthy volunteers, there was little or no demonstrable effect on their viability and nuclear morphology, indicating the relative specificity of this compound for cancer cells. Our previous results have shown that upon treatment with Pancratistatin, phosphatidyl-serine flips to the outer leaflet of the plasma membrane, caspase-3 is rapidly activated, and there is no occurrence of massive DNA damage at early time-points in cancerous cells. To better understand the selectivity of Pancratistatin, our recent work focuses on deducing the mechanism of action of this compound particularly in the acute human T-cell leukemia (Jurkat) cell line. This model of cancer was used in conjunction with normal, nucleated blood cells, obtained from healthy volunteers, as well as with samples from relapsed acute leukemia patients, prior to chemotherapeutic treatment at the Windsor Regional Cancer Centre. Our results suggest that Pancratistatin caused caspase-3 activation in association with the Fas-receptor within membranous lipid rafts, upstream of caspase-8 activation in Jurkat cells. Early events were followed by the loss of mitochondrial membrane potential, DNA degradation and subsequent cell death. Incubation of Jurkat cells with a caspase-3 inhibitor prior to treatment with Pancratistatin delayed these events dramatically, which implied that caspase-3 activation is an essential step in Pancratistatin-induced apoptosis. Interestingly, in contrast with these results, Pancratistatin caused none of these reactions in normal nucleated blood cells under identical treatment conditions. Taken together, our results indicate that Pancratistatin is a novel anti-cancer agent that specifically induces apoptosis in human leukemia cells through Fas-receptor associated caspase-3 activation, while leaving healthy white blood cells unscathed.

Chapter IV - Apoptosis is the result of an evolutionarily conserved program with common mechanisms for all euchariotic cells, in the same way that cell cycle does, thus creating a network of regulatory pathways that ensure the right development of cell populations included as essential parts of multicellular organisms. Cell death program is activated more easily and quickly in different cell types, and even in the same cells depending on their evolutionary stage. A single cell can also be sensitive to several types of death signals.

Physiological importance of apoptosis is enhanced by the fact that we are facing a phenomenon which, when disregulated, can be harmful for the organism, being in the core of a number of pathologies.Thus, the most relevant biological interest of apoptosis is the possibility of its modulation and so, the identification of inductive factors and their mechanisms of action appear as the most relevant challenges in apoptosis research.

Our research interest centers on neuroendocrine differentiation in prostatic carcinoma and its role on apoptosis modulation in the disease, with the development and applications of novel methodologies.

Neuroendocrine cells appear phisiologically inserted between epithelial cells, may be acting as trophyc stimuli in normal prostate. Of uncertain origin, the neuroendocrine cells are positive for neural markers and secrete neuropeptides with paracrine action on neighbouring cells. The presence of neuroendocrine differentiation -understood as an exaggerated ratio of neuroendocrine cells to normal or neoplastic epithelial cells- in the carcinoma is associated to worse prognostic, higher tumor progression and an androgen-independent status of the tumor, that becomes unresponsive to hormonal therapy.

Our research is based on three main goals. First: demonstrating the capability of prostate carcinoma epithelial cells in androgen sensitive and androgen insensitive stages to die by apoptosis after induction. That is, a preserved cell death machinery in these cells. Second, the role of representative neuropeptides in modulating the induction of cell death on the cells. Third, the study of intracellular alterations in both processes, i.e. ionic changes, mitochondrial permeability transition alterations. Novel methodologies are applied to develop working protocols for light and electron microscopy, electron probe analysis or confocal microscopy. As conclusions, we have demonstrated a role of neuropeptides bombesin and calcitonin on modulating etoposide-induced cell death, in vitro, on androgen sensitive and insensitive prostate cancer cells together with the fact that, in the presence of neuropeptides, apoptosis modulation is accompanied by reversion of ionic changes of apoptosis and of mitochondrial membrane potential alterations.

Chapter V - Prostate cancer is the most frequently diagnosed malignancy and the second leading cause of cancer death among men in the United States. The progression of prostate cancer from the androgen-dependent to the androgen-independent state is the main obstacle to improving the survival and quality of life in patients with advanced prostate cancer. Prostate cancer progression and the development of the androgen-independent state have been related to genetic abnormalities that influence not only the androgen receptor but also crucial molecules involved in apoptosis. Apoptosis is a programmed cell death executed by a family of intracellular cysteine proteases known as caspases. Activation of the caspase cascade is associated with proteolytic cleavage of diverse structural and regulatory proteins that contribute to apoptosis. Several proteins that inhibit apoptosis, including p53 and the Bcl-2 family, are involved in the promotion of tumorigenesis and drug-resistance in several cancers. In addition, the inhibitor of apoptosis proteins (IAPs) have been identified in baculovirus. The IAPs are a family of caspases inhibitors that block the apoptotic pathway via binding pro- and/or active forms of caspase-3, -7 and -9 and inhibiting their processing. To date, eight human IAPs have been recognized: X-chromosome-linked IAP (XIAP), human IAP-1 (hIAP-1), hIAP-2, survivin, neuronal apoptosis inhibitory protein (NAIP), apollon, livin, and IAP-like protein-2 (ILP-2). IAPs are overexpressed in prostate cancer and their expression correlates with tumor progression and drug-sensitivity, therefore therapies that target IAPs have the potential to improve outcomes for patients with prostate cancer. This review focuses on the experimental evidence that associates IAPs expression with prostate carcinogenesis, and analyzes the roles of IAPs in chemotherapy in order to develop novel therapeutic strategies.

Chapter VI - In the colon, signals from nutritional compounds (like dietary lipids) and endogenous factors regulating cell growth, differentiation and apoptosis are integrated within the cell and have a substantial impact determining the final phenotype, metabolism and kinetics of colon epithelial cell populations. Moreover, altered response of transformed cell to these signals is supposed during colon carcinogenesis.

Using human colon epithelial cells *in vitro*, derived either from normal human fetal colon tissue (FHC) or from adenocarcinoma (HT-29) the response to (i) TNF-α, anti-Fas antibody and TRAIL and (ii) polyunsaturated fatty acids of n-6 (arachidonic acid, AA) or n-3 (docosahexaenoic acid, DHA) series, short chain fatty acid - sodium butyrate (NaBt) and their mutual interactions have been investigated and compared. Generally, even if both cell lines showed some similar characteristics (morphology, extensive proliferation, not full differentiation), they differ in their ability to respond to endogenous inductors of apoptosis like TRAIL as well as to exogenous dietary factors like PUFAs or NaBt produced from fiber.

The response of HT-29 and FHC cells to TNF-α, anti-Fas antibody and TRAIL was compared. Both cell lines did not respond to TNF-α, they expressed a limited sensitivity to anti-Fas antibody and different response to TRAIL. Some molecular determinants of limited sensitivity of HT-29 and very low sensitivity of FHC cells to TRAIL were determined. This could be attributed to a lower activity of FHC cell mitochondrial pathway components followed by decreased caspase activities and PARP cleavage.

PUFAs and NaBt, working individually or together, initiated higher apoptotic response in FHC than in HT-29 cells. Moreover, NaBt also induced higher differentiation response in FHC cells, although the proliferation of both cell types was similarly affected. Combined treatment of cells with either AA or DHA and NaBt resulted in potentiation of apoptotic effects especially in FHC cells, DHA being more effective. The relationship between these effects and different lipid and oxidative metabolism of normal and cancer cells (changes of lipids in plasma as well as mitochondrial membrane, accumulation of lipid droplets in cytoplasm associated with changes of MMP and ROS production) is supposed.

Chapter VII - Developing brain is highly vulnerable to injury caused by radiation and chemotherapy leading to DNA damage. These stimuli activate the intrinsic apoptotic pathway that results in extensive neuronal loss over prolonged periods of time. High susceptibility of the developing neurons to apoptosis represents the major problem in treatment of childhood brain tumors by radiation and chemotherapy that cause significant loss of healthy neurons leading to development of severe cognitive and psychological deficits. Therefore, development of anti-cancer treatment strategies aimed at preservation of the developing brain is highly significant.

In this study, we made the first step in development of such a strategy using *in vitro* models of rat primary neurons and human tumor cell lines. Our results demonstrate that an inhibitor of cyclin-dependent kinases, flavopiridol potentiates death of tumor cells in cultures, while protecting cerebellar and cortical neurons from apoptosis after irradiation and treatment with the DNA damaging drug, etoposide. Taken together, our experimental data suggest that the neuroprotective effect of flavopiridol may be mediated by repression of specific BH3-only genes through the cyclin D1/CDK4/Rb/E2F-dependend pathway.

Future development and use of specific synthetic inhibitors of cyclin-dependent kinases in the clinic may have a significant impact on the improvement of the existing strategies to treat childhood brain tumors.

Chapter VIII - The ability to induce and execute apoptosis is essential for the maintenance and regulation of tissue and organ homeostasis. A common feature of malignant cells is their capability of evading apoptosis, which contributes to the high rate of resistance to radio- or chemotherapy of tumor cells. Although some progress has been made in inhibiting tumor growth by developing receptor-tyrosine-kinase antagonists that inhibit survival and growth promoting pathways, resistance against these and other compounds is mainly due to defects in apoptosis regulation. In contrast, low apoptotic thresholds may sensitize tumors to well tolerated and clinical established treatment regimens. Therefore, downstream targets of the apoptosis machinery, e.g. survivin, caspases or members of the bcl-2 family, seem to represent interesting new targets for cancer therapy. Modulating apoptotic thresholds can be achieved by reconstituting genetically lost or epigenetically silenced pro-apoptotic factors, e.g. by viral gene transfer or inducing re-expression with DNA methyltransferase or histone deacetylase inhibitors, or by inhibiting the activity or expression of anti-apoptotic factors. Here, RNAi-based approaches, e.g. against anti-apoptotic bcl-2, seem to hold great promise to overcome treatment resistance of a variety of malignant tumor types. Overall, restoring apoptosis sensitivity or lowering apoptotic thresholds will lead to better tolerated and more potent treatment strategies.

Chapter IX - Recent clinical studies show that dose-dense chemotherapy regimens offer the potential benefits of improved both response rate and survival for breast cancer patients. Pegfilgrastim is a covalent conjugate of filgrastim and polyethylene glycol with an increased elimination half-life due to decreased serum clearance. Few data are available on biological effects of pegylated granulocyte colony-stimulating factor. At our center two clinical trials are currently ongoing that will further evaluate pegfilgrastim utilized in dose-dense regimens for breast cancer patients both as neoadjuvant and adjuvant approach. Twenty patients were enrolled in two different dose-dense multicenter clinical trials. Four patients received four courses of concomitant Anthracycline and Taxane chemotherapy as primary systemic treatment and sixteen patients received eight courses of Anthracycline and Taxane on sequential scheme, both every 2 weeks (dose dense schedule), with pegfilgrastim administered from 24 to 72 hours after each chemotherapy course. On peripheral blood buffy coat smears obtained before starting treatment and immediately before each chemotherapy course we analyzed the following parameters in neutrophils: apoptosis by TUNEL technique, actin polymerization using phalloidin labeled with FITC, and alkaline phosphatase activity by cytochemistry. Our aim was to evaluate the influence of pegfilgrastim on these biological features in patients exposed to chemotherapy. After stimulation with pegfilgrastim in all patients we observed: stability of the absolute neutrophil count for the whole duration of treatment and no infectious events; a reduction of neutrophil apoptosis rate in comparison with that observed in control patients treated with standard chemotherapy courses without filgrastim support; persistent abnormalities of actin assembly in neutrophils, indicative of changes in cytoskeleton organization; a significant increase of the leukocyte alkaline phosphatase activity, that is a sensitive marker of myeloid differentiation. In conclusion, these results suggest that pegfilgrastim may improve the neutrophil function in patients with cancer exposed to chemotherapy by inhibiting the accelerated apoptosis and prolonging survival. This effect may, at least in part, be dependent on the influence of pegfilgrastim on actin cytoskeleton organization.

Chapter X - Suppression of apoptosis by survival signals is considered a hallmark of malignant transformation and resistance to anti-cancer therapy. The phosphoinositide-3 kinase (PI3k)/Akt pathway and NF-κB transcription factors are potent mediators of tumour cell survival. The carbocyclic lactone-lactam antibiotic rapamycin, a widely used immunosuppressant, inhibits the oncogenic transformation of human cells induced by PI3k or Akt by blocking the downstream mTOR kinase. However, inhibition of the PI3k/Akt/mTOR cascade may not be the only mechanism whereby rapamycin exerts anticancer effects. We previously demonstrated that rapamycin inhibits NF-κB by acting on FKBP51, a large immunophilin whose isomerase activity is essential for the functioning of the IKK kinase complex. This suggested that rapamycin may be effective also against neoplasias that express the tumour suppressor PTEN, which, by reducing cellular levels of phosphatidyl-inositol triphosphate, antagonizes the action of PI3k. To address this issue, we over-expressed PTEN in a human melanoma cell line characterized by high phospho-Akt and phospho-mTOR levels, and examined the effect of rapamycin on the apoptotic response to the NF-κB inducer doxorubicin versus cisplatin, which does not activate NF-κB. Rapamycin increased both cisplatin- and doxorubicin-induced apoptosis. Transient transfection of PTEN remarkably decreased phospho-mTOR levels and increased sensitivity to cisplatin's cytotoxic effect. Under these conditions, rapamycin failed to enhance cisplatin-induced apoptosis. This finding supports the notion that inhibition of a survival pathway increases the efficacy of cytotoxic drugs, and suggests that the pro-apoptotic effect of the rapamycin-cisplatin association requires activated mTOR. Rapamycin retained the capacity to enhance doxorubicin-induced apoptosis in cells over-expressing PTEN, which confirms our earlier observation that inhibition of the PI3k/Akt/mTOR pathway is not involved in the effect exerted by the rapamycin-doxorubicin association. These findings indicate that constitutive activation of mTOR is sufficient but not necessary for rapamycin's anti-cancer effect. Finally, we show that a decrease in FKBP51 expression levels, obtained with the small interfering

RNA technique in the leukemic cell line Jurkat, increased doxorubicin-induced apoptosis, suggesting that this rapamycin ligand is involved in resistance to chemotherapy-induced apoptosis.

In conclusion, rapamycin affects more than one signalling survival pathway and more than one target. Our data may impact on the synthesis of rapamycin derivatives. Thus far, rapamycin derivatives used in clinical trials have been tested for their mTOR-inhibiting effect. Our study opens the door to a novel class of anti-cancer drugs that specifically target immunophilins.

In: Cell Apoptosis Research Trends
Editor: Charles V. Zhang, pp. 1-39

ISBN: 1-60021-424-X
© 2007 Nova Science Publishers, Inc

Chapter I

The Role of BH3-Only Proteins in Cancer and Anti-Cancer Therapeutics

Junya Kuroda[1] and Shinya Kimura[2] [1]

1.Division of Hematology and Oncology, Department of Medicine, Kyoto Prefectural University of Medicine, 465 Kajii-cho, Kamigyo-ku, Kyoto, 602-8566, Japan
2.Department of Transfusion Medicine and Cell Therapy, Kyoto University Hospital, 54 Shogoin Kawahara-cho, Sakyo-ku, Kyoto, 606-8507, Japan

ABSTRACT

Apoptosis is a genetically conserved cell suicide program that is essential for proper development and the maintenance of tissue homeostasis. This cell death program removes unwanted cells, such as those harboring harmful mutations. Therefore, the genetic and functional integrity of molecules crucial for apoptosis is potentially significant for oncogenic processes. "Total cell kill" is essential in anti-cancer treatment, and apoptosis is the most extensively characterized mechanism for cancer cell killing. Recent studies reveal that apoptosis induced by anti-cancer agents is mostly the result of interactions between anti-apoptotic members of the Bcl-2 protein family and BH3-only proteins, a pro-apoptotic subgroup of the Bcl-2 family. In this context, BH3-only proteins act as initiators of cell death signals initiated by anti-cancer drugs by binding to anti-apoptotic Bcl-2 proteins, eventually leading to the activation of the mitochondria-dependent apoptosis pathway via the stimulation of Bax and Bak, disruption of mitochondrial outer membrane permeability and activation of the caspase cascade which causes dismantling of the cell. At least eight BH3-only proteins have been identified to date, most of which are involved in cell responses to anti-cancer agents. Since the expression profiles of BH3-only proteins are dependent on the type of cell, tissue, and nature of cellular stress, their degree of participation in cellular responses to various cancer treatments may also vary widely, according to the type of cancer and therapeutic

---

[1] Address correspondence to Kimura S. M.D., Ph.D.,Department of Transfusion Medicine and Cell Therapy, Kyoto University Hospital, 54 Shogoin Kawahara-cho, Sakyo-ku, Kyoto, 606-8507, Japan, Tel: +81 75 751 3630, Fax: +81-75-751 4283, Email: shkimu@kuhp.kyoto-u.ac.jp

approach. Here, we review current knowledge on the regulation of the Bcl-2 family, particularly the role of BH3-only proteins in apoptosis induced by anti-cancer agents, including recently developed molecular-based therapeutics, such as imatinib mesylate used to treat chronic myelogenous leukemia, and proteasome inhibitors as general anticancer drugs. Additionally, specific BH3-only proteins are likely to collaborate in specific cellular settings. Recent data suggest that the loss of function of crucial BH3-only proteins leads to resistance to certain anti-cancer drugs, confirming their significance in cancer treatment. We describe recent progress in translating these findings into novel therapeutic strategies, involving the direct manipulation of Bcl-2 proteins. These fundamental approaches directed towards understanding the diverse roles of BH3-only proteins in oncogenesis and cancer treatment should facilitate the development of more rational therapeutic approaches against cancer.

INTRODUCTION

Numerous studies have led to remarkable progress in our understanding of cancer biology in the past decades. One of the most important aspects in cancer biology is that deregulation of cell death, namely resistance to apoptosis, is closely associated with tumor development as well as the outcome of anti-cancer therapy [Brown JM, 2005; Cory S, 2003; Gerl R, 2005; Kasibhatla S, 2003; Strasser A, 2000; Vermeulen K, 2005]. Apoptosis is an evolutionarily conserved, genetically programmed cell death process essential for normal development, maintenance of tissue homeostasis and functioning of multicellular organisms. It eliminates redundant, damaged, or unwanted cells, such as infected cells or those harboring harmful mutations that potentially cause abnormal cell proliferation or aberrant immune responses [Adams JM, 2003; Cory S, 2003; Cory S, 2003; Daniel NN, 2004; Green DR, 2002]. Tight control of programmed cell death is crucial to prevent disease development, since excess cell death may cause degenerative disease, and inadequate apoptosis can lead to tumor development or autoimmune disease [Brunner T, 2003; Hughes P, 2006; Opfermann JT, 2003; Strasser A, 2005].

Mammals have two distinct apoptosis signaling pathways [Adams JM, 2003; Green DR, 2002; Strasser A, 2005] (Figure 1). One is called the 'intrinsic' (or 'mitochondrial' or 'Bcl-2 regulated') apoptosis pathway and it is regulated by the interplay of anti-apoptotic and pro-apoptotic members of the Bcl-2 family (Figure 2, 3). The other pathway is called the 'extrinsic' (or 'death receptor') pathway and it is activated by stimulation of death receptors (tumor necrosis factor (TNF) family receptors with intracellular death domains) by their cognate ligands, such as TNF, Fas or TRAIL. The intrinsic pathway converges upon mitochondria, where signals result in the release of pro-apoptotic factors that trigger activation of caspases (a family of cysteine proteases) and also caspase-independent processes [Chipuk JE, 2005; Green DR, 2004]. The death receptor pathway directly promotes activation of caspase-8 and downstream ('effector' caspases) [Hirata H, 1998]. The two pathways can be connected through an amplification loop involving caspase-mediated activation of the pro-apoptotic Bcl-2 family member Bid (Figure 1). Potent mitochondrial death factors include cytochrome c (cyt c), an integral part of the "Apoptosome" together with Apaf-1 and caspase-9 which causes activation of the caspase cascade [Kluck RM, 1997;

Yoshida H, 1998], and Diablo/Smac, which potentiates caspase activity by suppressing inhibitor of apoptosis proteins (IAPs) [Du C, 2000; Verhagen AM, 2000].

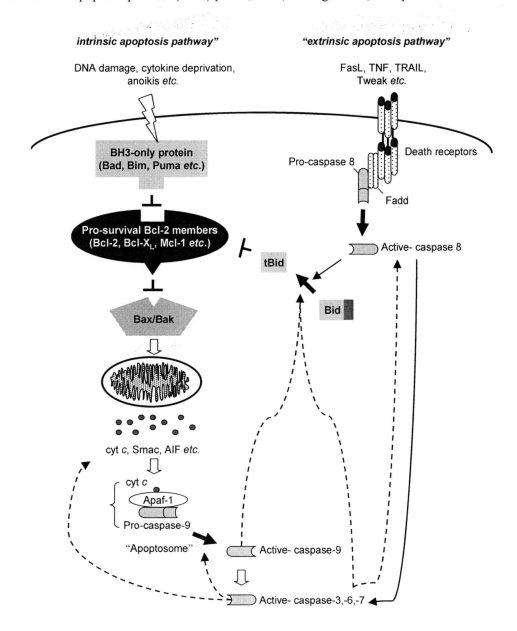

Figure 1. Simplified representation of the two main signaling cascades of apoptosis. Bcl-2 proteins regulate mitochondrial outer membrane integrity, and consequently the release of mitochondrial proteins, such as cytochrome *c* (cyt *c*), or Diablo/Smac (Smac). Stimulation of death receptors (members of the tumor necrosis factor receptor (TNF-R) family with an intra-cellular death domain by their cognate ligands (member of the tumor necrosis factor (TNF) family of ligands) induces caspase-8 activation and subsequent cleavage of Bid. Both pathways converge upon the activation of effector caspases, including caspase-3, -6, or -7. Once activated, most effector caspases can act as amplifiers of apoptosis signaling (dot lines).

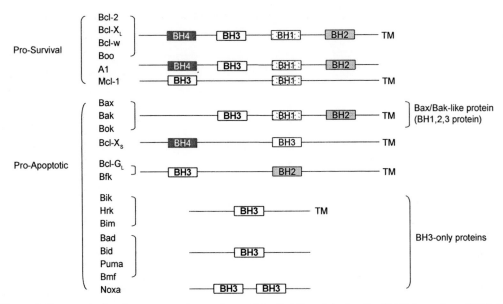

Figure 2. Pro-survival and pro-apoptotic members of the Bcl-2 protein family. Bcl-2 homology (BH1-4) and transmembrane regions (TM) are indicated.

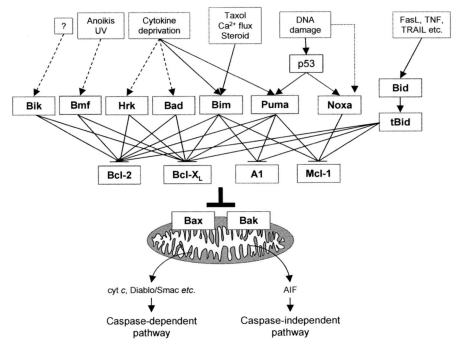

Figure 3. Apoptosis-initiating molecular cascades of the Bcl-2 family of proteins under cellular stress conditions. Apoptosis is initiated by the activation of certain sets of BH3-only proteins based on the nature of the death stimuli. This in turn leads to the inactivation of pro-survival proteins and subsequent activation of Bax/Bak on mitochondria.

Active caspases degrade vital proteins (e.g. Lamins) and activate latent enzymes that degrade DNA. This elicits the typical features of apoptosis, such as plasma membrane blebbing, chromatin condensation, internucleosomal DNA cleavage, surface exposure of "eat me signals" and phagocytosis [Kerr JF, 1972]. The Bcl-2 family of proteins regulates mitochondrial outer membrane permeability (MOMP) and release of these mitochondrial death factors [Susin SA, 1996; Susin SA, 1999; Spierings D, 2005; Green DR, 2004] (Figure 3). There are more than 20 Bcl-2 related proteins, comprising both pro-survival and pro-apoptotic members (Figure 2). Deregulation of Bcl-2 family members is frequently documented in various cancers. Both abnormal overexpression of pro-survival family members and loss of function of pro-apoptotic members can promote tumorigenesis, particularly in the context of activation of an oncogene (e.g. *c-myc*). These abnormalities can also confer upon cancer cells resistance to anti-cancer drugs or irradiation. BH3-only proteins have been shown to function as tumor suppressors, since even loss of a single allele of the BH3-only protein Bim accelerated *myc*-induced lymphoma development [Egle A, 2004]. Multiple lines of evidence demonstrate that activation of specific members of the proapoptotic Bcl-2 family, BH3-only proteins, is a prerequisite for apoptosis induction by anti-cancer drugs. As the expression of the BH3-only members is dependent on the cell/tissue-context and types of cellular stress [Bouillet P, 2002; Cory S, 2002; Huang DC, 2000], the types of these proteins utilized for apoptosis initiation vary depending on the type of cancer cell and the anti-cancer agents employed. Loss of function of crucial BH3-only proteins is thought to result in treatment failure. In certain settings, several BH3-only proteins act in cooperation to initiate apoptosis. Here we describe the current concepts on Bcl-2 family regulation in apoptosis, and review existing knowledge on the significance of BH3-only proteins in tumour development and the role of each member in cell death induced by anti-cancer therapeutics.

THE BCL-2 PROTEIN FAMILY

The Bcl-2 protein family is critical regulator of intrinsic, "mitochondria-dependent" apoptosis signaling. The Bcl-2 family includes at least 20 proteins that share between one and four conserved regions, designated "Bcl-2 homology domains (BH1, 2, 3 and BH4)". According to their activity, they are broadly classified into two groups: anti-apoptotic and pro-apoptotic members. Pro-apoptotic members are further subdivided into two groups according to their structure and function (Figure 2).

PRO-SURVIVAL BCL-2 FAMILY MEMBERS

The pro-survival Bcl-2 family members include Bcl-2, Bcl-X$_L$, Bcl-w (Bcl-2L2), A1 (also denoted Bcl-2A1 or Bfl-1), Mcl-1 (myeloid cell leukemia sequence 1) and Boo (Bcl-2 homolog of ovary, also known as Diva). These proteins contain three or four BH domains (BH1-4) (Figure 2). The BH1, BH2 and BH3 regions form a hydrophobic groove on the surface of pro-survival Bcl-2 family members where BH3 domains of BH3-only proteins can bind [Muchmore SW, 1996; Petros AM, 2000; Petros AM, 2001, Petros AM, 2004; Suzuki M, 2000].

Proper functioning of pro-survival Bcl-2 proteins is essential for organized tissue development. Bcl-2 and Mcl-1 are required for normal hematopoietic development [Opfermann JT, 2003; Opfermann JT, 2005; Nakayama K, 1994]. Bcl-X_L is essential for the survival of neurons and erythroid progenitor cells during embryogenesis [Chida D, 1999], and Bcl-w for spermatogenesis [Ross AJ, 1998]. A1 has been shown to be important for keeping granulocytes and mast cells alive [D'Sa-Eipper C, 1996]. However, there is so far no experimental evidence that Boo is critical for cell survival.

The founding member, *bcl-2* (B cell lymphoma/leukemia gene-2), was initially identified in 1985 as a proto-oncogene at the chromosomal breakpoint of t(14;18)(q32;q21) of human follicular lymphoma [Tsujimoto Y, 1988]. In contrast to the prevailing concept for oncogene function at that time, expression of Bcl-2 did not promote cellular proliferation, but blocked cell death triggered by a range of stress stimuli. In *bcl-2* transgenic mice, resting B cells accumulated as a result of extended cell survival, and developed polyclonal follicular hyperplasia [McDonnell TJ, 1989; Strasser A, 1991; Vaux DL 1988]. These mice later developed monoclonal lymphoma or plasmacytoma when cells acquired spontaneous rearrangement of the *c-myc* proto-oncogene [Strasser, 1993]. Furthermore, Strasser and colleagues showed that *bcl-2/myc* double transgenic mice developed undifferentiated leukemia much more rapidly than Eμ-*myc* singly transgenic mice, indicating cooperative interactions between Bcl-2 and Myc in oncogenesis [Strasser A, 1990]. This synergy was later also observed in other cell types, such as hepatocytes [DeoCampo ND, 2000] and breast epithelial cells [Jager R, 1997]. Bcl-2 also cooperates with the PML-RARα fusion oncoprotein (a result of the t(15;22)(q22;q21) chromosomal translocation) to induce acute promyelocytic leukemia [Kogan SC, 2001], and with the Bcr/Abl fusion tyrosine kinase, a hallmark of chronic myelogenous leukemia (CML), leading to more primitive myeloid leukemia [Jaiswal S, 2003]. Recently, the group of Letai showed that elimination of *bcl-2* by a conditional knockout strategy yields rapid loss of leukemic cells developing in constitutive *c-myc*/conditional *bcl-2* double transgenic mice [Letai A, 2004]. Collectively, the results mentioned above form the basis of the concept that a combination of deregulated proliferation and impaired apoptosis accelerates cancer development and that defects in apoptosis signaling are required to sustain tumour growth. Interestingly, expression of a *bcl-2* transgene in mice homozygous for the *lpr* loss of function mutation in Fas (CD95/APO-1) caused development of acute myelogenous leukemia (AML), indicating that dual blockade of the extrinsic and intrinsic apoptotic pathways can synergize in tumor development [Traver D, 1998].

Overexpression of pro-survival Bcl-2 proteins protects cells against a broad range of cytotoxic stimuli, including chemotherapeutic drugs. Therefore, deregulation of these proteins often correlates with cellular response to various chemotherapeutic drugs and/or γ-irradiation. Bcl-2 overexpression is frequently associated with lower chemosensitivity in most cancers [Reed JC, 1994; Real PJ, 2002; Groeger AM, 2004; Sartorius UA, 2002; Borner C, 1999]. As described above, the *bcl-2* gene is juxtaposed to Eμ enhancer elements of the immunoglobulin heavy chain gene by reciprocal chromosomal translocation t(14;18), and was found to be overexpressed in ~85% of human follicular lymphomas. Follicular lymphoma is classified as an indolent type lymphoma, but complete eradication of lymphoma cells solely using radiochemotherapy is still difficult to achieve [de Jong D, 2005]. This is possibly due to the potent inhibition to apoptosis afforded by the high levels of Bcl-2. A recent study revealed that Oct-2, a member of the POU homeodomain family of transcriptional regulators, is overexpressed and directly activates *bcl-2* transcription in t(14;18) lymphoma [Heckman CA, 2006]. Apart from chromosomal translocation, Bcl-2 transcription can be activated by various molecular mechanisms. One of the major transcription factors involved in Bcl-2 expression in some cancers is nuclear factor-κB (NF-κB) [Catz SD, 2001; Heckman CA, 2002; Kurland JF, 2001], a molecular target of anti-cancer drugs, such as proteasome inhibitors [Fahy BN, 2005; Mortenson MM, 2005], histone deacetylase inhibitors (HDACIs) [Fandy TE, 2005], thalidomide [Liu WN, 2004], and arsenic trioxide [Kerbauy DM, 2005; Mahieux R, 2001]. Another transcription factor that stimulates Bcl-2 expression is a family of signal transducers and activators of transcription (Stats) that are known mediators of cytokine signaling for cellular proliferation [Real PJ, 2002; Sakai I, 1997]. Despite the positive correlation between Bcl-2 expression and anti-apoptotic phenotype in cancers, the impact of the levels of this protein on patient prognosis varies among cancer types, and is not always a single prognostic indicator [Harada T, 2003; Kitada S, 1998]. Bcl-X_L is also potently anti-apoptotic and when overexpressed, it accelerates tumorigenesis (e.g. pancreatic β-cell tumors or leukemias) in combination with other oncogenes, such as *c-myc* [Linden M, 2004; Packham G, 1998; Pelengaris S, 2004; Swanson PJ, 2004]. Expression of Bcl-X_L correlates strongly with resistance to apoptosis induction by a panel of 122 chemotherapeutic drugs [Amundson SA, 2000]. Moreover, Bcl-X_L overexpression is associated with poor prognosis in childhood acute lymphoblastic leukemia [Addeo R, 2005], hepatocellular carcinoma [Watanabe J, 2004] and ovarian carcinoma [Williams J, 2005], but not in multiple myeloma [Peeters SD, 2005] or non-small cell lung cancers [Groeger AM, 2004]. Bcl-X_L is also transcriptionally regulated by NF-κB or Stats [Grad JM, 2000]. Mcl-1 overexpression is frequently observed in various cancer types [Fleischer B, 2006; Sano M, 2005; Erovic BM, 2005; Maeta Y, 2004]. Transgene of *mcl-1* leads to high incidence of B-cell lymphoma in mice [Zhou P, 2001], and Mcl-1 overexpression correlates with clinical grade and resistance to chemotherapy in B-cell non-Hodgkin lymphomas, chronic lymphocytic leukemia (CLL) and plasmacyte myeloma [Cho-Vega JH, 2004; Wuilleme-Toumi S, 2005]. Recently, Moshynska and colleagues revealed that mutations in the *mcl-1* gene, specifically, nucleotide insertion into the promoter sequence gene, induced high levels of Mcl-1 expression, which was closely associated with drug resistance and poor disease outcome in CLL [Moshynska O, 2004]. Finally, A1 is expressed abundantly in the bone marrow and at low levels in some other tissues [Choi SS, 1995], and possesses similar anti-apoptotic abilities to Bcl-2 [D'Sa-Eipper C, 1996]. A positive correlation was evident between the expression of A1 and development, progression, or resistance to apoptosis in some cancers [Choi SS, 1995; Kim JK, 2004; Morales AA, 2005].

A recent focus of research in cancer biology is cancer stem cells. These cells have been postulated to reside in the so-called 'stem cell niche', and produce proliferative daughter cancer cells as well as primitive stem cells that are quiescent and resistant to chemotherapeutic interventions [Huntly BJ, 2005; Wang JC, 2005; Konopleva M, 2002]. One of the crucial characteristics that protect cancer stem cells from cytotoxic stimuli or environmental stress is the acquisition of resistance to many apoptotic stimuli. Although neither Bcl-2 nor Bcl-X_L expression at initial presentation predicted achievement of complete remission in acute myelogenous leukemia (AML) patients, levels of these proteins were higher in the more primitive CD34-positive AML cells selected for survival after induction chemotherapy [Konopleva M, 2002]. The stem cell niche is a somewhat hypoxic area that stimulates specific subsets of genes for cellular survival and proliferation [Grayson WL, 2006]. One of the transcription factors critical for this process is a family of hypoxia-inducible factors (Hif). Previous studies have shown that Hif-1α induces Mcl-1 expression [Carmeliet P, 1998; Piret JP, 2005]. Collectively, these data indicate that anti-apoptotic Bcl-2 family members may play important roles in the maintenance of cancer stem cells.

BAX/BAK-LIKE MULTIDOMAIN PRO-APOPTOTIC MEMBERS

Pro-apoptotic Bax/Bak-like proteins include Bax (Bcl-2-associated X protein), Bak (Bcl-2 antagonist killer), Bok (Bcl-2-related ovarian killer) with three BH domains (BH1-3), and Bcl-X$_S$ possessing two BH domains (BH3 and BH4) (Figure 2). Bax and Bak are ubiquitously expressed in most tissues, while Bok has been reported to be more prevalent in reproductive tissues [Hsu SY, 1997]. In healthy cells, both Bax and Bak exist as their inactive forms (Figure 3). Bax preferentially localizes to the cytosol or attaches loosely to membranes as monomers [Guo B, 2003], while Bak translocates to the outer mitochondrial membrane as oligomers [Cheng EHY, 2003]. Upon overexpression, both Bax and Bak mainly act on mitochondria to trigger apoptosis. Gene knockout studies have demonstrated that deletion of either Bax or Bak does not significantly affect apoptosis. However, Bax/Bak double-deficient cells are highly resistant to apoptosis triggered by cytokine withdrawal or treatment with a broad range of cytotoxic drugs [Lindsten T, 2000]. Moreover, enforced expression of BH3-only proteins does not induce apoptosis in Bax/Bak double-deficient cells, indicating that BH3-only proteins initiate apoptosis signaling while Bax/Bak play a critical role downstream in apoptosis signaling [Kuwana T, 2005; Zong WX, 2001]. It is possible that Bax and Bak can function as tumor suppressors. Loss of either protein causes few tumors in mice, probably because Bax and Bak have largely overlapping functions. However, in conjunction with other oncogenic abnormalities, the loss of these proteins drives cells into a highly transformable phenotype. Bax deletion was shown to promote tumorigenesis in brain choroid plexus or mammary epithelium overexpressing a truncated version of SV40 large T antigen that blocks the Retinoblastoma protein [Shibata MA, 1999; Yin C, 1997], in B-lymphoid cells with *c-myc* overexpression [Eischen CM, 2001] and in primary mouse embryonic fibroblasts expressing adenovirus E1A [McCurrach ME, 1997]. Bax can be regulated by p53. The *bax* gene promoter region contains a p53-binding site, and is involved in a p53-regulated pathway for induction of apoptosis [Miyashita T, 1995]. Moreover, Bax deficiency accelerates tumorigenesis in p53$^{-/-}$ mice [McCurrach ME, 1997].

BH3-ONLY PROTEINS

BH3-only proteins, sharing only the short BH3 domain (9-16 amino acids), comprise the other subgroup of the pro-apoptotic Bcl-2 family. Experiments with gene targeted mice have shown that the BH3-only proteins are essential for connecting proximal death signals induced by cellular stress to the core apoptotic pathway executed by Bax and Bak [Adams JM, 2003; Cory S, 2003; Green DR, 2004; Huang DC, 2000; Strasser A, 2000; Strasser A, 2005; Willis SN, 2005]. Mammals have at least eight BH3-only proteins, including Bad (Bcl-2-anagonist of cell death), Bik (Bcl-2-interacting killer, also designated Nbk or Blk), Bid (BH3-interacting domain death agonist), Hrk (Harakiri, also called DP5), Bim (Bcl-2-interacting mediator of cell death, also called Bod), Noxa, Puma (p53-upregulated modulator of apoptosis, also called Bbc3) and Bmf (Bcl-2 modifying factor). All BH3-only proteins bind to pro-survival Bcl-2 members and inhibit their functions, but individual BH3-only proteins differ remarkably in their abilities to bind to different pro-survival Bcl-2 relatives. Bim and Puma bind to all pro-survival Bcl-2 family members with high affinity. Truncated Bid (tBid) binds with high affinity to Bcl-X_L, Bcl-w, Mcl-1 and A1 but only relatively weakly to Bcl-2. Bik and Hrk bind most potently to Bcl- X_L, Bcl-w and Mcl-1 but only poorly to Bcl-2 or A1. Bad and Bmf bind strongly to Bcl-2, Bcl-X_L and Bcl-w, but not to Mcl-1 or A1. Noxa binds only with high affinity to Mcl-1 and A1 [Certo M, 2006; Chen L, 2005; Letai A, 2002; Kuwana T, 2005] (Figure 3)(Table 1). Consistent with their binding specificities, overexpression of Bim or Puma, but not other BH3-only proteins, is sufficient to trigger apoptosis in the presence of Bax and Bak. Interestingly, in co-transfection assays, BH3-only proteins with different selective binding patterns (e.g. Bad plus Noxa) synergize in apoptosis induction. These results clearly indicate that BH3-only proteins can cooperate with each other to induce apoptosis [Certo M, 2006; Chen L, 2005]. In addition, Bim and Bid have been reported to bind directly to Bax and Bak to activate their pro-apoptotic functions [Certo M, 2006; Letai A, 2002], but this has not yet been shown to occur under physiological conditions. All these molecular interactions of BH3-only proteins require an intact BH3 domain that forms an amphipathic α-helix for insertion into the hydrophobic groove on the surface of pro-survival Bcl-2 family members [Chen L, 2005]. Individual BH3-only proteins are expressed in certain cell types and appear to be activated by specific types of cellular damage. In some situations, several BH3-only proteins share overlapping functions, which is also likely to be the case in apoptosis induced by anti-cancer drugs. As already mentioned above, at least certain BH3-only proteins can act as tumor suppressors [Egle A, 2004; Hemann MT, 2004; Hemann MT, 2005; Ranger AM, 2003; Zinkel SS, 2003]. The detailed regulatory pathways, physiologic functions and involvement of each BH3-only protein in cancer therapy are described in a later section.

Table 1. Affinities of BH3-only for pro-survival and Bax/Bak-like proteins. Binding affinities of BH3-ony proteins to prosurvival proteins: (+++) > (++) >(+), no affinity:(-). Binding affinity of BH3-only proteins to Bax/Bak: O: affinity (+), X: affinity (-).

	pro-survival members					Bax/Bak
	Bcl-2	Bcl-XL	Bcl-w	Mcl-1	A1	
Bad	+++	+++	+++	-	+	X
Bid	+	++	++	+	+	O
Bik	+	++	++	+	++	X
Bim	+++	+++	+++	+++	+++	O
Bmf	+++	+++	+++	+	+	X
Hrk	++	+++	++	++	++	X
Noxa	-	-	-	++	++	X
Puma	+++	+++	+++	+++	+++	X

HOW PRO-SURVIVAL AND PRO-APOPTOTIC BCL-2 PROTEINS INTERACT AND REGULATE APOPTOSIS: PROPOSED MODELS FOR INTRINSIC APOPTOSIS SIGNALING

Despite accumulating evidence, the mechanisms by which BH3-only proteins, Bax/Bak-like proteins and pro-survival Bcl-2 family members interact to mediate the cell life and death decision remain to be established. Here, we discuss several proposed models for the initiation and execution of intrinsic (mitochondria-mediated, Bcl-2 family-regulated) apoptosis based on current evidence.

Upon exposure to cell death stimuli, specific BH3-only proteins are induced and/or activated as the first step for initiating apoptosis. This event occurs upstream of Bax and Bak activation, since BH3-only proteins are unable to kill cells in the absence of these proteins [Kuwana T, 2005; Lindsten T, 2000; Zong WX, 2001]. Bax and Bak activation causes MOMP and leads to activation of the caspase cascade. There are several conceivable models for the interactions between the members of the Bcl-2 protein family, as shown below (Figure 4).

i) In healthy cells, pro-survival Bcl-2 family members bind and inhibit Bax and Bak to prevent their oligomerization on the mitochondrial outer membrane, while BH3-only proteins remain inactivated by transcriptional repression or post-translational sequestration. Once induced and/or activated by cytotoxic signals, BH3-only proteins bind to pro-survival Bcl-2 proteins thereby liberating Bax/Bak due to their stronger binding affinity. Released Bax and Bak form oligodimers on the mitochondrial outer membrane and induce MOMP and activation of the caspase cascade.

ii) In healthy cells, pro-survival Bcl-2 family members bind and inhibit both BH3-only proteins and Bax/Bak-like proteins to prevent apoptosis. Upon activation of a death stimulus, BH3-only proteins are either conformationally modified to facilitate removal from pro-survival relatives or increased in concentration, leading to excess "free" BH3-only proteins, which then directly activate Bax and Bak.

iii) The third model involves two classes of BH3-only proteins. One is able to both activate Bax/Bak directly and to inhibit pro-survival Bcl-2 proteins, while the other can only inhibit pro-survival Bcl-2 proteins. In healthy cells, pro-survival Bcl-2 family members bind to and inactivate Bax and Bak. Once activated and/or induced by cytotoxic stimuli, BH3-only proteins initially bind to pro-survival Bcl-2 family members to release Bax and Bak, and subsequently activate Bax/Bak also directly.

Pro-survival Bcl-2 family members have been postulated to inactivate Bax and Bak by direct binding or indirect mechanisms involving unknown adapter molecules. Activation of BH3-only proteins somehow (see above) causes activation of Bax/Bak. Once oligomerized on the mitochondrial outer membrane, Bax and Bak cause MOMP and release mitochondrial pro-apoptotic proteins to stimulate caspase-dependent and/or -independent processes [Green DR, 2004]. Thus, the balance between BH3-only proteins and pro-survival Bcl-2 family members is most likely to be the determinant of cell fate and activation of Bax and Bak via oligomerization on the mitochondrial outer membrane is the point of cell death commitment.

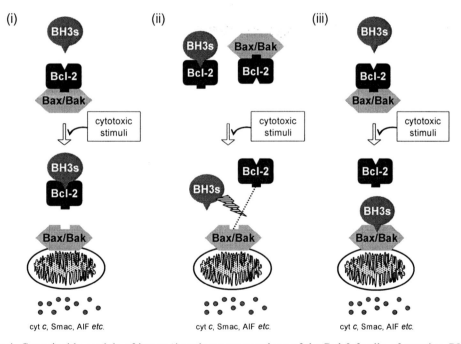

Figure 4. Conceivable models of interactions between members of the Bcl-2 family of proteins. BH3s: BH3-only proteins.

THE ROLES OF BH3-ONLY PROTEINS IN TUMOR DEVELOPMENT AND CANCER THERAPY

To execute apoptotic cell death, the initiation step requires the activation and proper functioning of BH3-only proteins (Figure 1, 3). Recent studies demonstrate that different BH3-only proteins are required for apoptosis induced by different anti-cancer drugs, and cell type-specific actions of BH3-only proteins have also been discovered. Here, we describe current knowledge on the role of individual BH3-only proteins in tumorigenesis and cancer therapy.

Bad (Bcl-2-Antagonist of Cell Death)

Bad is the first BH3-only family member identified. The protein was discovered as a binding partner of Bcl-X$_L$ by yeast two-hybrid library screening [Yang E, 1995]. In healthy cells, Bad is phosphorylated on several serine residues. Phosphorylated Bad is sequestered from Bcl-X$_L$ (and other pro-survival Bcl-2 family members) by binding to *14-3-3* scaffold proteins [Zha J, 1996]. Several kinases that mediate signaling by growth factors were shown to cause Bad phosphorylation. The serine/threonine kinases Akt/PKB and Raf, a downstream mediator of Ras, mediate phosphorylation at serine (Ser) 112 and Ser 136 [Fang X, 1999; Scheid MP, 1999], while protein kinase A (PKA) phosphorylates Ser 155 [Lizcano JM, 2000]. Shutdown of cytokine receptor signaling leads to Bad dephosphorylation, translocation to mitochondria, and interactions with Bcl-2 and Bcl-X$_L$. Pim-2, another serine/threonine kinase, also phosphorylates Bad at Ser112 [Yan B, 2003] (Figure 5). The finding that Bad-deficient mice are more prone to leukemia/lymphoma than their wild-type littermates, particularly when subjected to sub-lethal irradiation, supports the role of this protein as a tumor suppressor [Ranger AM, 2003].

Deregulation of Bad has been implicated in several cancers. Nicotine, a central component of cigarette smoke, potently induces Bad phosphorylation. Accordingly, nicotine-induced Bad sequestration from Bcl-X$_L$ has been postulated to lead to tumor development and chemoresistance in lung cancer cells [Jin Z, 2004]. Loss of Bad expression was observed in 51% of non-small lung carcinoma samples, with particularly high frequency in squamous cell carcinoma [Berrieman HK, 2005]. Bcr/Abl fusion tyrosine kinase, the primary cause of CML, constitutively activates multiple downstream signaling molecules, including PI3K/Akt and Ras/Raf/Erk pathways. These active pathways maintain Bad in the phosphorylated (inactive) state, which has been proposed to be responsible for the chemoresistance of CML [Neshat MS, 2000; Salomoni P, 2000]. We recently demonstrated that Bad dephosphorylation is involved in apoptotic initiation by the Bcr/Abl tyrosine kinase blocker, imatinib mesylate, in CML, possibly through Akt inhibition following Bcr/Abl inactivation. Bad is constitutively phosphorylated in AML, and the chemical inhibitor of PI3K, LY294002, induces dephosphorylation of Bad and increased apoptosis. Moreover, combined treatment with PI3K and MAPK inhibitors increases apoptotic cell death [Zhao S, 2004]. These data suggest that both PI3K/Akt and Ras/MAPK pathways may be suitable therapeutic targets for some cancers (Figure 5).

Bik (Bcl-2-Interacting Killer, Nbk: Natural Born Killer, Blk)

Bik, expressed predominantly in epithelial and hematopoietic cells, binds to some but not all pro-survival Bcl-2 family members, including Epstein-Barr virus BHRF1 and adenovirus E1B 19-kD [Boyd JM, 1995; Han J, 1998]. No striking abnormality or resistance to apoptosis was observed in response to B-cell receptor cross-linking, despite upregulation of the protein in *bik*-knockout (KO) mice [Coultas L, 2004]. However, marked synergy between loss of Bik and Bim was evident in double KO mice, suggesting functional overlap between these two (and perhaps also other combinations of) BH3-only proteins [Coultas L, 2005].

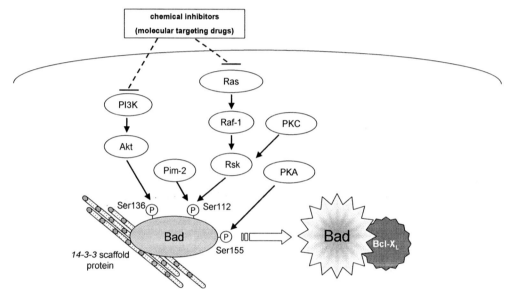

Figure 5. Regulatory pathways of Bad expression and chemotherapy. The pro-apoptotic activity of Bad is post-translationally regulated by signaling molecules that are often critical for cancer cell survival. Blockade of these signals dephosphorylate and activate Bad by releasing it from *14-3-3* scaffold proteins, and they are thus potentially therapeutic targets. P: phosphorylation, Ser: serine residue

Mutations in the *bik* gene have been reported in a substantial proportion of malignant lymphomas [Arena V, 2003], and epigenetic repression of Bik either through loss of heterozygosity (LOH) or DNA methylation was observed in a sizable fraction renal clear cell carcinoma [Sturm I, 2006]. Bik was shown to be activated during apoptosis induced by the proteasome inhibitor, bortezomib (PS-341) [Nikrad M, 2005; Zhu H, 2005] as well as in hormone treatment induced apoptosis in breast cancer cells [Hur J, 2004]. Enforced expression of the *bik* gene was shown to increase the number of apoptotic cells and enhance the apoptosis-inducing effects of certain chemotherapeutic drugs, such as etoposide or doxorubicin, and activation of death receptors by Fas ligand or TRAIL [Daniel PT, 1999].

Bid (BH3-Interacting Domain Death Agonist)

Wang and colleagues identified Bid as a death agonist that can heterodimerize with Bcl-2 as well as Bax by screening a λ-phage expression library [Wang K, 1996]. Bid is a unique amongst BH3-only proteins because it connects the death receptor pathway with the intrinsic apoptotic pathway. The *bid*-KO mice are resistant to fatal hepatocyte destruction mediated by injection with anti-Fas antibodies [Yin XM, 1999]. Full-length Bid (p22) is localized in the cytosol and has relatively poor pro-apoptotic activity. Following activation of the extrinsic apoptotic pathway, caspase-8 proteolytically cleaves Bid into a truncated fragment (tBid, p15) (Figure 1, 3). For this cleavage, full-length Bid has to be free of phosphorylation at Ser 61 and Ser 64, sites that can be phosporylated by casein kinases I and II [Dasagher S, 2001]. Once modified, tBid becomes a potent cell death inducer. The fragment displays enhanced ability to translocate to mitochondria by adapting N-terminal myristoylation, and binds pro-survival Bcl-2 members and reportedly also to Bax/Bak-like proteins (although not yet proven for physiological conditions) [Certo M, 2006]. *Bid*-KO mice are born with no developmental abnormalities. However, with age, a small fraction of these animals spontaneously develop myeloproliferative disorder, which progresses from myeloid hyperplasia to clonal malignancy resembling chronic myelomonocytic leukemia [Zinkel SS, 2001]. TRAIL (also designated as Apo-2L) is a member of the TNF family that promotes apoptosis by binding to the transmembrane death receptors DR4 and DR5. Its cytotoxic activity is relatively selective to the human tumor cell lines, but not to normal cells, and the apoptosis by TRAIL involves the cleavage of Bid into tBid by capspase-8 [Srivastava RK, 2001]. Apoptosis by some chemotherapeutic agents occasionally involve the activation of death receptor-mediated apoptotic pathway, therefore, also seem to involve Bid activation.

Hrk (Harakiri, DP5)

Hrk is mainly expressed in the neuronal system. Studies on *hrk*-KO mice confirm that Hrk is an important mediator of apoptosis in neurons when they are deprived of their essential survival factor, nerve growth factor (NGF) [Imaizumi K, 1997; Inohara N, 1997]. However, the influence of Hrk on neuronal tumors remains to be clarified.

Bim (Bcl-2-Interacting Mediator of Cell Death, Bod)

At present, Bim is the most intensively investigated BH3-only protein in association with cancers. The protein was originally isolated by O'Connor and colleagues [O'Connor L, 1998]. Bim is expressed in many tissues [O'Reilly LA, 2000], and has been reported to exist in at least three isoforms (Bim_{EL}, Bim_L and Bim_S), which are generated by alternative splicing [Bouillet P, 2001]. All isoforms bind to all pro-survival Bcl-2 family members [Chen L, 2005] (Figure 3)(Table 1). Studies with gene-targeted mice demonstrated that Bim is critical for embryonic development and hematopoietic cell homeostasis [Bouillet P, 1999]. *Bim*-KO mice show a marked increase in lymphoid and myeloid cell numbers, and disruption of T-cell development [Bouillet P, 2002; Hildeman DA, 2002]. Older mice accumulate excessive plasma cells, and (at least on a mixed C57BL/6x129SV background) many (~60% within a year) succumb to systemic lupus erythematosus-like autoimmune kidney disease [Bouillet P, 1999]. Lymphocytes and granulocytes from Bim-deficient mice are abnormally resistant to many apoptotic stimuli, such as cytokine deprivation, calcium ion flux (ionomycin), microtubule perturbation, or glucocorticosteroids and are partially resistant to etoposide [Bouillet P, 1999; Villunger A, 2003]. Bim is also critical for apoptosis of autoreactive thymocytes [Bouillet P, 2002] and immature B cells [Enders A, 2003] and for the death of activated T cells during shut-down of an immune response [Hildeman D, 2002, Pellegrini M, 2003].

The pro-apoptotic activity of Bim is regulated both transcriptionally and post-translationally (Figure 6). In healthy cells, Akt phosphorylates FoxO3a transcription factor, which in turn localizes to the cytosol. Once dephosphorylated as a result of Akt signaling shutdown, FoxO3a shuttles to the nucleus and acts as a transcription factor for Bim [Gilley J, 2003]. Therefore, chemical inhibitors for the PI3K/Akt pathway effectively enhance *bim* transcription [Stahl M, 2002]. There are three major mechanisms for post-translational Bim regulation. Bim_{EL} and Bim_{L} proteins can bind to LC8, the cytoplasmic dynein light chain, and are thereby sequestered to the microtubule-associated dynein motor complex. Certain apoptotic stimuli, such as the microtubule inhibitor paclitaxel, disrupt interactions between LC8 and the dynein motor complex, and thereby liberate Bim [Puthalakath H, 1999; Puthalakath H, 2002]. The second post-translational regulatory mechanism involves phosphorylation, ubiquitination and proteasomal degradation of Bim. In response to growth factor receptor stimulation, Erk-I/II can phosphorylate Bim_{EL} and $Bim_{L,}$ and thereby targets them for ubiquitination and proteasomal degradation, while inhibition of Erk leads to accumulation of dephosphorylated Bim_{EL} that then is no longer a target for ubiquitination. Therefore, inhibitors of the Ras/Erk pathway and the proteasome are capable of inducing Bim_{EL} and Bim_{L} accumulation [Ley R, 2003] (Figure 6). Additionally, Bim levels have been shown to be increased by Jnk, either via direct phosphorylation or by stimulation of the Jnk-activated transcription factor, Jun [Harris CA, 2001]. Thus, Bim is regulated by a combination of signaling pathways important for cell proliferation and survival in both normal and cancerous cells.

Bim has been shown to function as a tumor suppressor. Bim deficiency was shown to induce follicular lymphoma and to accelerate *myc*-induced lymphomagenesis [Egle A, 2004]. Interestingly, Tagawa and colleagues reported that 17% of cases of mantle cell lymphoma have loss of chromosome 2q13, the region where the human *bim* gene is located [Tagawa H, 2005]. In addition to genetic loss, Bim expression is frequently suppressed by various mechanisms in different cancers. Ras/Raf/Mek/Erk and PI3K/Akt pathways are frequently constitutively activated by various mechanisms. Through abnormal constitutive upregulation of tyrosine kinase activities, such as Her2 in breast cancer or EGFR in non-small lung cancers [Roskoski R Jr., 2004], Bcr/Abl fusion tyrosine kinase in CML [Steelman LS, 2004], in-frame internal tandem duplications in Flt3 receptor in AML [Stirewalt DL, 2003], or NPM/ALK fusion protein by t(2;5)(p23;q35) in anaplastic large-cell lymphoma [Gu TL, 2004], cancer cells acquire constitutive active Ras/Erk and PI3K/Akt that potentially promote proteasomal degradation of Bim and repress *bim* transcription by modulating FoxO3a, respectively. Consequently, Bim expression in primary CML cells is lower than that in normal hematopoietic cells. Treatment with imatinib mesylate, the Bcr/Abl tyrosine kinase inhibitor that is presently the first-line drug treatment for CML, substantially increases the levels of Bim in Bcr/Abl-positive leukemias. Gene knock-down studies using RNA interference technology indicated that Bim is the critical apoptotic initiator in imatinib-induced apoptosis [Aichberger KJ, 2005; Kuribara R, 2004; Kuroda J, 2006]. Recent experiments by our group have, however, demonstrated that Bim is the central, but not the sole BH3-only protein involved in apoptosis initiation by imatinib.

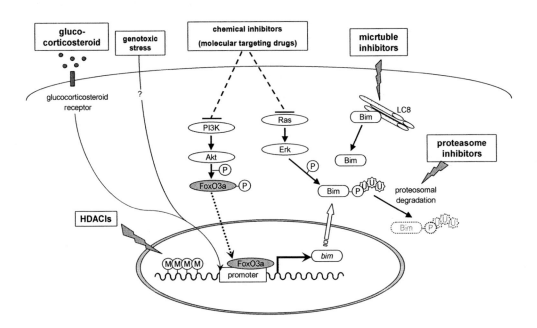

Figure 6. Regulation of Bim expression and chemotherapy. The most extensively characterized transcription factor for *bim* is FoxO3a; its activity is regulated by the PI3K/Akt pathway. The kinase Erk-I/II represses the pro-apoptotic activity of Bim by phosphorylating it, thereby targeting this protein for ubiquitination and proteasomal degradation. Hyper-methylation around the promoter region of the *bim* gene has been identified in specific cancers. Any molecular aberrations involved may be therapeutic targets. M: methylation, P: phosphorylation, U: ubiquitination

Our data show that imatinib could kill Bcr/Abl-transformed *bim*-KO hematopoietic cells, albeit with a delay compared to their wild-type counterparts. We found that imatinib also activated other BH3-only proteins, such as Bmf through transcriptional activation, and Bad through dephosphorylation. Combined loss of Bim and Bad nearly completely protected Bcr/Abl transformed leukemic cells from imatinib-induced apoptosis [Kuroda J, 2006]. These findings suggest that in CML cells, apoptosis induced by imatinib involves concerted interactions among several BH3-only proteins, specifically, Bim, Bad and, possibly Bmf. Cooperation between Bim and Bad was also observed during apoptosis induced by dexamethasone (Dex) plus PI3K inhibitor, LY294002, in a human follicular lymphoma cell line [Nuutinen U, 2006]. In multiple myeloma cells, Bim was shown to be negatively regulated by interleukin-6 signaling [Gomez-Bougie P, 2004], and dissociation of Bim from Mcl-1 is thought to be the prerequisite for melphalan-induced apoptosis [Gomez-Bougie P,

2005]. In cell lines of non-small lung and breast cancers, Bim expression correlated with sensitivity to paclitaxel-induced apoptosis. In this setting, paclitaxel activates Bim by dissociating it from LC8 and also increases the level of Bim via FoxO3a-mediated transcriptional upregulation [Li R, 2005]. Bim was also shown to be transcriptionally increased in colorectal cancer, gastric cancer and osteosarcoma cell lines after treatment with histone deacetylase inhibitors (HDACIs), such as trichostatin A or FK228 (depsipeptide) [Tan J, 2006; Zhao Y, 2005]. A recent report by Piazza and colleagues indicated that the *bim* promoter is methylated in certain leukemias and lymphomas. This methylation is associated with low levels of Bim expression and relative resistance to apoptotic stimuli [Piazza RG, 2005]. HDACI compounds, which can cause activation of silenced genes, may be a suitable approach to target such tumors. Finally, Bim is also thought to play a critical role in apoptosis of hematologic and epithelial neoplasms triggered by proteasome inhibitors, such as bortezomib, epoxomicin or MG132. Proteasome inhibitors cause accumulation of Bim protein by suppressing its proteasomal degradation [Aichberger KJ, 2005; Tan TT, 2005] (Figure 6). However, the role of the Bcl-2 family (both pro-survival and BH3-only proteins, including Bim) in apoptosis induced by proteasome inhibitors remains a subject of controversy. Despite the observed accumulation of Bim as well as other BH3-only proteins, such as Noxa or Bik [Fernandez Y, 2005; Nikrad M, 2005; Perez-Galan P, 2006], apoptosis induced by proteasome inhibitors is not significantly affected by Bcl-2 overexpression or knockdown of Bim [Herrmann JL, 1998; Mitsiades N, 2002; Fahy BN; 2005] (Kuroda J, in preparation). To complicate matters further, several reports support the involvement of the extrinsic pathway that involves caspase-8 activation in proteasome inhibitor-induced apoptosis [Servida F, 2005; Yeung BH, 2006] (Kuroda J, in preparation).

Noxa

Noxa was identified as a p53 responsive stress-induced gene, using the mRNA differential display method. *Noxa* mRNA is expressed at low levels in the brain, thymus, spleen, lung, kidney and testis of normal mice and the levels of Noxa were shown to be increased in several cell types following treatment with p53-dependent apoptotic stimuli [Oda E, 2000] (Figure 3, 7). *Noxa*-KO mice were born at the expected frequency, and became healthy adults. However, fibroblasts of these animals displayed abnormally decreased DNA damage-induced apoptosis [Shibue T, 2003; Villunger A, 2003]. Mutational analysis and expression profiling revealed that *noxa* mutations are rare in gastric, colorectal and lung cancers [Jansson AK, 2003; Lee SH, 2003]. Noxa was initially identified as an apoptosis mediator of the p53 pathway. Subsequently, p53-dependent and -independent activation of Noxa were shown to be critical for apoptosis induced by irinotecan, a topoisomerase-I inhibitor, HDACIs, or bortezomib [Fernandez Y, 2005; Qin JZ, 2004; Qin JZ, 2005; Sasaki T, 2004; Terui T, 2003] (Figure 3). *Noxa* mRNA was shown to increase in response to exposure to proteasome inhibitors and γ-secretase tripeptide, originally used in the treatment of Alzheimer's disease, although the precise mechanism is currently unclear [Qin JZ, 2005]. The γ-secretase tripeptide has the ability to block Notch receptor processing and activation that are thought to be important in the biology of cancer stem cells [Radtke F, 2005]. The relationship between Notch signaling and Noxa is an attractive research topic in cancer biology. Some studies have also demonstrated that *noxa* transcription is upregulated by the p53-related transcription factor p73 [Flinterman M, 2005].

Puma (p53-Upregulated Modulator of Apoptosis, Also Known As Bcl-2-Binding Component 3, Bbc3)

Several groups have identified a gene denoted Puma/Bbc3 either through global profiling of genes expressed downstream of p53 or by yeast 2-hybrid library screening for Bcl-2 interacting proteins. Puma is upregulated through p53 activation due to DNA damage or hypoxia [Nakano K, 2001; Villunger A, 2003; Yu J, 2001] (Figure 3, 7). Loss of Puma was shown to protect fibroblasts and lymphocytes from p53-mediated cell death caused by DNA damage, and, perhaps surprisingly, also from diverse p53-independent cytotoxic insults, including cytokine deprivation and exposure to glucocorticoids, staurosporine, or phorbol ester [Jeffers JR, 2003; Villunger A, 2003]. More recent study clarifies that FoxO3a regulates *puma* transcription under cytokine deprivation [You H, 2006]. The function of Puma as a p53 effector appears to be context-dependent, since Puma short hairpin RNAs (shRNAs) promoted oncogenic transformation of primary mouse fibroblasts, and dramatically accelerated Myc-induced lymphomagenesis, similar to p53 shRNAs, but did not cooperate with oncogenic Ras in transformation, distinct from p53 shRNAs [Hemann MT, 2004]. Low Puma expression is associated with disease progression, poor prognosis and poor response to DNA-damaging anti-cancer drugs in melanoma and CLL [Karst AM, 2005; Mackus WJ, 2005]. In addition to DNA-damaging agents, BMS-214662, a farnesyl transferase inhibitor (FTI) that suppresses the activities of various small GTPases by preventing protein prenylation, was shown to induce apoptosis through a p53/Puma/Bax/Bak/Mcl-1-regulated mechanism [Gomez-Benito M, 2005]. However, a number of reports demonstrated that FTIs trigger apoptosis in a p53-independent fashion [Crespo NC, 2002]. The involvement of the p53-Puma pathway in targeting protein prenylation of small GTPases during apoptosis is a subject of controversy. Accumulation of p53 due to DNA damage also leads to cell cycle arrest by inducing $p21^{Waf1/Cip1}$. The balance between Puma and $p21^{Waf1/Cip1}$ may therefore be pivotal in determining the cellular response (cell cycle arrest or cell death) to p53 activation. The mechanism by which either of the two pathways becomes dominant after DNA damage remains to be elucidated (Figure 7).

Bmf (Bcl-2 Modifying Factor)

Puthalakath and colleagues cloned *bmf*, and showed that the protein Bmf is sequestered into myosin V motors by association with dynein light chain-2. Specific damage signals, such as anoikis (loss of cellular attachment), were shown to trigger the translocation of Bmf and its binding to Bcl-2 and Bcl-X_L [Puthalakath H, 2001]. The *bmf* gene is located on chromosome 15q14, the site of a candidate tumor suppressor lost in many metastatic, but not primary carcinomas [Wick W, 1996]. Bmf is expressed in three isoforms, namely, Bmf-I, Bmf-II and

Bmf-III. Among these, only Bmf-I was shown to display pro-apoptotic activity [Morales AA, 2004]. *Bmf-I* is trancriptionally increased by serum starvation in leukemia or lymphoma cell lines, or treatment with imatinib in CML-derived cell lines [Kuroda J, in print]. FK228 HDACI acetylates histone H3 and H4 within the *bmf* promoter region, and requires Bmf, at least in part, to induce apoptosis [Zhang Y, 2006].

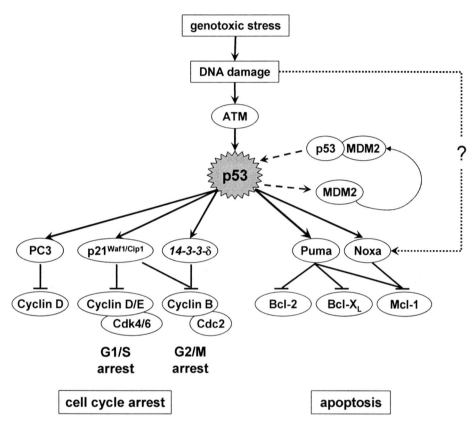

Figure 7. Simplified scheme of p53-mediated signaling cascades for cell cycle arrest and apoptosis. Genotoxic stress induces p53 accumulation that results in the activation of complex molecular cascades for cell cycle regulation, cell death or DNA repair. p21$^{Waf1/Cip1}$ is critical for p53-mediated cell cycle arrest and Puma is essential for p53-mediated apoptosis induction, respectively.

Other BH3-Only Proteins

Recent evidence shows that proteasomal degradation of Mcl-1 is required in an early stage of DNA damage-induced apoptosis [Cuconati A, 2003]. Mule/Arf-BP1 (Mule) is a unique BH3 domain-containing E3 ubiquitin ligase that polyubiquitinates Mcl-1 in a manner that requires its BH3-domain, and was reported to be critical for apoptosis induced by DNA-damaging stimuli [Zhong Q, 2005]. Spike is a unique BH3-only protein that does not translocate to mitochondria, but to endoplasmic reticulum. Reduced expression of Spike was observed in several epithelial carcinomas [Mund T, 2003].

PHARMACOLOGIC MANIPULATION
OF THE BCL-2 PROTEIN FAMILY

Given the critical roles of interactions between pro-survival and pro-apoptotic Bcl-2 family members in tumor development, disease progression and response to chemotherapy or γ-radiation, development of anti-cancer strategies through the manipulation of these proteins is an attractive goal. There are three approaches to manipulating Bcl-2 proteins for anti-cancer treatment. These involve the use of existing chemotherapeutic drugs that effectively modulate the levels or activity of Bcl-2 family members, therapeutic oligonucleotides to modulate expression of these proteins and synthetic peptides or small molecular compounds that mimic BH3-only proteins. The first strategy has been conventionally applied with or without intention in combination chemotherapy, as most cytotoxic drugs utilize BH3-only proteins for apoptosis initiation. One may expect that a combination of drugs that act to increase the levels of the same BH3-only protein have a combined/ synergistic effect to further enhance the expression of specific BH3-only protein and lead to enhanced apoptosis. However, this is not always the case. For example, both INNO-406 (formerly NS-187), a novel dual Bcr/Abl/Lyn tyrosine kinase inhibitor [Kimura S, 2005], and 17-(allylamino) -17-demethoxygeldamycin (17-AAG), a heat shock protein-90 inhibitor, trigger apoptosis in CML cells via Bim induction. However, a combination of these drugs does not increase either the Bim level or the number of apoptotic cells (Kuroda J, in preparation). Another example is combination chemotherapy, using several DNA-damaging agents for treatment of cancers with wild-type p53. If the mixture of DNA-damaging agents further increases Puma and/or Noxa levels via p53, conventional chemotherapy should always show at least additional apoptosis-inducing effects. However, this is not always the case. These results suggest boundaries in the levels of BH3-only proteins that can be induced by chemotherapeutic drugs.

Gene Therapy Targeting the Bcl-2 Family

Clinical trials of an antisense oligonucleotide (ODNs) targeting *bcl-2* mRNA, Oblimersen (Genasense, Genta Inc., Berkeley Heights, NJ), have been performed for lymphomas, AML, melanoma, multiple myeloma and CLL [Chanan-Khan A, 2004; Piro LD, 2004]. Oblimersen was potent in sensitizing cancer /leukemic cells to apoptosis by anti-cancer drugs both in studies *in vitro* and in preclinical models [Klasa RJ, 2002; Piro LD, 2004; Ramanarayanan J, 2004]. Phase I/II clinical trials for chemo-refractory B-cell malignancies and adult AML confirmed the bioactivity of this drug. However, a Phase III study on a combination of this compound with dacarbazine for melanoma and with Dex for multiple myeloma failed to meet the primary end-point of prolonging patient survival [Stein CA, 2005] (Table 2). It is possible that resistance to apoptosis by chemotherapy could not be overcome merely by Bcl-2 reduction in advanced refractory cancers that acquire a variety of other genetic mutations. Moreover, the anti-tumor activity of Oblimersen is not based solely on its ODN effect, but involves additional immunomodulatory effects of its CpG motifs triggering an inflammatory response through Toll-like receptor-9 [Gekeler V, 2006]. More recently, SPC2996, a new class of RNA antagonist targeting *bcl-2* mRNA with the locked nucleic acid modification (Santaris Pharma A/S, Horsholm, Denmark), was approved by FDA (U.S. Food and Drug Administration) for Phase I/II clinical trials for CLL.

Table 2. Therapeutic Inhibitors of pro-survival Bcl-2 proteins

Class	Compound	Developer	Status
Antisense ODNs	Oblimersen	Genta Inc.	Phase III in AML Failed in primary endpoint of Phase III studies in melanoma and myeloma
	SPC2996	Santa Pharma	Phase I in CLL
Small Molecule Compound	AT-101 (Gossypol)	Ascenta Therapeutics	Phase II in CLL, Prostate cancer, Lymphomas
	GX15-070	Gemin X Biotechnologies	Phase I/II in CLL
	ABT-737	Abbott Laboratories	Preclinical
	Antimycin A3	University of Washington	Preclinical
	Apogossypol	The Burnham Institute	Preclinical
	BH3I-1	Harvard University	Preclinical
	BH3I-2	Harvard University	Preclinical
	HA14-1	Raylight Chemokine Pharmaceuticals	Preclinical

Peptide-Based BH3 Mimetics

Nuclear magnetic resonance (NMR) structural analysis of the complex forms of Bcl-X$_L$ and Bad or Bak reveal that BH1, BH2 and BH3 domains of Bcl-X$_L$ form a hydrophobic groove on the molecular surface that envelopes the amphipathic α-helical BH3 domain of the BH3-only proteins [Muchmore SW, 1996; Petros AM, 2004]. Numerous studies have shown that synthetic peptides encoding amino acid sequences of the α-helical BH3 domain of BH3-only proteins bind pro-survival Bcl-2 proteins and can sensitize cells to apoptosis-inducing stimuli. Several attempts have been made to increase the cellular permeability and direct the peptide to the membrane. Treatment with a chimeric Bad peptide with fatty acids prolonged the survival of leukemia-xenografted mice [Wang JL, 2000]. Another approach to make BH3 peptides more accessible to cells is to generate a chimeric protein with cell surface proteins, such as human GM-CSF [Antignani A, 2005]. Walensky *et al.* developed the hydrocarbon stapling technique using α,α-disubstituted non-natural amino acids containing olefin-bearing tethers, and succeeded in improving the biologic activity of the Bid BH3 peptide by stabilizing its α-helical conformation, conferring protease resistance and increased cell permeability [Walensky LD, 2004]. The use of lactam bridges has also been reported to stabilize the α-helical conformation [Yang B, 2004].

Small Molecule BH3 Mimetics

Several groups have reported small molecule inhibitors that directly bind pro-survival Bcl-2 proteins, using structural-based design and high-throughput screening. Gossypol, a compound derived from cottonseeds, was originally used as a herbal medicine in China.

Subsequent studies showed that Gossypol inhibits Bcl-2, Bcl-X$_L$, Bcl-B and Bfl-1 [Reed JC, 2005]. Apogossypol (The Burham Institute, La Jolla, CA), a less toxic version of Gossypol, is currently under preclinical study [Becattini B, 2004]. Epigallocatechins (EGCGs) are also derived from natural products, green or black tea polyphenol. A part of the anticancer effect of EGCGs may be explained by their inhibitory effects on Bcl-2 [Leone M, 2003]. BH3I-1 and BH3I-2, a thiazolidin derivative and benzene sulfonyl derivative, respectively, are cell-permeable BH3 domain analogs that block Bcl-X$_L$ [Feng WY, 2003; Roa W, 2005]. Antimycin A3 was reported to mimic the BH3 peptide and to inhibit Bcl-X$_L$ and Bcl-2 [Tzung SP, 2001]. HA14-1, a small organic molecule, is thought to be a non-peptidic ligand of the Bcl-2 surface hydrophobic groove [Milanesi E, 2006]. In addition to its direct pro-apoptotic effect, HA14-1 is highly potent in enhancing the cell killing effects of a variety of other anti-cancer agents, including trastuzumab (Herceptin; humanized EGFR-2 antibody) in breast cancer cells [Milella M, 2004], bortezomib or flavopiridol (cyclin-dependent kinase inhibitor) in multiple myeloma [Pei XY, 2003; Pei XY, 2004], all-*trans*-retinoic acid in neuroblastoma [Niizuma H, 2006], and death-receptor mediated apoptosis by TRAIL [Hao JH, 2004; Sinicrope FA, 2004]. ABT-737 (Abbott Laboratories, Abbott Park, IL) and GX15-070 (Gemin X Biotechnologies, Montreal, Canada) are synthetic small-molecule inhibitors produced by NMR-guided structure-based drug design. The former is active against Bcl-2, Bcl-w and Bcl-X$_L$ [Oltersdorf T, 2005], while the latter is a broad inhibitor of several pro-survival Bcl-2 proteins [Beauparlant P, 2003; Reed JC, 2005]. ABT-737 is a potent apoptosis inducer in primary B-cell malignant cells and enhances the killing effect of paclitaxel in preclinical models of non-small lung cancer [Oltersdolf T, 2005]. GX15-070 is currently under study in a Phase I/II clinical trial for the treatment of CLL and a separate Phase I trial for solid tumors. Phase I clinical data demonstrated that GX15-070 is well tolerated and generates biological and clinical improvements in CLL, as evident from the increase in histone-oligonucleosomal DNA levels (a hallmark of apoptosis) 1-6 h after infusion, and an average reduction of 29% in peripheral lymphocyte counts in 18/25 patients [Gemin X, 2005 News]. To our knowledge, no specific inhibitor for Mcl-1 or A1 has been reported to date.

Conclusions

Here, we review emerging evidence on the significance of apoptosis regulation, focusing on the roles of BH3-only proteins in tumor development and cancer therapy. According to recent reports on molecular therapeutics for cancers, the concurrent blockade of several signaling molecules appears to be necessary for tumor eradication [Komarova NL, 2005], and the concurrent inhibition of signalings responsible for cancer cell proliferation and survival seems to be rational. We here show that approaches targeting the molecular network of the Bcl-2 protein family are rational and promising strategies for cancer therapy. Based on the knowledge on the regulation of the Bcl-2 family, several drugs targeting these proteins have already been put forward for analysis in clinical trials. However, considerable more work is required to effectively improve pharmacologic properties, optimize formulations, and decrease toxicities for further advancement of anti-cancer therapy.

Acknowledgements

We are grateful to Professors Andreas Strasser (Molecular Genetics of Cancer Division, The Walter and Eliza Hall Institute of Medical Research, Parkville, Victoria, Australia) and Taira Maekawa (Department of Transfusion Medicine and Cell Therapy, Kyoto University Hospital, Kyoto, Japan) for scientific advice and critical review of the article.

REFERENCES

[1]   Adams, JM (2003). Ways of dying: multiple pathways to apoptosis. *Genes Dev, 17,* 2481-2495.

[2]   Addeo, R; Caraglia, M; Baldi, A; D'Angelo Casale, F; Crisci, S; Abbruzzese, A; Vincenze, B; Campioni, M; Di Tullio, MT; Indolfi, P (2005). Prognostic role of bcl-xL and p53 in childhood acute lymphoblastic leukemia (ALL). *Cancer Biol Ther, 4,* 32-38.

[3]   Aichberger, KJ; Mayerhofer, M; Krauth, MT; Vales, A; Kondo, R; Derdak, S; Pickl, WF; Selzer, E; Deininger, M; Druker, BJ; Sillaber, C; Esterbauer, H; Valent, P (2005). Low-level expression of proapoptotic Bcl-2-interacting mediator in leukemic cells in patients with chronic myeloid leukemia: role of BCR/ABL, characterization of underlying signaling pathways, and reexpression by novel pharmacologic compounds. *Cancer Res, 65,* 9436-9444.

[4]   Amundson, SA; Myers, TG; Scudiero, D; Kitada, S; Reed, JC; Fornace, AJ Jr. (2000). An informatics approach identifying markers of chemosensitivity in human cancer cell lines. *Cancer Res, 60,* 6101-6110.

[5]   Antignani, A; Youle, RJ (2005). A chimeric protein induces tumor cell apoptosis by delivering the human Bcl-2 family BH3-only protein Bad. *Biochemistry, 44,* 4074-4082.

[6]   Arena, V; Martini, M; Luongo, M; Capelli, A; Larocca, LM (2003). Mutations of the BIK gene in human peripheral B-cell lymphomas. *Genes Chromosomes Cancer, 38,* 91-96.

[7]   Beauparlant, P; Shore, GC (2003). Therapeutic activation of caspases in cancer: a question of selectivity. *Curr Opin Drug Discov Devel, 6,* 179-187.

[8]   Becattini, B; Kitada, S; Leone, M; Monosov, E; Chandler, S; Zhai, D; Kipps, TJ; Reed, JC; Pellecchia M (2004). Rational design and real time, in-cell detection of the proapoptotic activity of a novel compound targeting Bcl-X(L). *Chem Biol, 11,* 389-395.

[9]   Berrieman, HK; Smith, L; O'Kane, SL; Campbell, A; Lind, MJ; Cawkwell, L (2005). The expression of Bcl-2 family proteins differs between nonsmall cell lung carcinoma subtypes. *Cancer, 103,* 1415-1419.

[10]   Borner, C; Schlagbauer Wadl, H; Fellay, I; Selzer, E; Polterauer P; Jansen, B (1999). Mutated N-ras upregulates Bcl-2 in human melanoma in vitro and in SCID mice. *Melanoma Res, 9,* 347-350.

[11]   Bouillet, P; Strasser, A (2002). BH3-only proteins - evolutionarily conserved proapoptotic Bcl-2 family members essential for initiating programmed cell death. *J Cell Sci, 115,* 1567-1574.

[12] Bouillet, P; Zhang, LC; Huang, DC; Webb, GC; Bottema, CD; Shore, P; Eyre, HJ; Sutherland, GR; Adams, JM (2001). Gene structure alternative splicing, and chromosomal localization of pro-apoptotic Bcl-2 relative Bim. *Mamm Genome, 12,* 163-168.

[13] Bouillet, P; Metcalf D; Huang, DC; Tarlinton, DM; Kay, TW; Kontgen, F; Adams, JM; Strasser, A (1999). Proapoptotic Bcl-2 relative Bim required for certain apoptotic responses, leukocyte homeostasis, and to preclude autoimmunity. *Science, 286,*1735-1738.

[14] Bouillet, P; Purton JF; Godfrey, DI; Zhang, LC; Coultas L; Puthalakath, H; Pellegrini M; Cory, S; Adams, JM; Strasser, A (2002). BH3-only Bcl-2 family member Bim is required for apoptosis of autoreactive thymocytes. *Nature, 415,* 922-926.

[15] Boyd, JM; Gallo, GJ; Elangovan, B; Houghton, AB; Malstrom, S; Avery, BJ; Ebb, RG; Subramanian, T; Chittenden T; Lutz, RJ; Chinnadurai, G (1995). Bik, a novel death-inducing protein shares a distinct sequence motif with Bcl-2 family proteins and interacts with viral and cellular survival-promoting proteins. *Oncogene, 11,* 1921-1928.

[16] Brown, JM; Attardi, LD (2005). The role of apoptosis in cancer development and treatment response. *Nat Rev Cancer, 5,* 231-237.

[17] Brunner T; Mueller, C (2003). Apoptosis in disease: about shortage and excess. *Essays Biochem, 39,* 119-130.

[18] Carmeliet, P; Dor, Y; Herbert, JM; Fukumura, D; Brusselmans, K; Dewerchin, M; Neeman M; Bono, F; Abramovitch, R; Maxwell, P; Koch, CJ; Ratcliffe, P; Moons, L; Jain, RK; Collen, D; Keshert, E (1998). Role of HIF-1alpha in hypoxia-mediated apoptosis, cell proliferation and tumour angiogenesis. *Nature, 394,* 485-490.

[19] Catz, SD; Johnson, JL (2001). Transcriptional regulation of bcl-2 by nuclear factor kappa B and its significance in prostate cancer. *Oncogene, 20,* 7342-7351.

[20] Certo, M; Del Gaizo Moore, V; Nishino, M; Wei, G; Korsmeyer, S; Armstrong, SA; Letai, A (2006) Mitochondria primed by death signal determine cellular addiction to antiapoptotic BCL-2 family members. *Cancer Cell, 9,* 351-365.

[21] Chanan-Khan, A (2004). Bcl-2 antisense therapy in hematologic malignancies. *Curr Opin Oncol, 16,* 581-585.

[22] Chen, L; Willis, SN; Wei A; Smith, BJ; Fletcher, JI; Hinds, MG; Colman, PM; Day, CL; Adams, JM; Huang, DC (2005). Differential targeting of prosurvival Bcl-2 proteins by their BH3-only ligands allows complementary apoptotic function. *Mol Cell, 17,* 393-403.

[23] Cheng, EHY; Sheiko, TV; Fisher, JK; Craigen, WJ; Korsmeyer, SJ (2003). VDAC2 inhibits BAK activation and mitochondrial apoptosis. *Science, 301,* 513-517.

[24] Chida, D; Miura, O; Yoshimura, A; Miyajima, A (1999). Role of cytokine signaling molecules in erythroid differentiation of mouse fetal liver hematopoietic cells: functional analysis of signaling molecules by retrovirus-mediated expression. *Blood,* 93, 1567-1578.

[25] Chipuk, JE; Green, DR (2005). Do inducers of apoptosis trigger caspase-independent cell death? *Nat Rev Mol Cell Biol, 6,* 268-275.

[26] Choi, SS; Park, IC; Yun, JW; Sung, YC; Hong SI; Shin HS (1995). A novel Bcl-2 related gene, Bfl-1, is overexpressed in stomach cancer and preferentially expressed in bone marrow. *Oncogene, 11,* 1693-1698.

[27] Cho-Vega, JH; Rassidakis GZ; Admirand, JH; Oyarzo, M; Ramalingam P; Paraguya, A; McDonnell, TJ; Amin, HM; Medeiros, LJ (2004). MCL-1 expression in B-cell non-Hodgkin's lymphomas. *Hum Pathol, 35,* 1095-1100.

[28] Cory, S; Huang, DC; Adams, JM (2003). The Bcl-2 family: roles in cell survival and oncogenesis. *Oncogene, 22,* 8590-8607.

[29] Cory, S; Adams, JM (2002). The Bcl2 family: regulators of the cellular life-or-death switch. *Nat Rev Cancer, 2,* 647-656.

[30] Coultas, L; Bouillet P; Stanley EG; Brodnicki TC; Adams JM; Strasser, A (2004). Proapoptotic BH3-only Bcl-2 family member Bik/Blk/Nbk is expressed in hemopoietic and endothelial cells but is redundant for their programmed death. *Mol Cell Biol, 24,* 1570-1581.

[31] Coultas, L; Bouillet, P; Loveland, KL; Meachem, S; Perlman, H; Adams, JM; Strasser, A (2005). Concomitant loss of proapoptotic BH3-only Bcl-2 antagonists Bik and Bim arrests spermatogenesis. *EMBO J, 24,* 3963-3973.

[32] Crespo, NC; Delarue F; Ohkanda J; Carrico, D; Hamilton AD; Sebti, SM (2002). The farnesyltransferase inhibitor, FTI-2153, inhibits bipolar spindle formation during mitosis independently of transformation and Ras and p53 mutation status. *Cell Death Differ, 9,* 702-709.

[33] Cuconati, A; Mukherjee, C; Perez, D; White, E (2003). DNA damage response and MCL-1 destruction initiate apoptosis in adenovirus-infected cells. *Genes Dev, 17,* 2922-2932.

[34] Danial, NN; Korsmeyer, SJ (2004). Cell death: critical control points. *Cell, 116,* 205-219.

[35] Daniel, PT; Pun, KT; Ritschel, S; Sturm, I; Holler, J; Dorken, B; Brown, R (1999). Expression of the death gene Bik/Nbk promotes sensitivity to drug-induced apoptosis in corticosteroid-resistant T-cell lymphoma and prevents tumor growth in severe combined immunodeficient mice. *Blood, 94,* 1100-1107.

[36] DeoCampo, ND; Wilson, MR; Trosko, JE (2000). Cooperation of bcl-2 and myc in the neoplastic transformation of normal rat liver epithelial cells is related to the down-regulation of gap junction-mediated intercellular communication. *Carcinogenesis, 21,* 1501-1506.

[37] Desagher, S; Osen-Sand A; Montessuit, S; Magnenat, E; Vilbois, F; Hochmann, A; Journot, L; Antonsson, B; Martinou, JC (2001). Phosphorylation of bid by casein kinases I and II regulates its cleavage by caspase 8. *Mol Cell, 8,* 601-611.

[38] de Jong, D (2005). Molecular pathogenesis of follicular lymphoma: a cross talk of genetic and immunologic factors. *J Clin Oncol, 23,* 6358-6363.

[39] D'Sa-Eipper, C; Subramanian, T; Chinnadurai, G (1996). bfl-1, a bcl-2 homologue, suppresses p53-induced apoptosis and exhibits potent cooperative transforming activity. *Cancer Res, 56,* 3879-3882.

[40] Du, C; Fang, M; Li, Y; Li, L; Wang, X (2000). Smac, a mitochondrial protein that promotes cytochrome c-dependent caspase activation by eliminating IAP inhibition. *Cell, 102,* 33-42.

[41] Egle, A; Harris, AW; Bouillet, P; Cory, S (2004). Bim is a suppressor of Myc-induced mouse B cell leukemia. *Proc Natl Acad Sci USA, 101,* 6164-6169.

[42] Eischen, CM; Roussel, MF; Korsmeyer, SJ; Cleveland, JL (2001). Bax loss impairs Myc-induced apoptosis and circumvents the selection of p53 mutations during Myc-mediated lymphomagenesis. *Mol Cell Biol, 21,* 7653-7662.

[43] Enders, A; Bouillet, P; Puthalakath, H; Xu, Y; Tarlinton, DM; Strasser A (2003). Loss of the pro-apoptotic BH3-only Bcl-2 family member Bim inhibits BCR stimulation-induced apoptosis and deletion of autoreactive B cells. *J Exp Med, 198,* 1119-1126.

[44] Erovic, BM; Pelzmann, M; Grasl MCH; Pammer, J; Kornek, G; Brannath, W; Selzer, E; Thurnher, D (2005). Mcl-1, vascular endothelial growth factor-R2, and 14-3-3sigma expression might predict primary response against radiotherapy and chemotherapy in patients with locally advanced squamous cell carcinomas of the head and neck. *Clin Cancer Res, 11,* 8632-8636.

[45] Fahy, BN; Schlieman, MG; Mortenson, MM; Virudachalam, S; Bold, RJ (2005). Targeting BCL-2 overexpression in various human malignancies through NF-kappaB inhibition by the proteasome inhibitor bortezomib. *Cancer Chemother Pharmacol, 56,* 46-54.

[46] Fandy, TE; Shankar, S; Ross, DD; Sausville, E; Srivastava, RK (2005). Interactive effects of HDAC inhibitors and TRAIL on apoptosis are associated with changes in mitochondrial functions and expressions of cell cycle regulatory genes in multiple myeloma. *Neoplasia, 7,* 646-657.

[47] Fang, X; Yu, S; Eder, A; Mao, M: Bast, RC Jr; Boyd, D; Mills, GB (1999). Regulation of BAD phosphorylation at serine 112 by the Ras-mitogen-activated protein kinase pathway. *Oncogene, 18,* 6635-6640.

[48] Feng, WY; Liu, FT; Patwari, Y; Agrawal, SG; Newland, AC; Jia, L (2003). BH3-domain mimetic compound BH3I-2' induces rapid damage to the inner mitochondrial membrane prior to the cytochrome c release from mitochondria. *Br J Haematol, 121,* 332-340.

[49] Fernandez, Y; Verhaegen, M; Miller, TP; Rush, JL; Steiner, P; Opipari AW Jr; Lowe, SW; Soengas, MS (2005). Differential regulation of noxa in normal melanocytes and melanoma cells by proteasome inhibition: therapeutic implications. *Cancer Res, 65,* 6294-6304.

[50] Fleischer, B; Schulze-Bergkamen, H; Schuchmann, M; Weber, A; Biesterfeld, S; Muller, M; Krammer, PH; Galle, PR (2006). Mcl-1 is an anti-apoptotic factor for human hepatocellular carcinoma. *Int J Oncol, 28,* 25-32.

[51] Flinterman, M; Guelen, L; Ezzati-Nik, S; Killick, R; Melino, G; Tominaga, K; Mymryk, JS; Gaken, J; Tavassoli, M (2005). E1A activates transcription of p73 and Noxa to induce apoptosis. *J Biol Chem, 280,* 5945-5959.

[52] Gekeler, V; Gimmnich, P; Hofmann, HP; Grebe, C; Rommele, M; Leja, A; Baudler, M; Benimetskaya, L; Gonser, B; Pieles, U; Maier, T; Wagner, T; Sanders, K; Beck, JF; Hanauer, G; Stein, CA (2006). G3139 and Other CpG-Containing Immunostimulatory

Phosphorothioate Oligodeoxynucleotides Are Potent Suppressors of the Growth of Human Tumor Xenografts in Nude Mice. *Oligonucleotides, 16,* 83-93.

[53] Gemin X. News. 2005: Available from: http://www.geminx.com/en/news/20051212-clin.php

[54] Gerl, R; Vaux, DL (2005). Apoptosis in the development and treatment of cancer. *Carcinogenesis, 26,* 263-270.

[55] Gilley, J; Coffer, PJ; Ham, J (2003). FOXO transcription factors directly activate bim gene expression and promote apoptosis in sympathetic neurons. *J Cell Biol, 162,* 613-622.

[56] Gomez-Benito, M; Marzo, I; Anel, A; Naval, J (2005). Farnesyltransferase inhibitor BMS-214662 induces apoptosis in myeloma cells through PUMA up-regulation, Bax and Bak activation, and Mcl-1 elimination. *Mol Pharmacol, 67,*1991-1998.

[57] Gomez-Bougie, P; Bataille, R; Amiot, M (2004). The imbalance between Bim and Mcl-1 expression controls the survival of human myeloma cells. *Eur J Immunol, 34,* 3156-3164.

[58] Gomez-Bougie, P; Oliver, L; Le Gouill, S; Bataille, R; Amiot, M (2005). Melphalan-induced apoptosis in multiple myeloma cells is associated with a cleavage of Mcl-1 and Bim and a decrease in the Mcl-1/Bim complex. *Oncogene, 24,* 8076-8079.

[59] Grad, JM; Zeng, XR; Boise LH (2000). Regulation of Bcl-xL: a little bit of this and a little bit of STAT. *Curr Opin Oncol, 12,* 543-549.

[60] Grayson, WL; Zhao, F; Izadpanah, R; Bunnell, B; Ma, T (2006). Effects of hypoxia on human mesenchymal stem cell expansion and plasticity in 3D constructs. *J Cell Physiol, 207,* 331-339.

[61] Green, DR; Evan, GI (2002). A matter of life and death. *Cancer Cell, 1,* 19-30.

[62] Green DR, Kroemer G (2004). The pathophysiology of mitochondrial cell death. *Science, 305,* 626-629.

[63] Groeger, AM; Esposito, V; De Luca, A; Cassandro, R; Tonini, G; Ambrogi, V; Baldi, F; Goldfarb, R; Mineo, TC; Baldi, A; Wolner, E (2004). Prognostic value of immunohistochemical expression of p53, bax, Bcl-2 and Bcl-xL in resected non-small-cell lung cancers. *Histopathology, 44,* 54-63.

[64] Gu, TL; Tothova, Z; Scheijen, B; Griffin, JD; Gilliland, DG; Sternberg, DW (2004). NPM-ALK fusion kinase of anaplastic large-cell lymphoma regulates survival and proliferative signaling through modulation of FOXO3a. *Blood, 103,* 4622-9.

[65] Guo, B; Zhai, D; Cabezas, E; Welsh, K; Nouraini, S; Satterthwait, AC; Reed, JC (2003). Humanin peptide suppresses apoptosis by interfering with Bax activation. *Nature, 423,* 456-461.

[66] Han, J; Modha, D; White, E (1998). Interaction of E1B 19K with Bax is required to block Bax-induced loss of mitochondrial membrane potential and apoptosis. *Oncogene, 17,* 2993-3005.

[67] Hao, JH; Yu, M; Liu, FT; Newland, AC; Jia, L (2004). Bcl-2 inhibitors sensitize tumor necrosis factor-related apoptosis-inducing ligand-induced apoptosis by uncoupling of mitochondrial respiration in human leukemic CEM cells. *Cancer Res, 64,* 3607-3616.

[68] Harada, T; Ogura, S; Yamazaki, K; Kinoshita, I; Itoh, T; Isobe, H; Yamashiro, K; Dosaka-Akita, H; Nishimura, M (2003). Predictive value of expression of P53, Bcl-2

and lung resistance-related protein for response to chemotherapy in non-small cell lung cancers. *Cancer Sci, 94*, 394-399.

[69] Harris, CA; Johnson, EM Jr. (2001) BH3-only Bcl-2 family members are coordinately regulated by the JNK pathway and require Bax to induce apoptosis in neurons. *J Biol Chem, 276*, 37754-37760.

[70] Heckman, CA; Mehew, JW; Boxer, LM (2002). NF-kappaB activates Bcl-2 expression in t(14;18) lymphoma cells. *Oncogene, 21*, 3898-3908.

[71] Heckman, CA; Duan H; Garcia, PB; Boxer, LM (2006). Oct transcription factors mediate t(14;18) lymphoma cell survival by directly regulating bcl-2 expression. *Oncogene, 25*, 888-98.

[72] Hemann, MT; Bric, A; Teruya-Feldstein, J; Herbst, A; Nilsson, JA; Cordon-Cardo, C; Cleveland, JL; Tansey, WP; Lowe, SW (2005). Evasion of the p53 tumour surveillance network by tumour-derived MYC mutants. *Nature, 436*, 807-811.

[73] Hemann, MT; Zilfou, JT; Zhao, Z; Burgess, DJ; Hannon, GJ; Lowe, SW (2004). Suppression of tumorigenesis by the p53 target PUMA. *Proc Natl Acad Sci USA, 101*, 9333-9338.

[74] Herrmann, JL; Briones, F Jr.; Brisbay, S; Logothetis, CJ; McDonnell, TJ (1998). Prostate carcinoma cell death resulting from inhibition of proteasome activity is independent of functional Bcl-2 and p53. *Oncogene, 17*, 2889-2899.

[75] Hildeman, DA; Zhu, Y; Mitchell, TC; Bouillet, P; Strasser, A; Kappler, J; Marrack, P (2002). Activated T cell death in vivo mediated by proapoptotic bcl-2 family member bim. *Immunity, 16*, 759-67.

[76] Hirata, H; Takahashi, A; Kobayashi, S; Yonehara, S; Sawai, H; Okazaki, T; Yamamoto, K; Sasada, M (1998). Caspases are activated in a branched protease cascade and control distinct downstream processes in Fas-induced apoptosis. *J Exp Med, 187*, 587-600.

[77] Hsu, SY; Kaipia, A; McGee, E; Lomeli, M; Hsueh, AJ (1997). Bok is a pro-apoptotic Bcl-2 protein with restricted expression in reproductive tissues and heterodimerizes with selective anti-apoptotic Bcl-2 family members. *Proc Natl Acad Sci USA, 94*,12401-12406.

[78] Huang, DC; Strasser, A (2000). BH3-Only proteins-essential initiators of apoptotic cell death. *Cell, 103*, 839-842.

[79] Hughes, P; Bouillet, P; Strasser, A (2006). Role of Bim and other Bcl-2 family members in autoimmune and degenerative diseases. *Curr Dir Autoimmun, 9*, 74-94.

[80] Huntly, BJ; Gilliland, DG (2005). Leukaemia stem cells and the evolution of cancer-stem-cell research. *Nat Rev Cancer, 5*, 311-321.

[81] Hur, J; Chesnes, J; Coser, KR; Lee, RS; Geck P; Isselbacher, KJ; Shioda, T (2004). The Bik BH3-only protein is induced in estrogen-starved and antiestrogen-exposed breast cancer cells and provokes apoptosis. *Proc Natl Acad Sci USA, 101*, 2351-2356.

[82] Imaizumi, K; Tsuda, M; Imai, Y; Wanaka, A; Takagi, T; Tohyama, M (1997). Molecular cloning of a novel polypeptide, DP5, induced during programmed neuronal death. *J Biol Chem, 272*, 18842-18848.

[83]   Inohara, N; Ding, L; Chen, S; Nunez, G (1997). harakiri, a novel regulator of cell death, encodes a protein that activates apoptosis and interacts selectively with survival-promoting proteins Bcl-2 and Bcl-X(L). *EMBO J, 16*, 1686-1694.

[84]   Jager, R; Herzer, U; Schenkel, J; Weiher, H (1997). Overexpression of Bcl-2 inhibits alveolar cell apoptosis during involution and accelerates c-myc-induced tumorigenesis of the mammary gland in transgenic mice. *Oncogene, 15,* 1787-1795.

[85]   Jaiswal, S; Traver, D; Miyamoto, T; Akashi, K; Lagasse, E; Weissman, IL (2003). Expression of BCR/ABL and BCL-2 in myeloid progenitors leads to myeloid leukemias. *Proc Natl Acad Sci USA, 100,* 10002-10007.

[86]   Jansson, AK; Emterling, AM; Arbman, G; Sun, XF (2003). Noxa in colorectal cancer: a study on DNA, mRNA and protein expression. *Oncogene, 22,* 4675-4678.

[87]   Jeffers, JR; Parganas, E; Lee, Y; Yang, C; Wang, J; Brennan, J; MacLean, KH; Han, J; Chittenden, T; Ihle, JN; McKinnon, PJ; Cleveland, JL; Zambetti, GP (2003). Puma is an essential mediator of p53-dependent and -independent apoptotic pathways. *Cancer Cell,* 4, 321-328.

[88]   Jin, Z; Gao, F; Flagg, T; Deng, X (2004). Nicotine induces multi-site phosphorylation of Bad in association with suppression of apoptosis. *J Biol Chem, 279,* 23837-23844.

[89]   Karst, AM; Dai, DL; Martinka, M; Li, G (2005). PUMA expression is significantly reduced in human cutaneous melanomas. *Oncogene, 24,* 1111-1116.

[90]   Kasibhatla, S (2003), Tseng B. Why target apoptosis in cancer treatment? *Mol Cancer Ther, 2,* 573-580.

[91]   Kerbauy, DM; Lesnikov, V; Abbasi, N; Seal, S; Scott, B; Deeg, HJ (2005). NF-kappaB and FLIP in arsenic trioxide (ATO)-induced apoptosis in myelodysplastic syndromes (MDSs). *Blood, 106,* 3917-3925.

[92]   Kerr, JF; Wyllie, AH; Currie, AR (1972). Apoptosis: a basic biological phenomenon with wide-ranging implications in tissue kinetics. *Br J Cancer, 26,* 239-257.

[93]   Kim, JK; Kim, KD; Lee, E; Lim, JS; Cho, HJ; Yoon, HK; Cho, MY; Baek, KE; Park, YP; Paik, SG; Choe, YK; Lee, HG (2004). Up-regulation of Bfl-1/A1 via NF-kappaB activation in cisplatin-resistant human bladder cancer cell line. *Cancer Lett, 212,* 61-70.

[94]   Kimura, S; Naito, H; Segawa, H; Kuroda, J; Yuasa, T; Sato, K; Yokota, A; Kamitsuji, Y; Kawata, E; Ashihara, E; Nakaya, Y; Naruoka, H; Wakayama, T; Nasu, K; Asaki, T; Niwa, T; Hirabayashi, K; Maekawa, T (2005). NS-187, a potent and selective dual Bcr-Abl/Lyn tyrosine kinase inhibitor, is a novel agent for imatinib-resistant leukemia. *Blood, 106,* 3948-3954.

[95]   Kitada, S; Andersen, J; Akar, S; Zapata, JM; Takayama, S; Krajewski, S; Wang, HG; Zhang, X; Bullrich, F; Croce, CM; Rai, K; Hines, J; Reed, JC (1998). Expression of apoptosis-regulating proteins in chronic lymphocytic leukemia: correlations with In vitro and In vivo chemoresponses. *Blood, 91,* 3379-3389.

[96]   Klasa, RJ; Gillum, AM; Klem, RE; Frankel, SR (2002). Oblimersen Bcl-2 antisense: facilitating apoptosis in anticancer treatment. *Antisense Nucleic Acid Drug Dev, 12,*193-213.

[97]   Kluck, RM; Bossy-Wetzel, E; Green, DR; Newmeyer, DD (1997). The release of cytochrome c from mitochondria: a primary site for Bcl-2 regulation of apoptosis. *Science, 275,* 1132-1136.

[98] Kogan, SC; Brown, DE; Shultz, DB; Truong, BT; Lallemand-Breitenbach, V; Guillemin, MC; Lagasse, E; Weissman, IL; Bishop, JM (2001). BCL-2 cooperates with promyelocytic leukemia retinoic acid receptor alpha chimeric protein (PMLRARalpha) to block neutrophil differentiation and initiate acute leukemia. *J Exp Med, 193,* 531-543.

[99] Komarova, NL; Wodarz, D (2005). Drug resistance in cancer: Principle of emergence and prevention. *Proc Natl Acad Sci USA, 102,* 9714-9719.

[100] Konopleva, M; Zhao, S; Hu, W; Jiang, S; Snell, V; Weidner, D; Jackson, CE; Zhang, X; Champlin, R; Estey, E; Reed, JC; Andreeff, M (2002). The anti-apoptotic genes Bcl-X(L) and Bcl-2 are over-expressed and contribute to chemoresistance of non-proliferating leukaemic CD34+ cells. *Br J Haematol, 118,* 521-534.

[101] Konopleva, M; Konoplev, S; Hu, W; Zaritskey, AY; Afanasiev, BV; Andreeff, M (2002). Stromal cells prevent apoptosis of AML cells by up-regulation of anti-apoptotic proteins. *Leukemia, 16,* 1713-1724.

[102] Kuribara, R; Honda, H; Matsui, H; Shinjyo, T; Inukai, T; Sugita, K; Nakazawa, S; Hirai, H; Ozawa, K; Inaba T (2004). Roles of Bim in apoptosis of normal and Bcr-Abl-expressing hematopoietic progenitors. *Mol Cell Biol, 24,* 6172-6183.

[103] Kurland, JF; Kodym, R; Story, MD; Spurgers, KB; McDonnell, TJ; Meyn, RE (2001). NF-kappaB1 (p50) homodimers contribute to transcription of the bcl-2 oncogene. *J Biol Chem, 276,*45380-45386.

[104] Kuroda, J; Puthalakath, H; Cragg, MS; Kelly, PN; Bouillet, P; Huang, DCS; Kimura, S; Ottmann, OG; Druker, BJ; Villunger, A; Roberts, AW; Strasser, A (2006). Bim and Bad mediate imatinib-induced killing of Bcr/Abl[+] leukemic cells and resistance due to their loss is overcome by a BH3 mimetic. *Proc Natl Acad Sci USA,* 103, 14907-14912

[105] Kuwana, T; Bouchier-Hayes, L; Chipuk, JE; Bonzon, C; Sullivan, BA; Green, DR; Newmeyer, DD (2005). BH3 domains of BH3-only proteins differentially regulate Bax-mediated mitochondrial membrane permeabilization both directly and indirectly. *Mol Cell, 17,* 525-535.

[106] Lee, SH; Soung, YH; Lee, JW; Kim, HS; Lee, JH; Kim, HS; Lee, JH; Park, JY; Cho, YG; Kim, CJ; Kim, SY; Park, WS; Kim, SH; Lee, JY; Yoo, NJ (2003). Mutational analysis of Noxa gene in human cancers. *APMIS, 111,* 599-604.

[107] Leone, M; Zhai, D; Sareth, S; Kitada, S; Reed, JC; Pellecchia, M (2003). Cancer prevention by tea polyphenols is linked to their direct inhibition of antiapoptotic Bcl-2-family proteins. *Cancer Res, 6,* 8118-8121.

[108] Letai, A; Bassik, MC; Walensky, LD, Sorcinelli, MD; Weiler, S; Korsmeyer, SJ (2002). Distinct BH3 domains either sensitize or activate mitochondrial apoptosis, serving as prototype cancer therapeutics. *Cancer Cell, 2,* 183-192.

[109] Letai, A; Sorcinelli, MD; Beard, C; Korsmeyer, SJ (2004). Antiapoptotic BCL-2 is required for maintenance of a model leukemia. *Cancer Cell, 6,* 241-249.

[110] Ley, R; Balmanno, K; Hadfield, K; Weston, C; Cook, SJ (2003). Activation of the ERK1/2 signaling pathway promotes phosphorylation and proteasome-dependent degradation of the BH3-only protein, Bim. *J Biol Chem, 278,* 18811-18816.

[111] Li, R; Moudgil, T; Ross, HJ; Hu, HM (2005). Apoptosis of non-small-cell lung cancer cell lines after paclitaxel treatment involves the BH3-only proapoptotic protein Bim. *Cell Death Differ, 12*, 292-303.

[112] Linden, M; Kirchhof, N; Carlson, C; Van Ness, B (2004). Targeted overexpression of Bcl-XL in B-lymphoid cells results in lymphoproliferative disease and plasma cell malignancies. *Blood, 103*, 2779-2786.

[113] Lindsten, T; Ross, AJ; King, A; Zong, WX; Rathmell, JC; Shiels, HA; Ulrich, E; Waymire, KG; Mahar, P; Frauwirth, K; Chen, Y; Wei, M; Eng, VM; Adelman, DM; Simon, MC; Ma, A; Golden, JA; Evan, G; Korsmeyer, SJ; MacGregor, GR; Thompson, CB (2000). The combined functions of proapoptotic Bcl-2 family members bak and bax are essential for normal development of multiple tissues. *Mol Cell, 6*, 1389-1399.

[114] Liu, WM; Strauss, SJ; Chaplin, T; Shahin, S; Propper, DJ; Young, BD; Joel, SP; Malpas, JS (2004). s-thalidomide has a greater effect on apoptosis than angiogenesis in a multiple myeloma cell line. *Hematol J, 5*, 247-254.

[115] Lizcano, JM; Morrice, N; Cohen, P (2000). Regulation of BAD by cAMP-dependent protein kinase is mediated via phosphorylation of a novel site, Ser155. *Biochem J, 349*, 547-557.

[116] Mackus, WJ; Kater, AP; Grummels, A; Evers, LM; Hooijbrink, B; Kramer, MH; Castro, JE; Kipps, TJ; van Lier, RA; van Oers, MH; Eldering, E (2005). Chronic lymphocytic leukemia cells display p53-dependent drug-induced Puma upregulation. *Leukemia, 19*, 427-434.

[117] McCurrach, ME; Connor, TM; Knudson, CM; Korsmeyer, SJ; Lowe, SW (1997). bax-deficiency promotes drug resistance and oncogenic transformation by attenuating p53-dependent apoptosis. *Proc Natl Acad Sci USA, 94*, 2345-2349.

[118] Maeta, Y; Tsujitani, S; Matsumoto, S; Yamaguchi, K; Tatebe, S; Kondo, A; Ikeguchi, M; Kaibara, N (2004). Expression of Mcl-1 and p53 proteins predicts the survival of patients with T3 gastric carcinoma. *Gastric Cancer, 7*, 78-84.

[119] Mahieux, R; Pise-Masison, C; Gessain, A; Brady, JN; Olivier, R; Perret, E; Misteli, T; Nicot, C (2001). Arsenic trioxide induces apoptosis in human T-cell leukemia virus type 1- and type 2-infected cells by a caspase-3-dependent mechanism involving Bcl-2 cleavage. *Blood, 98*, 3762-3769.

[120] McDonnell, TJ; Deane, N; Platt, FM; Nunez, G; Jaeger, U; McKearn, JP; Korsmeyer, SJ (1989). bcl-2-immunoglobulin transgenic mice demonstrate extended B cell survival and follicular lymphoproliferation. *Cell, 57*, 79-88.

[121] Milanesi, E; Costantini, P; Gambalunga, A; Colonna, R; Petronilli, V; Cabrelle, A; Semenzato, G; Cesura, AM; Pinard, E; Bernardi, P (2006). The Mitochondrial Effects of Small Organic Ligands of BCL-2: Sensitizarion of BCL-2-Overexpressing Cells to Apoptosis by a Pyrimidine-2,4,6-Trione Derivative. *J Biol Chem, 281*,10066-10072.

[122] Milella, M; Trisciuoglio, D; Bruno, T; Ciuffreda, L; Mottolese, M; Cianciulli, A; Cognetti, F; Zangemeister-Wittke, U; Del Bufalo, D; Zupi, G (2004). Trastuzumab down-regulates Bcl-2 expression and potentiates apoptosis induction by Bcl-2/Bcl-XL bispecific antisense oligonucleotides in HER-2 gene--amplified breast cancer cells. *Clin Cancer Res, 10*, 7747-7756.

[123] Mitsiades, N; Mitsiades, CS; Poulaki, V; Chauhan, D; Fanourakis, G; Gu, X; Bailey, C; Joseph, M; Libermann, TA; Treon, SP; Munshi, NC; Richardson, PG; Hideshima, T; Anderson, KC (2002). Molecular sequelae of proteasome inhibition in human multiple myeloma cells. *Proc Natl Acad Sci USA, 99,* 14374-14379.

[124] Miyashita, T; Reed, JC (1995). Tumor suppressor p53 is a direct transcriptional activator of the human bax gene. *Cell, 80,* 293-299.

[125] Morales, AA; Olsson, A; Celsing, F; Osterborg, A; Jondal, M; Osorio, LM (2005). High expression of bfl-1 contributes to the apoptosis resistant phenotype in B-cell chronic lymphocytic leukemia. *Int J Cancer, 113,* 730-737.

[126] Morales, AA; Olsson, A; Celsing, F; Osterborg, A; Jondal, M; Osorio, LM (2004). Expression and transcriptional regulation of functionally distinct Bmf isoforms in B-chronic lymphocytic leukemia cells. *Leukemia, 18,* 41-47.

[127] Mortenson, MM; Schlieman, MG; Virudachalam, S; Lara, PN; Gandara, DG; Davies, AM; Bold, RJ (2005). Reduction in BCL-2 levels by 26S proteasome inhibition with bortezomib is associated with induction of apoptosis in small cell lung cancer. *Lung Cancer, 49,*163-170.

[128] Moshynska, O; Sankaran, K; Pahwa, P; Saxena, A (2004). Prognostic significance of a short sequence insertion in the MCL-1 promoter in chronic lymphocytic leukemia. *J Natl Cancer Inst, 96,* 673-682.

[129] Muchmore, SW; Sattler, M; Liang, H; Meadows, RP; Harlan, JE; Yoon, HS; Nettesheim, D; Chang, BS; Thompson, CB; Wong, SL; Ng, SL; Fesik, SW (1996). X-ray and NMR structure of human Bcl-xL, an inhibitor of programmed cell death. *Nature, 381,* 335-341.

[130] Mund, T; Gewies, A; Schoenfeld, N; Bauer, MK; Grimm, S (2003). Spike, a novel BH3-only protein, regulates apoptosis at the endoplasmic reticulum. *FASEB J, 17,* 696-698.

[131] Nakano, K; Vousden, KH (2001). PUMA, a novel proapoptotic gene, is induced by p53. *Mol Cell, 7,* 683-694.

[132] Nakayama, K; Nakayama, K; Negishi, I; Kuida, K; Sawa, H; Loh, DY (1994). Targeted disruption of Bcl-2 alpha beta in mice: occurrence of gray hair, polycystic kidney disease, and lymphocytopenia. *Proc Natl Acad Sci USA, 91,* 3700-3704.

[133] Neshat, MS; Raitano, AB; Wang, H; Reed, JC; Sawyers, CL (2000). The survival function of the Bcr-Abl oncogene is mediated by Bad-dependent and -independent pathways: Roles for phosphatidylinositol 3-kinase and Raf. *Mol Cell Biol, 20,* 1179-1186.

[134] Niizuma, H; Nakamura, Y; Ozaki, T; Nakanishi, H; Ohira, M; Isogai, E; Kageyama, H; Imaizumi, M; Nakagawara, A (2006). Bcl-2 is a key regulator for the retinoic acid-induced apoptotic cell death in neuroblastoma. *Oncogene, 25,* 5046-5055.

[135] Nikrad, M; Johnson, T; Puthalalath, H; Coultas, L; Adams, J; Kraft, AS (2005). The proteasome inhibitor bortezomib sensitizes cells to killing by death receptor ligand TRAIL via BH3-only proteins Bik and Bim. *Mol Cancer Ther, 4,* 443-449.

[136] Nuutinen, U; Postila, V; Matto, M; Eeva, J; Ropponen, A; Eray, M; Riikonen, P; Pelkonen, J (2006). Inhibition of PI3-kinase-Akt pathway enhances dexamethasone-

induced apoptosis in a human follicular lymphoma cell line. *Exp Cell Res, 312,* 322-330.

[137] O'Connor, L; Strasser, A; O'Reilly, LA; Hausmann, G; Adams, JM; Cory, S; Huang, DC (1998). Bim: a novel member of the Bcl-2 family that promotes apoptosis. *EMBO J, 17,* 384-395.

[138] Oda, E; Ohki, R; Murasawa, H; Nemoto, J; Shibue, T; Yamashita, T; Tokino, T; Taniguchi, T; Tanaka, N (2000). Noxa, a BH3-only member of the Bcl-2 family and candidate mediator of p53-induced apoptosis. *Science, 288,* 1053-1058.

[139] Oltersdorf, T; Elmore, SW; Shoemaker, AR; Armstrong, RC; Augeri, DJ; Belli, BA; Bruncko, M; Deckwerth, TL; Dinges, J; Hajduk, PJ; Joseph, MK; Kitada, S; Korsmeyer, SJ; Kunzer, AR; Letai, A; Li, C; Mitten, MJ; Nettesheim, DG; Ng, S; Nimmer, PM; O'Connor, JM; Oleksijew, A; Petros, AM; Reed, JC; Shen, W; Tahir, SK; Thompson, CB; Tomaselli, KJ; Wang, B; Wendt, MD; Zhang, H; Fesik, SW; Rosenberg, SH (2005). An inhibitor of Bcl-2 family proteins induces regression of solid tumours. *Nature, 435,* 677-681.

[140] O'Reilly, LA; Cullen, L; Visvader, J; Lindeman, GJ; Print, C; Bath, ML; Huang, DC; Strasser, A (2000). The proapoptotic BH3-only protein bim is expressed in hematopoietic, epithelial, neuronal, and germ cells. *Am J Pathol, 157,* 449-461.

[141] Opferman, JT; Korsmeyer, SJ (2003). Apoptosis in the development and maintenance of the immune system. *Nat Immunol, 4,* 10-15.

[142] Opferman, JT; Iwasaki, H; Ong, CC; Suh, H; Mizuno, S; Akashi, K; Korsmeyer, SJ (2005). Obligate role of anti-apoptotic MCL-1 in the survival of hematopoietic stem cells. *Science, 307,* 1101-1104.

[143] Opferman, JT; Letai, A; Beard, C; Sorcinelli, MD; Ong, CC; Korsmeyer, SJ (2003). Development and maintenance of B and T lymphocytes requires antiapoptotic MCL-1. *Nature, 426,* 671-676.

[144] Packham, G; White, EL; Eischen, CM; Yang, H; Parganas, E; Ihle, JN; Grillot, DA; Zambetti, GP; Nunez, G; Cleveland, JL (1998). Selective regulation of Bcl-XL by a Jak kinase-dependent pathway is bypassed in murine hematopoietic malignancies. *Genes Dev, 12,* 2475-2487.

[145] Peeters, SD; Hovenga, S; Rosati, S; Vellenga, E (2005). Bcl-xl expression in multiple myeloma. *Med Oncol, 22,* 183-190.

[146] Pei, XY; Dai, Y; Grant, S (2003). The proteasome inhibitor bortezomib promotes mitochondrial injury and apoptosis induced by the small molecule Bcl-2 inhibitor HA14-1 in multiple myeloma cells. *Leukemia, 17,* 2036-2045.

[147] Pei, XY; Dai, Y; Grant, S (2004). The small-molecule Bcl-2 inhibitor HA14-1 interacts synergistically with flavopiridol to induce mitochondrial injury and apoptosis in human myeloma cells through a free radical-dependent and Jun NH2-terminal kinase-dependent mechanism. *Mol Cancer Ther, 3,* 1513-1524.

[148] Pelengaris, S; Abouna, S; Cheung, L; Ifandi, V; Zervou, S; Khan, M (2004). Brief inactivation of c-Myc is not sufficient for sustained regression of c-Myc-induced tumours of pancreatic islets and skin epidermis. *BMC Biol, 2,* 26.

[149] Pellegrini, M ; Belz, G ; Bouillet, P; Strasser, A (2003). Shutdown of an acute T cell immune response to viral infection is mediated by the proapoptotic Bcl-2 homology 3-only protein Bim. *Proc Natl Acad Sci USA, 100,* 14175-14180.

[150] Perez-Galan, P; Roue, G; Villamor, N; Montserrat, E; Campo, E; Colomer, D (2006). The proteasome inhibitor bortezomib induces apoptosis in mantle-cell lymphoma through generation of ROS and Noxa activation independent of p53 status. *Blood, 107,* 257-264.

[151] Petros, AM; Olejniczak, ET; Fesik, SW (2004). Structural biology of the Bcl-2 family of proteins. *Biochim Biophys Acta, 1644,* 83-94.

[152] Petros, AM; Nettesheim, DG; Wang, Y; Olejniczak, ET; Meadows, RP; Mack, J; Swift, K; Matayoshi, ED; Zhang, H; Thompson, CB; Fesik, SW (2000). Rationale for Bcl-xL/Bad peptide complex formation from structure, mutagenesis, and biophysical studies. *Protein Sci, 9,* 2528-2534.

[153] Petros, AM; Medek, A; Nettesheim, DG; Kim, DH; Yoon, HS; Swift, K; Matayoshi, ED; Oltersdorf, T; Fesik, SW (2001). Solution structure of the antiapoptotic protein bcl-2. *Proc Natl Acad Sci USA, 98,* 3012-3017.

[154] Piazza, RG; Magistroni, V; Andreoni, F; Franceschino, A; Gambacorti, C (2005). Bim promoter is highly methylated in malignant lymphoid cell lines, leading to downregulation of Bim expression and protection from apoptosis. *Blood, 106,* 737A, 2625.

[155] Piret, JP; Minet, E; Cosse, JP; Ninane, N; Debacq, C; Raes, M; Michiels, C (2005). Hypoxia-inducible factor-1-dependent overexpression of myeloid cell factor-1 protects hypoxic cells against tert-butyl hydroperoxide-induced apoptosis. *J Biol Chem, 280,* 9336-9344.

[156] Piro, LD (2004). Apoptosis, Bcl-2 antisense, and cancer therapy. *Oncology (Williston Park), 18,* 5-10.

[157] Puthalakath, H; Strasser, A (2002). Keeping killers on a tight leash: transcriptional and post-translational control of the pro-apoptotic activity of BH3-only proteins. *Cell Death Differ, 9,* 505-512.

[158] Puthalakath, H; Huang, DC; O'Reilly, LA; King, SM; Strasser, A (1999). The proapoptotic activity of the Bcl-2 family member Bim is regulated by interaction with the dynein motor complex. *Mol Cell, 3,* 287-296.

[159] Puthalakath, H; Villunger, A; O'Reilly, LA; Beaumont, JG; Coultas, L; Cheney, RE; Huang, DC; Strasser, A (2001). Bmf: a proapoptotic BH3-only protein regulated by interaction with the myosin V actin motor complex, activated by anoikis. *Science, 293,* 1829-1832

[160] Qin, JZ; Ziffra, J; Stennett, L; Bodner, B; Bonish, BK; Chaturvedi, V; Bennett, F; Pollock, PM; Trent, JM; Hendrix, MJ; Rizzo, P; Miele, L; Nickoloff, BJ (2005). Proteasome inhibitors trigger NOXA-mediated apoptosis in melanoma and myeloma cells. *Cancer Res, 65,* 6282-6293.

[161] Qin, JZ; Stennett, L; Bacon, P; Bodner, B; Hendrix, MJ; Seftor, RE; Seftor, EA; Margaryan, NV; Pollock, PM; Curtis, A; Trent, JM; Bennett, F; Miele, L; Nickoloff, BJ (2004). p53-independent NOXA induction overcomes apoptotic resistance of malignant melanomas. *Mol Cancer Ther, 3,* 895-902.

[162] Radtke, F; Clevers, H (2005). Self-renewal and cancer of the gut: two sides of a coin. *Science, 307,* 1904-1909.

[163] Ramanarayanan, J; Hernandez-Ilizaliturri, FJ; Chanan-Khan, A; Czuczman MS (2004). Pro-apoptotic therapy with the oligonucleotide Genasense (oblimersen sodium) targeting Bcl-2 protein expression enhances the biological anti-tumour activity of rituximab. *Br J Haematol, 127,* 519-530.

[164] Ranger, AM; Zha, J; Harada, H; Datta, SR; Danial, NN; Gilmore, AP; Kutok, JL; Le Beau, MM; Greenberg, ME; Korsmeyer, SJ (2003). Bad-deficient mice develop diffuse large B cell lymphoma. *Proc Natl Acad Sci USA, 100,* 9324-9329.

[165] Real, PJ; Sierra, A; De Juan, A; Segovia, JC; Lopez-Vega, JM; Fernandez-Luna, JL (2002). Resistance to chemotherapy via Stat3-dependent overexpression of Bcl-2 in metastatic breast cancer cells. *Oncogene, 21,* 7611-7618.

[166] Reed, JC; Kitada, S; Takayama, S; Miyashita, T (1994). Regulation of chemoresistance by the bcl-2 oncoprotein in non-Hodgkin's lymphoma and lymphocytic leukemia cell lines. *Ann Oncol, 5,* 61-65.

[167] Reed, JC; Pellecchia, M (2005). Apoptosis-based therapies for hematologic malignancies. *Blood, 106,* 408-418.

[168] Roa, W; Chen, H; Alexander, A; Gulavita, S; Thng, J; Sun, XJ; Petruk, K; Moore, R (2005). Enhancement of radiation sensitivity with BH3I-1 in non-small cell lung cancer. *Clin Invest Med, 28,* 55-63.

[169] Roskoski, R Jr (2004). The ErbB/HER receptor protein-tyrosine kinases and cancer. *Biochem Biophys Res Commun, 319,* 1-11.

[170] Ross, AJ; Waymire, KG; Moss, JE; Parlow, AF; Skinner, MK; Russell, LD; MacGregor, GR (1998). Testicular degeneration in Bclw-deficient mice. *Nat Genet, 18,* 251-256.

[171] Sakai, I; Kraft, AS (1997). The kinase domain of Jak2 mediates induction of bcl-2 and delays cell death in hematopoietic cells. *J Biol Chem, 272,* 12350-12358.

[172] Salomoni, P; Condorelli, F; Sweeney, SM; Calabretta, B (2000). Versatility of BCR/ABL-expressing leukemic cells in circumventing proapoptotic BAD effects. *Blood, 96,* 676-684.

[173] Sano, M; Nakanishi, Y; Yagasaki, H; Honma, T; Oinuma, T; Obana, Y; Suzuki, A; Nemoto, N (2005). Overexpression of anti-apoptotic Mcl-1 in testicular germ cell tumours. *Histopathology, 46,* 532-539.

[174] Sartorius, UA; Krammer, PH (2002). Upregulation of Bcl-2 is involved in the mediation of chemotherapy resistance in human small cell lung cancer cell lines. *Int J Cancer, 97,* 584-592.

[175] Sasaki, T; Sasahira, T; Shimura, H; Ikeda, S; Kuniyasu, H (2004). Effect of human noxa on irinotecan-induced apoptosis in human gastric carcinoma cell lines. *Hepatogastroenterology, 51,* 912-915.

[176] Scheid, MP; Schubert, KM; Duronio, V (1999). Regulation of bad phosphorylation and association with Bcl-x(L) by the MAPK/Erk kinase. *J Biol Chem, 274,* 31108-31113.

[177] Servida, F; Soligo, D; Delia, D; Henderson, C; Brancolini, C; Lombardi, L; Deliliers, GL (2005). Sensitivity of human multiple myelomas and myeloid leukemias to the proteasome inhibitor I. *Leukemia, 19,* 2324-2331.

[178] Shibata, MA; Liu, ML; Knudson, MC; Shibata, E; Yoshidome, K; Bandey, T; Korsmeyer, SJ; Green, JE (1999). Haploid loss of bax leads to accelerated mammary tumor development in C3(1)/SV40-TAg transgenic mice: reduction in protective apoptotic response at the preneoplastic stage. *EMBO J, 18,* 2692-2701.

[179] Shibue, T; Takeda, K; Oda, E; Tanaka, H; Murasawa, H; Takaoka, A; Morishita, Y; Akira, S; Taniguchi, T; Tanaka, N (2003). Integral role of Noxa in p53-mediated apoptotic response. *Genes Dev, 17,* 2233-2238.

[180] Sinicrope, FA; Penington, RC; Tang, XM (2004). Tumor necrosis factor-related apoptosis-inducing ligand-induced apoptosis is inhibited by Bcl-2 but restored by the small molecule Bcl-2 inhibitor, HA 14-1, in human colon cancer cells. *Clin Cancer Res, 10,* 8284-8292.

[181] Spierings, D; McStay, G; Saleh, M; Bender, C; Chipuk, J; Maurer, U; Green, DR (2005). Connected to death: the (unexpurgated) mitochondrial pathway of apoptosis. *Science, 310,* 66-67.

[182] Srivastava, RK (2001). TRAIL/Apo-2L: mechanisms and clinical applications in cancer. *Neoplasia, 3,* 535-546.

[183] Stahl, M; Dijkers, PF; Kops, GJ; Lens, SM; Coffer, PJ; Burgering, BM; Medema, RH (2002). The forkhead transcription factor FoxO regulates transcription of p27Kip1 and Bim in response to IL-2. *J Immunol, 168,* 5024-5031.

[184] Steelman, LS; Pohnert, SC; Shelton, JG; Franklin, RA; Bertrand, FE; McCubrey, JA (2004). JAK/STAT, Raf/MEK/ERK, PI3K/Akt and BCR-ABL in cell cycle progression and leukemogenesis. *Leukemia, 18,* 189-218.

[185] Stein, CA; Benimetskaya, L; Mani, S (2005). Antisense strategies for oncogene inactivation. *Semin Oncol, 32,* 563-572.

[186] Stirewalt, DL; Radich, JP (2003). The role of FLT3 in haematopoietic malignancies. *Nat Rev Cancer, 3,* 650-665.

[187] Strasser, A (2005). The role of BH3-only proteins in the immune system. *Nat Rev Immunol, 5,* 189-200.

[188] Strasser, A; Harris, AW; Cory, S (1993). E mu-bcl-2 transgene facilitates spontaneous transformation of early pre-B and immunoglobulin-secreting cells but not T cells. *Oncogene, 8,* 1-9.

[189] Strasser, A; Whittingham, S; Vaux, DL; Bath, ML; Adams, JM; Cory, S; Harris, AW (1991). Enforced BCL2 expression in B-lymphoid cells prolongs antibody responses and elicits autoimmune disease. *Proc Natl Acad Sci USA, 88,* 8661-8665.

[190] Strasser, A; O'Connor, L; Dixit, VM (2000). Apoptosis signaling. *Annu Rev Biochem, 69,* 217-245.

[191] Strasser, A; Harris, AW; Bath, ML; Cory, S (1990). Novel primitive lymphoid tumours induced in transgenic mice by cooperation between myc and bcl-2. *Nature, 348,* 331-333.

[192] Sturm, I; Stephan, C; Gillissen, B; Siebert, R; Janz, M; Radetzki, S; Jung, K; Loening, S; Dorken, B; Daniel, PT (2006). Loss of the tissue-specific proapoptotic BH3-only protein Nbk/Bik is a unifying feature of renal cell carcinoma. *Cell Death Differ, 13,* 619-627.

[193] Susin, SA; Lorenzo, HK; Zamzami, N; Marzo, I; Snow, BE; Brothers, GM; Mangion, J; Jacotot, E; Costantini, P; Loeffler, M; Larochette, N; Goodlett, DR; Aebersold, R; Siderovski, DP; Penninger, JM; Kroemer, G (1999). Molecular characterization of mitochondrial apoptosis-inducing factor. *Nature, 397,* 441-446.

[194] Susin, SA; Zamzami, N; Castedo, M; Hirsch, T; Marchetti, P; Macho, A; Daugas, E; Geuskens, M; Kroemer, G (1996). Bcl-2 inhibits the mitochondrial release of an apoptogenic protease. *J Exp Med, 184,*1331-1341.

[195] Suzuki, M; Youle, RJ; Tjandra, N (2000). Structure of Bax: coregulation of dimer formation and intracellular localization. *Cell, 103,* 645-654.

[196] Swanson, PJ; Kuslak, SL; Fang, W; Tze, L; Gaffney, P; Selby, S; Hippen, KL; Nunez, G; Sidman, CL; Behrens, TW (2004). Fatal acute lymphoblastic leukemia in mice transgenic for B cell-restricted bcl-xL and c-myc. *J Immunol, 172,* 6684-6691.

[197] Tagawa, H; Karnan, S; Suzuki, R; Matsuo, K; Zhang, X; Ota, A; Morishima, Y; Nakamura, S; Seto, M (2005). Genome-wide array-based CGH for mantle cell lymphoma: identification of homozygous deletions of the proapoptotic gene BIM. *Oncogene, 24,*1348-1358.

[198] Tan, J; Zhuang, L; Jiang, X; Yang, KK; Karuturi, KM; Yu, Q (2006). Apoptosis Signal-regulating Kinase 1 Is a Direct Target of E2F1 and Contributes to Histone Deacetylase Inhibitor induced Apoptosis through Positive Feedback Regulation of E2F1 Apoptotic Activity. *J Biol Chem, 281,* 10508-10515.

[199] Tan, TT; Degenhardt, K; Nelson, DA; Beaudoin, B; Nieves-Neira, W; Bouillet, P; Villunger, A; Adams, JM; White, E (2005). Key roles of BIM-driven apoptosis in epithelial tumors and rational chemotherapy. *Cancer Cell, 7,* 227-238.

[200] Terui, T; Murakami, K; Takimoto, R; Takahashi, M; Takada, K; Murakami, T; Minami, S; Matsunaga, T; Takayama, T; Kato, J; Niitsu, Y (2003). Induction of PIG3 and NOXA through acetylation of p53 at 320 and 373 lysine residues as a mechanism for apoptotic cell death by histone deacetylase inhibitors. *Cancer Res, 63,* 8948-8954.

[201] Traver, D; Akashi, K; Weissman, IL; Lagasse, E (1998). Mice defective in two apoptosis pathways in the myeloid lineage develop acute myeloblastic leukemia. *Immunity, 9,* 47-57.

[202] Tsujimoto, Y; Cossman, J; Jaffe, E; Croce, CM (1985). Involvement of the bcl-2 gene in human follicular lymphoma. *Science, 228,*1440-1443.

[203] Tzung, SP; Kim, KM; Basanez, G; Giedt, CD; Simon, J; Zimmerberg, J; Zhang, KY; Hockenbery, DM (2001). Antimycin A mimics a cell-death-inducing Bcl-2 homology domain 3. *Nat Cell Biol, 3,* 183-191.

[204] Vaux, DL; Cory, S; Adams, JM (1988). Bcl-2 gene promotes haemopoietic cell survival and cooperates with c-myc to immortalize pre-B cells. *Nature, 335,* 440-442.

[205] Verhagen, AM; Ekert, PG; Pakusch, M; Silke, J; Connolly, LM; Reid, GE; Moritz, RL; Simpson, RJ; Vaux, DL (2000). Identification of DIABLO, a mammalian protein that promotes apoptosis by binding to and antagonizing IAP proteins. *Cell, 102,* 43-53.

[206] Vermeulen, K; Van Bockstaele, DR; Berneman, ZN (2005). Apoptosis: mechanisms and relevance in cancer. *Ann Hematol, 84,* 627-639.

[207] Villunger, A; Scott, C; Bouillet, P; Strasser, A (2003). Essential role for the BH3-only protein Bim but redundant roles for Bax, Bcl-2, and Bcl-w in the control of granulocyte survival. *Blood, 101,* 2393-2400.

[208] Villunger, A; Michalak, EM; Coultas, L Mullauer; F; Bock, G; Ausserlechner, MJ; Adams, JM; Strasser, A (2003). p53- and drug-induced apoptotic responses mediated by BH3-only proteins puma and noxa. *Science, 302,* 1036-1038.

[209] Walensky, LD; Kung, AL; Escher, I; Malia, TJ; Barbuto, S; Wright, RD; Wagner, G; Verdine, GL; Korsmeyer, SJ (2004). Activation of apoptosis in vivo by a hydrocarbon-stapled BH3 helix. *Science, 305,* 1466-1470.

[210] Wang, JC; Dick, JE (2005). Cancer stem cells: lessons from leukemia. *Trends Cell Biol, 15,* 494-501.

[211] Wang, JL; Zhang, ZJ; Choksi, S; Shan, S; Lu, Z; Croce, CM; Alnemri, ES; Korngold, R; Huang, Z (2000). Cell permeable Bcl-2 binding peptides: a chemical approach to apoptosis induction in tumor cells. *Cancer Res, 60,* 1498-1502.

[212] Wang, K; Yin, XM; Chao, DT; Milliman, CL; Korsmeyer, SJ (1996). BID: a novel BH3 domain-only death agonist. *Genes Dev, 10,* 2859-2869.

[213] Watanabe, J; Kushihata, F; Honda, K; Sugita, A; Tateishi, N; Mominoki, K; Matsuda, S; Kobayashi, N (2004). Prognostic significance of Bcl-xL in human hepatocellular carcinoma. *Surgery, 135,* 604-612.

[214] Wick, W; Petersen, I; Schmutzler, RK; Wolfarth, B; Lenartz D; Bierhoff, E; Hummerich, J; Muller, DJ; Stangl, AP; Schramm, J; Wiestler, OD; von Deimling, A (1996). Evidence for a novel tumor suppressor gene on chromosome 15 associated with progression to a metastatic stage in breast cancer. *Oncogene, 12,* 973-978.

[215] Williams, J; Lucas, PC; Griffith, KA; Choi, M; Fogoros, S; Hu, YY; Liu, JR (2005). Expression of Bcl-xL in ovarian carcinoma is associated with chemoresistance and recurrent disease. *Gynecol Oncol, 96,* 287-295.

[216] Willis, SN; Adams, JM (2005). Life in the balance: how BH3-only proteins induce apoptosis. *Curr Opin Cell Biol, 17,* 617-625.

[217] Wuilleme-Toumi, S; Robillard, N; Gomez, P; Moreau, P; Le Gouill, S; Avet-Loiseau, H; Harousseau, JL; Amiot, M; Bataille, R (2005). Mcl-1 is overexpressed in multiple myeloma and associated with relapse and shorter survival. *Leukemia, 19,* 1248-1252.

[218] Yan, B; Zemskova, M; Holder, S; Chin, V; Kraft, A; Koskinen, PJ; Lilly, M (2003). The PIM-2 kinase phosphorylates BAD on serine 112 and reverses BAD-induced cell death. *J Biol Chem, 278,* 45358-45367.

[219] Yang, B; Liu, D; Huang, Z (2004). Synthesis and helical structure of lactam bridged BH3 peptides derived from pro-apoptotic Bcl-2 family proteins. *Bioorg Med Chem Lett, 14,* 1403-1406.

[220] Yang, E; Zha, J; Jockel, J; Boise, LH; Thompson, CB; Korsmeyer, SJ (1995). Bad, a heterodimeric partner for Bcl-XL and Bcl-2, displaces Bax and promotes cell death. *Cell, 80,* 285-291.

[221] Yeung, BH; Huang, DC; Sinicrope, FA (2006). PS-341 (Bortezomib) induces lysosomal cathepsin B release and a caspase-2-dependent mitochondrial permeabilization and apoptosis in human pancreatic cancer cells. *J Biol Chem, 281,* 11923-11932.

[222] Yin, C; Knudson, CM; Korsmeyer, SJ; Van Dyke, T (1997). Bax suppresses tumorigenesis and stimulates apoptosis in vivo. *Nature, 385,* 637-640.

[223] Yin, XM; Wang, K; Gross, A; Zhao, Y; Zinkel, S; Klocke, B; Roth, KA; Korsmeyer, SJ (1999). Bid-deficient mice are resistant to Fas-induced hepatocellular apoptosis. *Nature, 400, 886-891.*

[224] Yoshida, H; Kong, YY; Yoshida, R; Elia, AJ; Hakem, A; Hakem, R; Penninger, JM; Mak, TW (1998). Apaf1 is required for mitochondrial pathways of apoptosis and brain development. *Cell, 94,* 739-750.

[225] You, H ; Pellegrini, M; Tsuchihara, K; Yamamoto, K; Hacker, G; Erlacher, M; Villunger, A; Mak, TW (2006). FOXO3a-dependent regulation of Puma in response to cytokine/growth factor withdrawal. *J Exp Med, 203,* 1657-1663.

[226] Yu, J; Zhang, L; Hwang, PM; Kinzler, KW; Vogelstein, B (2001). PUMA induces the rapid apoptosis of colorectal cancer cells. *Mol Cell, 7,* 673-682.

[227] Zha, J; Harada, H; Yang, E; Jockel, J; Korsmeyer, SJ (1996). Serine phosphorylation of death agonist BAD in response to survival factor results in binding to 14-3-3 not BCL-X(L) *Cell, 87,* 619-628.

[228] Zhang, Y ; Adachi, M ; Kawamura, R ; Imai, K (2006). Bmf is a possible mediator in histone deacetylase inhibitors FK228 and CBHA-induced apoptosis. *Cell Death Differ, 13,* 129-140.

[229] Zhao, S; Konopleva, M; Cabreira-Hansen, M; Xie, Z; Hu, W; Milella, M; Estrov, Z; Mills, GB; Andreeff, M (2004). Inhibition of phosphatidylinositol 3-kinase dephosphorylates BAD and promotes apoptosis in myeloid leukemias. *Leukemia, 18, 267-275.*

[230] Zhao, Y; Tan, J; Zhuang, L; Jiang, X; Liu, ET; Yu, Q (2005). Inhibitors of histone deacetylases target the Rb-E2F1 pathway for apoptosis induction through activation of proapoptotic protein Bim. *Proc Natl Acad Sci USA, 102,* 16090-16095.

[231] Zhong, Q ; Gao, W ; Du, F ; Wang, X (2005). Mule/ARF-BP1, a BH3-only E3 ubiquitin ligase, catalyzes the polyubiquitination of Mcl-1 and regulates apoptosis. *Cell, 121,* 1085-1095.

[232] Zhou, P; Levy, NB; Xie, H; Qian, L; Lee, CY; Gascoyne, RD; Craig, RW (2001). MCL1 transgenic mice exhibit a high incidence of B-cell lymphoma manifested as a spectrum of histologic subtypes. *Blood, 97,* 3902-3909.

[233] Zhu, H; Zhang, L; Dong, F; Guo, W; Wu, S; Teraishi, F; Davis, JJ; Chiao, PJ; Fang, B (2005). Bik/NBK accumulation correlates with apoptosis-induction by bortezomib (PS-341, Velcade) and other proteasome inhibitors. *Oncogene, 24,* 4993-4999.

[234] Zinkel, SS; Ong, CC; Ferguson, DO; Iwasaki, H; Akashi, K; Bronson, RT; Kutok, JL; Alt, FW; Korsmeyer, SJ (2003). Proapoptotic BID is required for myeloid homeostasis and tumor suppression. *Genes Dev, 17,* 229-239.

[235] Zong, WX; Lindsten, T; Ross, AJ; MacGregor, GR; Thompson, CB (2001). BH3-only proteins that bind pro-survival Bcl-2 family members fail to induce apoptosis in the absence of Bax and Bak. *Genes Dev, 15,* 1481-1486.

In: Cell Apoptosis Research Trends
Editor: Charles V. Zhang, pp. 41-91

ISBN: 1-60021-424-X
© 2007 Nova Science Publishers, Inc

Chapter II

Tumor Resistance to Cell Death and its Relationship with Protein Prion

Maryam Mehrpour[1,22], Franck Meslin[2] and Ahmed Hamaï[2]

[1]Chinese Academy of Sciences, The Laboratory of Apoptosis and Cancer Biology, The National Key Laboratory of Biomembrane and Membrane Biotechnology, Institute of Zoology, Beijing 100080, P.R. China.
[2]INSERM, U 753, Laboratoire d'Immunologie des Tumeurs Humaines : Interaction effecteurs cytotoxiques-système tumoral, Institut Gustave Roussy PR1 and IFR 54, 94805 Villejuif, France.

ABSTRACT

Cancer is an insidious disease, in which virtually every aspect of cellular control can be subverted to allow the uncontrolled, invasive cellular growth that defeats multicellular cooperation and kills an organism. In testimony to the essential role for proper execution of cell death in tumor suppression, apoptosis is widely recognized as an essential tumor-suppressor system. Indeed, defects in apoptosis are considered a hallmark of cancer, and are known to render the tumor resistant to immunosurveillance and therapy. Prion infections represent a fascinating biological phenomenon which has elicited at the interface between neuroscience and immunology. Although Prion protein (PrPc) is well known for its implication in transmissible spongiform encephalopathy, recent data indicated that PrPc may participate in program cell death regulation. PrPc would be correlated to the acquisition of a resistance phenotype by tumor cells to cytotoxic effectors or antitumor drugs. This review revisits the molecular mechanisms of tumor resistance to apoptosis and the implication of the PrPc in this phenomenon.

Keywords: prion protein, tumor resistance to cell death, breast cancer, normal human myoepithelial breast cells.

---

[2] Correspondence concerning this article should be addressed to Maryam Mehrpour, E-mail: mehrpour@ioz.ac.cn.
Tel : (8610) – 62521552.

ABBREVIATIONS

ABC transporter	ATP-Binding Cassette transporter
AIF	Apoptosis-Inducing Factor
AKT	v-akt murine thymoma viral oncogene homolog
ALPS	Autoimmune lymphoproliferative system
ANT	Adenine Nucleotide Translocator
Apaf	Apoptosis protease activation factor
ARF	Alternate Open Reading Frame
A-Smase	acid sphingomyelinase
ASPP	Apoptosis-Stimuling Protein of p53
ATM	Ataxia-Telangiectasia Mutated
ATR	ATM- and Rad3-related
BAD	BCL2-antagonist of cell death
BAK	BCL2-antagonist/killer
BAX	BCL2-associated X protein
BCL2	B-cell leukemia/lymphoma 2
BCL-X$_L$/BCL2L1	BCL2-like 1
BCL-w/BCL2L2	BCL2-like 2
BFL1/BCL2A1	BCL2-related protein A1
BH	BCL2 Homology
BID	BH3 interacting domain death agonist
BIM/BCL2L11	BCL2-like 11
BIR	Baculoviral IAP repeat
BOK	BCL2-related Ovarian Killer
BOO/DIVA/BCL 2L10	BCL2-like 10
CD95L	CD95 Ligand
CDKN2a	Cyclin-Dependent Kinase Inhibitor 2a
CTSD	endolysosomal aspartate protease cathepsin D
DED	Death Effector Domain
DD	Death Domain
DcR	decoy receptor
DISC	Death-inducing signaling complex
DR	Death Receptor
EBV	Epstein-Barr Virus
EGFR	Epidermial Growth Factor
ERK	Extracellular signal-Regulated Kinase
FADD	Fas-associated DD Kinase
FLICE	FADD-like interleukin-1 β-converting enzyme
FLIP	FLICE-like inhibitory Protein
GPI	Glycosyl-phosphatidylinositol
HR	Homologous Recombination
IAP	Inhibitor of Apoptosis protein

c-IAP	Inhibitor of cellular Apoptosis
I-κB	Inhibitor of κB
JMY	Junction-meditaing and regulatory protein
MALT	Mucosa-Associated Lymphoid Tissue
MCL1	Myeloid Cell Leukemia sequence 1
MDM2	Double Minute 2 protein
MOMP	Mitochondrial Outer Membrane Permeabilization
MRP	Multidrug resistance-associated protein
MSH2	MutS homolog 2
NAIP	Neuronal Apoptosis Inhibitory Protein
NHEJ	Non-Homologous End Joining
NF-κB	nuclear factor-κB
NK	Natural Killer Cell
PI-9/SPI-9	Protease Inhibitor 9
PIDD	P53-induced protein with a death domain
PIPLC	phosphoinositol phospholipase C
PrPc	Cellular Prion Protein
PrP^{Sc}	Scrapie Prion Protein
PTP	Permeability Transition Pore
PI3K	Phosphoinositide 3-Kinase
PIP	Phosphoinoisitol Phosphate
PTEN	Phosphatase and Tensin homolog
RAIDD	receptor-interacting protein (RIP)-associated ICH-1/CED-3 homologous protein with a death domain
RIP	Receptor-Interacting Protein
ROS	reactive oxygen species
SAPK	Stress Activated Protein Kinase
Smac/Diablo	Second Mitochondria-derived Activator of Caspase/direct IAP binding protein
STAT3	Signal Transducer and Activator of Transcription 3
TNF	Tumor Necrosis Factor
TNFR	TNF receptor
TRADD	TNFR-associated DD
TRAF	TNFR-associating factor
TRAIL	Tumor Necrosis Factor-Related Apoptosis-Inducing Ligand
TRAIL-R	TRAIL Receptor
TRID	TNF-R1 internalization domain
VDAC	Voltage-dependent anion channel
XAF 1	XIAP-Associated Factor.

INTRODUCTION

New genetic and biochemical approaches have fostered remarkable progress in our understanding of cancer biology during the past decade [1]. One of the most important advances has been the recognition that resistance to cell death, particularly apoptotic cell death, is an important aspect of both tumorigenesis and the development of resistance to anticancer drugs [1-3]. In addition to apoptosis, cells can be effectively eliminated by necrosis, mitotic catastrophe or autophagic cell death. In addition, premature senescence causes reversible arrests of cell division. Some of the characteristics of these different modes of cell death are summarized in Table 1. Much recent research on new cancer therapies has therefore focused on devising ways to overcome this resistance and to trigger the cell death of tumor cells. Although the detailed mechanisms underlying tumor cell's resistance to apoptosis remain to be characterized, some important components and steps in this process have already been elucidated. A simple look at the wide range of antineoplastic treatments that are ineffective at killing cancer cells (Table 2) implies that the tumor resistance mechanisms are complex. For decades, clinicians and basic scientists have been puzzled by the fact that tumor cells simultaneously acquire the capability to escape immune surveillance mechanisms and evade the cytotoxic action of diverse cytotoxic insults, for example, DNA damage (e.g. by irradiation, alkylation, methylation or crosslinking), microtubule destabilization or topoisomerase inhibition. Complete treatment responses after chemotherapeutic regimens rarely benefit more than 20% of the melanoma patients, and the term 'remission' is rarely used with the melanoma. Moreover, new genetic and biochemical approaches in breast cancer indicate also that drug resistance is likely not a primary consequence of acquired genetic alterations selected *during* or *after* therapy, but rather inherent to the malignant behavior of cancer cells at diagnosis. Various mechanisms including genetic instability, oncogene overexpression, tumor suppressor downregulation, epigenetic modifications, loss of cell cycle control and impact of tumor microenvironnement result in development of tumor resistance to cell death. Understanding these mechanisms at the molecular level provides deeper insight into carcinogenesis, influences therapeutic strategy and might, ultimately, lead to new therapeutic approaches based on modulation of apoptosis sensitivity.

Prion diseases are a group of transmissible neurodegenerative disorders, including Creutzfeldt-Jakob disease, Gerstmann-Sträussler syndrome, kuru, and fatal familial insomnia in humans as well as scrapie and bovine spongiform encephalopathy in animals [4]. These diseases are caused by the conversion of PrPc (cellular Prion protein), a normal cell surface glycoprotein, into PrP^{Sc}, a β-sheet-rich conformer that is infectious in the absence of nucleic acid [5-6]. Although a great deal is known about the role of PrP^{Sc} in the disease process, the normal function of PrPc has remained elusive. A variety of functions have been proposed for PrPc, including roles in metal ion trafficking [7], cell adhesion [8], and transmembrane signaling. Identifying the function of PrPc may provide important clues to the pathogenesis of prion diseases, since there is evidence that PrPc plays an essential role in mediating the neurotoxic effects of PrP^{Sc} [9]. Several intriguing lines of evidence have emerged recently indicating that PrPc may function to protect cells from various kinds of internal or environmental stress. PrPc overexpression rescues not only cultured neurons but also tumor cell lines from pro-apoptotic stimuli, including BAX expression, serum withdrawal, DNA damage, cytokine and anti-cancer drug treatments [10-17]. This review revisits the molecular mechanisms of tumor resistance to apoptosis in cancer and the implication of the cellular prion protein PrPc in this phenomenon.

Table 1. Characteristics of different types of cell death

Type of cell death	General characteristics of death				Detection methods
	Morphological changes			Cellular and Biochemical features	
	Nucleus	Cell membrane	Cytoplasm		
Apoptosis	Chromatin condensation, nuclear margination, DNA laddering	Blebbing is often seen	Formation of apoptotic bodies	Caspase dependent	Annexin-V staining; TUNEL staining; DNA laddering; caspase activation flow cytommetry to detect cells in sub G1 content; detection of change in mitochondrial membrane potential, detection of ROS levels electron microscopy.
Autophagy	Partial chromatin condensation, No DNA laddering	Blebbing	Formation of double membrane vacuoles which sequester mitochondria and ribosomes	Caspase and p53 independent. The celldigest itself	Lack of marginated condensed nuclear chromatin by electron microscopy; protein degradation assays; exclusion of vital dyes until late stages. Prominent cytoplasmic vacuoles detected with monodansylcadavenine.
Mitotic catastrophe	Multiple micronuclei, nuclear fragmentation	-	-	Occurs after or during mitosis and is probably caused by mis-segregation of chromosomes and /or cell fusion. Giant cell formation. Caspase independent (at early stage). Abnormal CDK1/cyclin B activation. Can lead to apoptosis and is p53-independent	Multinucleated cells detected by light or electron microscopy
Necrosis	Clumping and random degradation of nuclear DNA	Swelling, rupture	Increased vacuolation; organelle degeneration, mitochondrial swelling	Typically not genetically determined	Early permeability to vital dyes such as trypan blue, flow cytometry for vital dye staining.
Senescence	Distinct heterochromatic structure	-	Flattening and increased granularity	Cells are metabolically active but non-dividing and show an increase in cell size; Cells express senescence-associated b-galactosidase and this process is p53-dependent.	Electron microscopy, Staining for senescence-associated b-galactosidase

Apoptosis can be initiated by two alternative pathways: either through death receptors on the cell surface (extrinsic pathway) or through mitochondria (intrinsic pathway). In both pathways (Figure 1), cysteine aspartyl-specific proteases (caspases) are activated that cleave cellular substrates, and this leads to the biochemical and morphological changes that are characteristic of apoptosis (Table 1). Finally, the contents of dead cells are packaged into apoptotic bodies, which are recognized by neighboring cells or macrophages and cleared by phagocytosis.

Table 2. Wide spectrum of chemotherapeutic drugs cancer cells show resistance against
in vivo

Alkylating agents Cyclophosphamide Triazenes *Dacarbazine* *Temozolomide*	Alkylation and methylation of nucleic acids	Inhibition of nucleic acid and protein synthesis
Nitosoureas *Carmustine* *Lomustine* *Semustine*	Alkylation of nucleic acids and protein	ssDNA Breaks DNA crosslinking Carbamoylation of proteins
Nitrogen mustard	Alkylation of nucleic acids	DNA crosslinking
Antibiotics Anthracyclines *Adriamycin* *Doxorubicin* *Epirubicin*	DNA intercalating agent Free radicals	ssDNA Breaks DNA crosslinking Inhibition of DNA and RNA synthesis
Plant-derived products Epipodophyllotoxins *Etoposide*	Inhibition of topoisomerase II	DNA breaks
Taxanes *Taxol* *Paclitaxel* *Docetaxel*	Microtubule disruption (prevent depolymerisation)	Altered cell division, motility Intracellular transport
Vinca alkaloids *Vincristine* *Vinblastine*	Microtubule disruption (prevent assembly)	Altered cell division, motility Intracellular transport
Hormonal analogs Antiestrogen *Tamoxifen*	Competitive inhibitor of endogenous estrogens	Altered estrogen signaling
Platinum Drugs Cisplatin Carboplatin	DNA and protein crosslink	ssDNA and dsDNA breaks Changes in DNA structure Inhibition of DNA and RNA synthesis

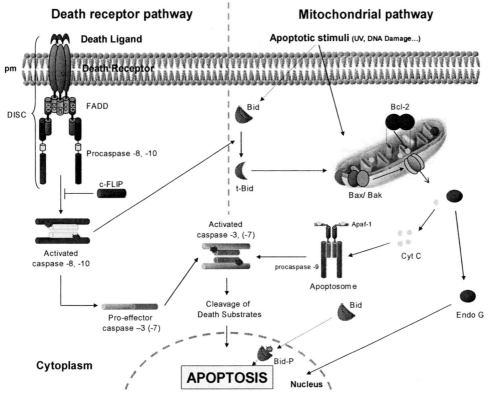

Figure 1. The two main apoptotic signalling pathways. Death receptorsare members of the tumor-necrosis factor (TNF) receptor superfamily and comprise a subfamily that is characterized by an intracellular domain — the death domain (DD). Death receptors are activated by their natural ligands. When ligands bind to their respective death receptors the death domains attract the intracellular adaptor protein FADD (Fas-associated death domain protein, also known as MORT1), which, in turn, recruits the inactive proforms of certain members of the caspase protease family. The caspases that are recruited to this death-inducing signalling complex (DISC) — caspase-8 and caspase-10 — function as 'initiator' caspases. At the DISC, procaspase-8 and procaspase-10 are cleaved and yield active initiator caspases. In some cells — known as type I cells — the amount of active caspase-8 formed at the DISC is sufficient to initiate apoptosis directly, but in type II cells, the amount is too small and mitochondria are used as 'amplifiers' of the apoptotic signal. Activation of mitochondria is mediated by the BCL2 family member BID. BID is cleaved by active caspase-8 and translocates to the mitochondria. Then cytochrome c is released from mitochondria and interacts with Apaf-1, procaspase- 9 and dATP to form the apoptosome. This causes the caspase- 9 activation, which then activates effector caspase- 3. Active executioner caspases (caspase- 3, -7) cleave the death substrates, which eventually results in apoptosis and causes morphological changes. There is crosstalk between these two pathways.

I. APOPTOSIS PATHWAYS

The main effector cells against tumors are cytotoxic T cells and Natural killer (NK) cells. These immune cells use two mains mechanisms to kill tumor cells: the granule exocytosis (perforin/granzymes) pathway and the death receptors pathways.

1. The Death Receptor Family: Extrinsic Pathway

Death receptors are members of the Tumor Necrosis Factor receptor (TNFR) superfamily, which consists of 32 members and 19 ligands (a downloadable table can be accessed at http://www.niams.nih.gov/rtbc/ImageStore/Test/WORD/AB/IRG/tnfchart.doc.) with a broad range of biological functions including the regulation of cell death, survival, differentiation or immune regulation [18-22]. Members of the TNF receptor family share similar cysteine-rich extracellular domains. In addition, death receptors are defined by a cytoplasmic domain of about 80 amino acids called 'death domain', which plays a crucial role in transmitting the death signal from the cell's surface to intracellular signaling pathways. The best-characterized death receptors comprise CD95 (APO-1Fas) TNF receptor 1 (TNFRI), the two agonistic-receptors TRAIL-R1 (DR4) and TRAIL-R2 (DR5), while the role of three antagonistic decoy receptors TRAIL-R3 (DcR1), TRAIL-4 (DcR2), and osteoprotegrin has not exactly been defined.

The corresponding ligands comprise death receptor ligands such as CD95 ligand, TNF, lymphotoxin (the latter two bind to TNFRI), and Tumor Necrosis Factor-Related Apoptosis-Inducing Ligand (TRAIL). With the exception of lymphotoxin, all ligands are type II transmembrane proteins, which also exist as soluble molecules after cleavage by metalloproteases present in the microenvironment. Death receptors are activated upon oligomerization in response to ligand binding.

1.1. CD95 Receptor-CD95 Ligand System

The CD95 receptor-CD95 ligand system is a key signal pathway involved in the regulation of apoptosis in different cell types [21-23]. CD95, a 48 kDa type I transmembrane receptor, is expressed in activated lymphocytes, in a variety of tissues of lymphoid or nonlymphoid origin, as well as in tumor cells. CD95L, a 40 kDa type II transmembrane molecule, occurs in a membrane-bound and in a soluble form, generated through cleavage by metalloproteases. CD95L is produced by activated T cells and plays a crucial role in the regulation of the immune system by triggering autocrine suicide or paracrine death in neighboring lymphocytes or other target cells. Also, CD95L is constitutively expressed in several tissues and has been implied in immune privilege of certain organs such as the testis or the eye [25]. Moreover, CD95L expression on cancer cells has been implicated in immune escape of tumors [26]. By constitutive expression of death receptor ligands such as CD95L, tumors may adopt a killing mechanism from cytotoxic lymphocytes to delete the attacking antitumor T cells through the induction of apoptosis via CD95/CD95L interaction [26]. However, this model of tumor counterattack has also been challenged, since no study has so far conclusively demonstrated that tumor counterattack is a relevant immune escape mechanism *in vivo*.

Links between the receptor and the mitochondrial pathway exist at different levels [27]. Upon death receptor triggering, activation of caspase-8 may result in cleavage of BID, which in turn translocates to mitochondria to release cytochrome *c*, thereby initiating a mitochondrial amplification loop [27]. The activated death receptors recruit and activate an adaptor protein called Fas-associated death domain (FADD) through interactions between the death domain (DD) on the death receptors and FADD. The death effector domain (DED) of

FADD recruits and activates caspase-8, leading to the formation of the death-inducing signaling complex (DISC). In type I cells, the presence of activated caspase-8, a so-called initiator caspase, is sufficient to induce activation of one or more effector caspases (e.g., caspase-3 or -7), which then act on final death substrates in apoptosis [28-29]. However, in type II cells, even a small amount of activated caspase-8, although not enough to activate the effector caspases, is sufficient to trigger a mitochondria-dependent apoptotic amplification loop by activating BID, which induces the accumulation of BAX in mitochondria, the release of cytochrome c from mitochondria, the activation of caspase-9, caspase-3, and caspase-7, and finally, programmed cell death In addition, cleavage of caspase-6 downstream of mitochondria may positively feed-back to amplify the receptor pathway by cleaving caspase-8 [30].

Recently, an unexpected role for BID was found in the ATM pathway that responds to DNA damage [31-32] and BID can be a target for ATM/ATR. Although BID clearly induces apoptosis following activation of death receptors, its role in controlling apoptosis following DNA damage need to be clarified.

1.2. TRAIL Receptor and TRAIL/Apo-2 Ligand System

TRAIL/Apo-2L was identified based on its sequence homology to other members of the TNF superfamily [18, 33]. Similar to CD95L, TRAIL is a type II transmembrane protein, but can be proteolytically cleaved from the cell surface. TRAIL is constitutively expressed in a wide range of tissues. Interestingly, the TRAIL receptor system is rather complex [34-35]. TRAIL-R1 (DR4) and TRAIL-R2 (DR5), the two agonistic TRAIL receptors, contain a conserved cytoplasmic death domain motif, which engages the apoptosis machinery upon ligand binding [23, 36-42] (reviewed in Refs [43 ,36]). TRAIL-R3 (DcR1), and TRAIL-R4 (DcR2) are antagonistic decoy receptors that bind TRAIL but do not transmit a death signal [35, 39, 44-45]. DcR1 is a glycosyl-phosphatidylinositol GPI-anchored cell surface protein, which lacks a cytoplasmic tail, while DcR2 harbors a substantially truncated cytoplasmic death domain. In addition to these four membrane-associated receptors, osteoprotegerin is a soluble decoy receptor, which is involved in the regulation of osteoclastogenesis [46]. Similar to CD95L, TRAIL rapidly triggers apoptosis in many tumor cells [18,35]. Ligation of the agonistic TRAIL receptors DR4 and DR5 by TRAIL or agonistic antibodies results in receptor trimerization and clustering of the receptors' death domains [47].

In addition to inducing apoptosis by DISC formation (Figure 2), TRAIL binding to its receptors also leads to activation of the transcription factor nuclear factor-kappa B (NF-κB) (Figure 2). TRAIL death receptors, like the TNF receptor 1 (TNFR1) [48], activate NF-κB through the TNFR1-associated death domain protein (TRADD) [37]. Activated TRADD recruits the DD-containing protein RIP and TNF receptor-associated factor-2 (TRAF2), leading to activation of the NF-κB pathway; in contrast, dominant-negative TRADD can block the NF-κB activation induced by TRAIL receptors [37, 49]. In TNFR1 signaling, TRADD is believed to be upstream of FADD. However, in TRAIL signaling, TRADD may be mediated by FADD, because TRADD recruitment to the DISC is observed only in the presence of FADD [49].

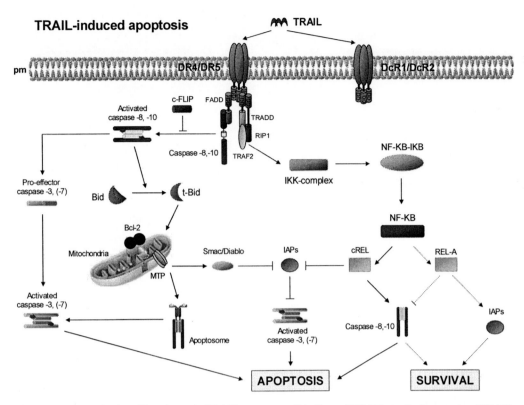

Figure 2. Apoptosis signalling through TRAIL receptors. Binding of TRAIL to their receptor (TRAIL-R1/DR4 and TRAIL-R2/DR5) may transduce via activated caspase-8 directly by activating effector caspase-3 or indirectly by activating BID, which in turn translocated into mitochondria and cause cytochrome *c* and Smac release to cytosol. The released cytochrome *c* interacts with Apaf-1 and cause caspase-9 activation, which then activates effector caspase-3. The released Smac inhibits IAPs by preventing their binding to caspase-9 and caspase-3. Flip bind to the DISC and prevent the activation of caspase-8. Apoptosis pathways induced by TRAIL can also activate NF-κB. Different subunit activation of NF-κB determines whether NF-κB facors apoptosis or survival.

The death-inducing signaling complexes of both TRAIL and Fas receptors contain FADD and caspase-8. In addition to caspase-8, also caspase-10 is recruited to the TRAIL DISC. However, the importance of caspase-10 in the TRAIL DISC for apoptosis induction has been controversially discussed [18]. The TRAIL system has attracted considerable attention for cancer therapy since TRAIL has been shown to kill cancer cells predominantly, while sparing normal cells (see below). The underlying mechanisms for the differential sensitivity of malignant *vs* nonmalignant cells is not well understood. Screening of various different tumor cell types in normal cells did not reveal a consistent association between TRAIL sensitivity and expression of agonistic or antagonistic TRAIL receptors as suggested [50]. Instead, susceptibility for TRAIL-induced cytotoxicity has been suggested to be regulated intracellularly.

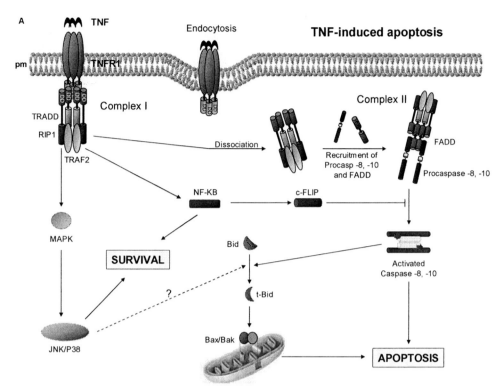

Figure 3A. Signalling through TNF receptor. First model of TNFR1-Mediated Apoptosis. After binding of TNF to TNFR1, rapid recruitment of TRADD, RIP1, and TRAF2 occurs (complex I). Subsequently TNFR1, TRADD, and RIP1 become modified and dissociate from TNFR1. The liberated death domain (DD) of TRADD (and/or RIP1) now binds to FADD, resulting in caspase-8/10 recruitment (forming complex II) and resulting in apoptosis. If NF-κB activation triggered by complex I is successful, cellular FLIP$_L$ levels are sufficiently elevated to block apoptosis and cells survive.

1.3. TNF Receptor and TNF Ligand System

Tumor necrosis factor alpha (TNF) is a highly pleiotropic cytokine that elicits diverse cellular responses ranging from proliferation and differentiation to activation of apoptosis [19, 51]. The different biological activities are mediated by two distinct cell surface receptors: TNF receptor 1 (TNF-R1) and TNF-R2. TNF-R1 appears to be the key mediator of TNF signaling. Recent reports indicated that one possible clue to the basic mechanisms of TNF-R1 signal diversification may lie in a compartmentalization of TNF signaling (see comment in [52]). Based on pharmacological inhibitor studies, authors proposed that induction of apoptosis requires internalization of TNF-R1 [53]. In addition, TNF-mediated NF-κB activation was recently linked to the recruitment of TNF-R1 to lipid rafts on the cell surface [54]. Along this line, Micheau and Tschopp [48] have proposed a model describing induction of TNF-R1-mediated apoptosis via two sequential signaling complexes: a plasma membrane-located NF-κB signaling complex I consisting of TNF-R1, TRADD, RIP-1, and TRAF-2 and a cytosolic apoptosis signaling complex II, which is composed of TRADD, FADD, and caspase-8 but is not associated with TNF-R1 (Figure 3A). Since the first complex can activate survival signals and influence the activity of the second complex, this mechanism provides a checkpoint to control the execution of apoptosis. In contrast to this model, recent report from Schütze's group [55] shows existence of a ligand-induced establishment of a TNF-R1-associated DISC within the first minutes after receptor internalization on endocytic vesicles (Figure 3B). In this model, within minutes internalized TNF-R1 (TNF receptosomes) recruits TRADD, FADD, and caspase-8 to establish the DISC. In addition, the authors identified the TNF-R1 internalization domain (TRID) required for receptor endocytosis and showed that TNF-R1 internalization, DISC formation, and apoptosis are inseparable events. Furthermore, fusion of TNF receptosomes with trans-Golgi vesicles results in activation of acid sphinomylinase and cathepsin D (Figure 3B). In addition to inducing apoptosis by receptosome trafficking TNF-receptor binding also leads to NF-κB activation. This is initiated by TRADD-dependent recruitment of RIP-1 and TRAF-2. This conclusion is based on the fact that when the authors used the TNF-R1 deleted TRID cells, they obtained the complex containing of RIP1 and TRAF2. If NF-κB activation is successful cellular factors are sufficiently elevated to block apoptosis and cells survive.

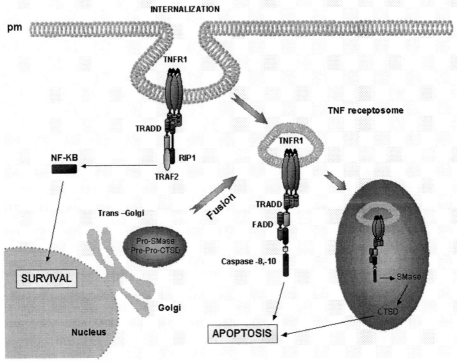

Figure 3B. Signalling through TNF receptor. Second model of TNFR1-Mediated signaling: After TNF binding internalization of TNF-R1 (TNF receptosome) precedes formation of TNF-R1-associated DISC containing TRADD, FADD, and caspase-8. Caspase-8 is autocatalytically cleaved during receptosome trafficking. TNF receptosomes fuse with trans-Golgi vesicles containing pro-A-SMase and pre-pro-CTSD to form multivesicular endosomes in which A-SMase and CTSD (Cathepsin D) activation is induced trigger apoptosis. This differs from the situation with FAS, where the DISC assembles rapidly and receptor internalization can be prevented by caspase inhibitors. In addition to inducing apoptosis by receptosome trafficking TNF receptor binding also leads to NF-κB activation. This is initiated by TRADD-dependent recruitment of RIP-1 and TRAF-2. If NF-κB activation is successful and cellular factors are sufficiently elevated to block apoptosis cells survive.

2. Mitochondria Intrinsic Pathway

Chemotherapy, irradiation and other stimuli can initiate apoptosis through the mitochondrial intrinsic pathway. The regulation of this pathway is complex. The inducers include oncogene, low oxygen (hypoxia), growth factor deprivation, cell-cell detachment (anoïkis) and other stress signals. Pro-apoptotic BCL2 family proteins (reviewed in Refs [56-57]) — for example, BAX, BID, BAD and BIM — are important mediators of these signals. The pro-apoptotic members can be subdivided into the BAX subfamily (BAX, BAK, BOK) and the BH3-only proteins (for example, BID, BAD and BIM) [58]. Activation of mitochondria leads to release of apoptogenic factors such as cytochrome c, Smac/Diablo (second mitochondria-derived activator of caspase/direct IAP binding protein), Omi/HtrA2, endonuclease G and AIF (Apoptosis-inducing factor) from the mitochondrial intermembrane space [59-64]. The release of cytochrome c into the cytosol triggers caspase-3 activation through the formation of the cytochrome c/Apaf-1/caspase-9-containing apoptosome complex, while Smac/Diablo and Omi/HtrA2 promote caspase activation through neutralizing the inhibitory effects to inhibitor of apoptosis proteins (IAPs). The concomitant events are the mitochondrial outer membrane permeabilization (MOMP), disruption of electron transport, loss of mitochondrial transmembrane potential ($\Delta\psi$m, DeltaPsim), decline in ATP levels, production of reactive oxygen species (ROS), and loss of mitochondrial structural integrity. In addition, some of the mitochondrial proteins (AIF, HtrA2/Omi, endonuclease G) released as a result of MOMP can promote caspase-independent cell death through mechanisms that are relatively poorly defined. Caspase-independent cell death can also result from stimuli that cause lysosomal membrane permeabilization with the consequent release of cathepsin proteases [64-66]. Although the impact of BCL2 family members on apoptosis is well known, the biochemical mechanism of their function is not entirely clear. Currently, there are different nonexclusive models that have been proposed to explain how cytohrome c release is regulated. One potential mechanism involved mitochondrial swelling, either due to the opening of mitochondrial permeability transition pore (mPTP), or to mitochondrial hyperpolarization, both of which result in the rupture of the outer mitochondrial membrane, thereby the release of cytochrome c and the loss of mitochondrial membrane potential. This is questionable since there are evidence showing that cytochrome c release and apoptosis occur in the absence of mitochondrial swelling. Recent report from Tsujimoto's group suggests that mPTP is more related to necrosis than to apoptosis [67]. According to one model, mitochondrial membrane permeabilization involves the permeability transition pore complex (PTPC), a multiprotein complex that consists of the adenine nucleotide translocator (ANT) of the inner membrane, the voltage-dependent anion channel (VDAC) of the outer membrane and various other proteins [56]. BCL2 proteins might interact with the PTPC and regulate its permeability. Accoding to the another model the pro-apoptotic molecules in BCL2 family proteins are activated to create discontinuity or pore at the outer membrane of mitochondria to mediate cytochrome c release [68-70]. In this model, BH3-only proapoptotic proteins (such as BID, Bim, Noxa etc.) induce mitochondrial membrane permeabilization either through activation of BAX/BAK or directly neutralize anti-apoptotic molecules such as BCl2/BCL-X_L, two structurally and functionally identical molecules [71-73].

II. Tumor Resistance Mechanisms

Tumor cells can acquire resistance to apoptosis by various mechanisms that interfere at different levels of apoptosis signaling (summarized in Table 3).

Table 3. Mechanisms of tumor resistance to apoptosis

Molecules		*References*
Death receptors		
Expression of soluble receptors for death ligands	Soluble CD95 DcR3	[85-88] [89-91]
TRAIL-R1/ TRAIL-R2 ratio		[107]
Deficient receptor redistribution		[19]
Downregulation and mutation of Death receptors	CD95 TNF TRAIL-R1 TRAIL-2	[92-95] [96] [97-100] [74-77] [78]
Expression of anti-apoptotic molecules		
BCL2 family members	BCL2/ BCL-Xl MCL1	[138-140,142-146,259] [140,147-150] [151,260]
Flip		[112,114,116-119,261-267]
IAPs	Survivin CIAP2 ML-IAP	[133,134,268,269] [135, 136]
PI-9/SPI-6		[137,270,271]
Prion protein		[15,16,272]
Downregulation and mutation of pro-apoptotic		
Bax		[153-156,273-275]
APAF1		[214]
Caspase 8		[152]
XAF1		[130]
Alterations of p53 pathway		
P53		[164-166,168,169,276-280]
INK4A/ARF		[172]
ASPP		[173, 281]
Further mechanisms		
Alterations of the survival pathways		
PI3K		[177]
PTEN		[179-187]
AKT		[178]
NF-kB activity		[188-190] [191, 192]-[194]
MAP kinases		[146], [11]
Chemoresistance mechanisms		
Expression of transporters	MDR1/PglycoproteinMRP	[200]; [196]
Expression of nuclear enzyme DNA repair mechanisms	topoisomerase II MSH2	[205, 206] [210]- [211]
Extracellular matrix		[215]

1. Defective Death Receptor Signaling

1.1. Loss of Function of Receptor Genes by Mutations or Methylation

Death receptors expression may vary between different cell types and can be downregulated or absent in resistant tumor cells. Their inactivation by, e.g. mutations or gene methylation, will probably result in less immune surveillance-induced apoptosis and, consequently, contribute to a malignant phenotype. Deletions and mutations of the death receptors TNF receptors, Fas and TRAIL-R1 and TRAIL-R2 have also been observed in tumors [74-78]. The fact that loss of chromosome 8p21–22, where the TRAIL-receptor genes are mapped [44, 79-82] is a frequent event in various cancers.

Methylation of the Fas promotor and enhancer region was found in colon cancer cell lines and in colon carcinomas. It contributed to the down-regulation of Fas expression and subsequent loss of sensitivity to Fas-induced apoptosis in colon carcinoma cells [83-84].

1.2. Expression of Soluble Receptors

One mechanism by which tumors interfere with death-receptor-mediated apoptosis might be the expression of soluble receptors that act as decoys for death ligands. Two distinct soluble receptors soluble CD95 (sCD95) and decoy receptor 3 (DcR3) have been shown to competitively inhibit CD95 signaling. sCD95 is expressed in various malignancies, and elevated levels can be found in the sera of cancer patients. High sCD95 serum levels were associated with poor prognosis in melanoma patients [85-88]. DcR3 binds to CD95L and the TNF family member LIGHT and inhibits CD95L-induced apoptosis. It is genetically amplified in several lung and *colon carcinomas* and is overexpressed in several adenocarcinomas, glioma cell lines and glioblastomas [89-91]. Ectopic expression of DcR3 in a rat glioma model resulted in decreased immune-cell infiltration, which indicates that DcR3 is involved in immune evasion of malignant glioma [91]. Moreover, death receptors are downregulated or inactivated in many tumors. The expression of the death receptor CD95 is reduced in some tumor cells for example, in *hepatocellular carcinomas*, neoplastic colon epithelium, melanomas and other tumors [92-95] compared with their normal counterparts. Loss of CD95, probably by downregulation of transcription, might contribute to chemoresistance and immune evasion. Oncogenic *RAS* seems to downregulate CD95 [96], and in hepatocellular carcinomas loss of CD95 expression is accompanied by p53 aberrations [95].

Several *CD95* gene mutations have been reported in primary samples of myeloma and T-cell leukaemia [97-99]. The mutations include point mutations in the cytoplasmic death domain of CD95 and a deletion that leads to a truncated form of the death receptor. These mutated forms of CD95 might interfere in a dominant-negative way with apoptosis induction by CD95. In families with germ-line *CD95* mutations, which usually result in *autoimmune lymphoproliferative syndrome* (ALPS), the risk of developing lymphomas is increased [100].

1.3. The Importance of TRAIL-R1/ TRAIL-R2 Ratio

Originally, the regulation of TRAIL induced apoptosis was suggested to be especially controlled by the membrane expression of TRAIL-receptors. Although membrane expression of TRAIL-R1 and TRAIL-R2 is clearly necessary to induce the death signal of TRAIL, membrane expression of DcR1 or DcR2, however, does not necessarily correlate with cell sensitivity to TRAIL [101-102].

Besides the importance of the balance of death receptors and decoy receptor membrane expression, the ratio between may play a role in determining TRAIL-sensitivity. Although this has not been studied very well, there are some indications that in some situations or cell lines, one of the two death receptors is more efficient in death signaling than the other. TRAIL can induce apoptosis through TRAIL-R1 and TRAIL-R2 independently [103-104], but at physiological conditions (37 °C) it binds with a higher affinity to TRAIL-R2 than to TRAIL-R1 [105]. Additionally, in order to induce apoptosis, TRAIL-R1 and TRAIL-R2 have distinct cross-linking requirements. TRAIL-R1 equally responds to cross-linked (e.g. membrane bound) TRAIL and non-cross-linked (soluble) rhTRAIL, whereas TRAIL-R2 signals only in response to cross-linked soluble rhTRAIL [106]. Thus, depending on the type of TRAIL used for apoptosis induction, TRAIL-R1/TRAIL-R2 ratio can determine TRAIL-sensitivity [107].

1.4. Deficient Receptor Redistribution

Studies in melanoma cells have demonstrated that binding of TRAIL to the TRAIL-receptors on the membrane causes a redistribution of the TRAIL-receptors within the cell [108]. The movement of decoy-receptors was dependent on signals from TRAIL-R1 and TRAIL-R2. Reduced decoy receptor relocalization may, therefore, contribute to sensitivity of melanoma cells (and possibly other cancer cells) to TRAIL-induced apoptosis [109].

In colon cancer cells, it was found that a deficient TRAIL death receptor transport to the cell surface resulted in TRAIL-resistant cells. TRAIL-resistant clones were isolated after exposure of a TRAIL-sensitive cell line to TRAIL. Although total cellular mRNA and protein levels were similar in TRAIL-sensitive parental and TRAIL-resistant clones, TRAIL-R1 surface expression and TRAIL-R1 recruitment to the DISC were not detected in the TRAIL-resistant clones. Glycosylation inhibitor tunicamycin increased TRAIL-R1 (and TRAIL-R2) expression and resensitized the resistant clones to TRAIL [110].

Redistribution of death receptors in lipid rafts, which are plasma membrane microdomains enriched in cholesterol and glycosphingolipids, can regulate the efficacy of signaling by death receptors [19]. In colon cancer cells, resveratrol induced TRAIL-R1 and TRAIL-R2 (and FAS) redistribution in lipid rafts and this sensitized the cells to TRAIL-induced apoptosis [111]. This indicates that not only the basic expression levels on the cell surface, but also redistribution/relocalization of TRAIL-receptors to the cell membrane and formation of lipid rafts determine TRAIL-sensitivity.

2. Expression of Anti-Apoptotic Proteins

Various proteins inhibit the apoptotic process at different levels (see Figure 1 and 2)

FLIPs (FADD-like interleukin-1 β-converting enzyme-like protease (FLICE/caspase-8)-inhibitory proteins) interfere with the initiation of apoptosis directly at the level of death receptors [112]. Two splice variants -a long form (cFLIP$_L$) and a short form (cFLIP$_S$)- have been identified in human cells. Although the role of c-FLIP$_S$ as an inhibitor of death receptor-mediated apoptosis is well understood and resembles the activity of viral analogues of FLIP, called viral FLIPs proteins (v-FLIPs) [112, 116], the specific role of cFLIP$_L$ in death receptor-mediated apoptosis remains unclear. The overexpression of cFLIP$_L$ may either induce or inhibit apoptosis, depending on protein levels and cell type. Both forms share structural homology with procaspase-8, but lack its catalytic site. This structure allows them to bind to the DISC, thereby inhibiting the processing and activation of the initiator caspase-8. High FLIP expression that has been found in many tumor cells has been correlated with resistance to CD95- and TRAIL-induced apoptosis [113]. In addition, FLIP expression was associated with tumor escape from T-cell immunity and enhanced tumor progression in experimental studies *in vivo* pointing to a role of FLIP as a tumor progression factor [113]. However, the impact of FLIP on apoptosis sensitivity towards cytotoxic drugs may vary between cell types. Thus, overexpression of FLIP did not confer protection against cytotoxic drugs in T-cell leukemia cells, while it inhibits chemotherapy-induced cell death in colorectal cancer [114]. FLIP antisense oligonucleotides or downregulation of FLIP expression by metabolic inhibitors sensitized various tumor cells for death receptor-induced apoptosis [104, 114-115]. v-FLIPs are encoded by some tumorigenic viruses, including HHV8 [116-118]. In cells that are latently infected with HHV8, v-FLIP is expressed at low levels, but its expression is increased in advanced *Kaposi's sarcomas* or on serum withdrawal from lymphoma cells in culture [119]. Therefore, v-FLIPs might contribute to the persistence and oncogenicity of v-FLIP-encoding viruses.

The IAPs (inhibitor of apoptosis proteins) are cellular caspase inhibitors, which are characterized by the presence of one to three baculoviral IAP repeats (BIRs). Through these BIR domains the IAP proteins bind to and inhibit caspases. IAPs inhibit directly active caspase-3 and -7 and to block caspase-9 activation. Members of the IAP-family include *XIAP, cIAP1, cIAP2, NAIP,* livin, BRUCE and *survivin* [120]. Increasing evidence demonstrates that IAPs, especially survivin and XIAP, are upregulated in several tumor types. Recently, it is demonstrated that survivin binds to Smac/Diablo, which is released from the mitochondria. By binding to Smac/Diablo, survivin prevents binding of Smac/Diablo to IAPs. Free IAPs, such as XIAP may then directly bind to caspases to prevent apoptosis. In addition to the regulation of apoptosis, IAP members such as survivin have been found to be involved in the regulation of mitosis [121-122]. Inhibition of apoptosis by IAPs in response to cytotoxic therapy has been suggested by several experimental studies to be an important mechanism of resistance [64, 123-125]. A mutation in the promotor region of survivin was detected in several cancer cell lines and causes (at least in part) overexpression of survivin [126]. Although it is clear from studies in several cancer cell lines that down-regulation of survivin enhances TRAIL-induced apoptosis [127-128]. Since caspase 3 is regulated by XIAP, upregulation of XIAP is a likely mechanism of TRAIL-resistance. Recently, it was shown that disruption of the gene encoding XIAP in human colon cancer cells did not interfere with basal proliferation, but indeed caused a remarkable sensitivity to TRAIL [129]. This suggests that XIAP (and other IAPs) may be important for the modulation of TRAIL-sensitivity in many tumor types. Recently, a XIAP-interacting protein named XIAP-associated factor 1 (XAF1) has been identified [130]. XAF1 antagonizes the ability of XIAP to suppress caspase activity and cell death *in vitro*. It is not known how XAF1 interacts with XIAP to inhibit its activity. In contrast to Smac/Diablo, XAF1 does not need to be processed and seems to be constitutively able to interact with and inhibit XIAP. However, unlike XIAP which is found primarily in the cytoplasm, endogenous XAF1 is localized in the nucleus. The different compartmentalization between XIAP and XAF1 raises the question of how these two proteins can interact: does XAF1 translocate into the cytoplasm where it inhibits XIAP, or does XIAP enter the nucleus where it is sequestered by XAF1? It is conceivable that XIAP could be transported into the nucleus within the caspase-3 protein complex, where it would continue to inhibit caspase-3 activity. This caspase-inhibiting activity could then be relieved by nuclear XAF1, in a fashion similar to the cytoplasmic inhibition of XIAP by Smac/Diablo.

Expression of survivin is highly tumor specific [131,132]. It is found in most human tumors but not in normal adult tissues [131-132]. In *neuroblastoma*, expression correlates with a more aggressive and unfavorable disease [123]. Survivin inhibited UVB-induced apoptosis *in vitro* and *in vivo*, whereas it did not affect CD95-induced cell death [123]. Expression of a non-phosphorylatable mutant of survivin induces cytochrome *c* release and cell death. In xenograft tumor models, this mutant suppressed tumor growth and reduced intra-peritoneal tumor dissemination [133-134]. The frequent translocation t(11;18)(q21;q21) that is found in about 50% of marginal cell lymphomas of the mucosa-associated lymphoid tissue (MALT) affects cIAP2 gene [135]. This indicates a role for cIAP2 in the development of MALT lymphoma. ML-IAP is expressed at high levels in melanoma cell lines but not in primary melanocytes. Melanoma cell lines that express ML-IAP are significantly more resistant to drug-induced apoptosis than those that do not express ML-IAP [136].

PI-9/SPI-6 expression of this serine protease inhibitor, which inhibits granzyme B, results in the resistance of tumor cells to cytotoxic lymphocytes, leading to immune escape [137].

Anti-apoptotic BCL2 family members (BCL2, BCL-X$_L$, *BCL-w,* MCL1, A1/*BFL1*, *BOO*/DIVA, NRH) can inhibit apoptosis through mitochondria on different levels. Most anti-apoptotic members contain the BCL2 homology (BH) domains 1, 2 and 4, whereas the BH3 domain seems to be crucial for apoptosis induction. Moreover, it has been shown *in vitro* and *in vivo* models that BCL2 expression confers resistance to many kinds of chemotherapeutic drugs and irradiation [138-140]. A characteristic feature of BCL2 proteins is the formation of homodimers and heterodimers with other members of the family. For example, BCL2 can form heterodimers with BAX, BAD or other proapoptotic molecules, and may thus block their proapoptotic activity. In some types of tumors, a high level of BCL2 expression is associated with a poor response to chemotherapy and seems to be predictive of shorter, disease-free survival [140-143]. The tumor-associated viruses Epstein–Barr virus (*EBV*) and human herpesvirus 8 (*HHV8* or Kaposi's sarcoma-associated herpesvirus) encode proteins that are homologues of BCL2. Both proteins BHRF1 from EBV and KSBCL2(vBCL2) from HHV8 have an anti-apoptotic function and enhance survival of the infected cells [144-146]. In this way, they might contribute to tumor formation after virus infection, and to resistance of these tumors to therapy. In addition, other anti-apoptotic BCL2 family members also seem to be involved in resistance of tumors to apoptosis. For example, BCL-X$_L$ can confer resistance to multiple apoptosis-inducing pathways in cell lines and seems to be upregulated by a constitutively active mutant epidermal growth factor receptor (*EGFR*) *in vitro* [147-150]. *MCL1* (myeloid cell leukaemia sequence 1) can also render cell lines resistant to chemotherapy [151]. In some leukaemia patients, MCL1 expression was increased at the time of relapse, which indicates that some anticancer drugs might select for leukaemia cells that have elevated MCL1 levels.

3. Inactivation of Pro-Apoptotic Genes

Caspase-8

Convincing evidence is accumulating which shows caspase-8 to be a key and irreplaceable molecule in TRAIL-induced as well as Fas-L and TNF alpha-induced apoptosis. Caspase-8 is deleted or silenced preferentially in childhood neuroblastomas with amplification of MYCN [152]. In these tumors, the gene for the initiator caspase-8 is frequently inactivated by gene deletion or methylation. Caspase-8-deficient neuroblastoma cells are resistant to death receptor- and doxorubicin -mediated apoptosis.

BAX and BAK

Besides overexpression of anti-apoptotic genes, tumors can acquire apoptosis resistance by down-regulating or mutating pro-apoptotic molecules. In certain types of cancer, the pro-apoptotic BCL2 family member BAX or BAK are mutated [153-155]. Reduced *BAX* expression is associated with a poor response rate to chemotherapy and shorter survival in some situations [156]. Several studies in mice have confirmed the function of BAX as a tumor suppressor. The importance of BAX, BAK or both in death receptor signaling or chemotherapy drug-induced apoptosis seems to depend on the cell type. However this is the ratio between BAX or BAK/BCL2 or BC-X_L may play a role in determining cell death sensitivity.

APAF-1

Metastatic melanomas have found another way to escape mitochondria-dependent apoptosis. These tumors often do not express APAF-1, which forms an integral part of the apoptosome [157], and the *APAF-1* locus shows a high rate of allelic loss. The remaining allele is transcriptionally inactivated by gene methylation. *APAF1*-negative melanomas fail to respond to chemotherapy a situation that is commonly found in this type of tumor.

4. p53-Dependent Pathways

Defects in DNA repair including defects in non-homologous end joining (NHEJ) and homologous recombination (HR), in DNA-damage checkpoints or in telomere maintenance lead to genomic instability and a predisposition to cancer (Figure 4) [1]. This genetic instability provides a way in which a normal cell can accumulate sufficient mutations to become malignant, and is the basis for the so-called 'mutator hypothesis' of cancer [158]. However, the cell has important mechanisms to protect against genomic instability driven by the loss of cell-cycle checkpoints, persistent DNA damage or telomere dysfunction. Central to this mechanism is the tumor-suppressor protein p53, which acts as a 'guardian of the genome' in protecting cells against cancer. *TP53* is the most commonly mutated gene in human cancer, a finding that reflects its crucial anticancer activity [159]. p53 acts to obstruct tumorigenesis by serving as a cellular-stress and DNA-damage sentinel.

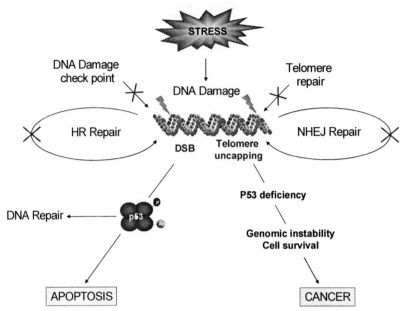

Figure 4. p53-dependent apoptosis, malignancy and resistance to cancer treatments. Defects in several processes, including DNA-damage checkpoint function, telomere maintenance, and DNA double-strand break (DSB) repair including non-homologous end joining (NHEJ) and homologous recombination (HR), result in DSBs or exposed chromosomal ends. In the presence of p53, cells with DSBs or uncapped telomeres are predisposed to apoptosis, which limits tumorigenesis. If, however, p53 is absent, cells with DSBs or eroded telomeres can survive inappropriately, creating a permissive environment for the generation of genomic instability that can drive carcinogenesis. Defects in apoptosis underlie not only tumorigenesis, but also resistance to cancer treatments.

In response to a variety of stress signals such as DNA damage, oncogene activation, hypoxia or nucleotide depletion, p53 undergoes post translational modification (phosphorylation, acetylation and sumoylation) that allows its stabilization and accumulation in the nucleus. Then, p53 activates the transcription of genes involved in cell cycle arrest, apoptosis, and DNA repair [160]. p53 is inhibited by MDM2, a ubiquitin ligase that targets p53 for its destruction by the proteasome. MDM2 is inactivated by binding to ARF (alternate open reading frame). Cellular stress, including that induced by chemotherapy or γ-irradiation, activates p53 either directly, by inhibition of MDM2, or indirectly by activation of ARF. ARF can also be induced by proliferative oncogenes such as RAS. p53 can induce the expression of numerous apoptotic genes such as CD95, TRAIL-R2, BAX, Bam, Puma and Noxa that can contribute to the activation of both death-receptor and mitochondrial apoptotic pathways. The choice of response to p53 activation is determined, in part, by differential regulation of p53 activity in normal and tumor cells. In this model, activation of p53 in normal cells leads to the selective expression of cell-cycle-arrest target genes (such as *CDKN1A*, which encodes WAF1), resulting in a reversible or permanent inhibition of cell proliferation. In tumor cells, phosphorylation of p53 at Ser46 (through activation of kinases, expression of co-activators such as p53DINP1 or repression of phosphatases such as WIP1) and/or functional interaction with apoptotic cofactors, such as ASPP (apoptosis-stimulating protein of p53), JMY (Junction-mediating and regulatory protein) and the other p53-family members p63 and p73 allows for the activation of apoptotic target genes. These cofactors can bind p53 (directly or indirectly) as shown for ASPP and JMY or as shown for p63 and p73 assist p53 DNA binding by directly interacting with p53-responsive promoters.

In parallel, p53 can possess a non nuclear proapoptotic function that is independent of its transcriptional activity. This pathway occur through transcriptional upregulation of PIDD protein (P53-induced protein with a death domain) [161]. PIDD can promote assembly of a complex between itself, RAIDD (receptor-interacting protein (RIP)-associated ICH-1/CED-3 homologous protein with a death domain) and caspase- 2 ('the piddosome'). It remains unclear how assembly of the piddosome can promote cell death, but this may involve caspase 2−dependent MOMP [66]. More recent evidence has also revealed a function for p53 as a binding partner of antiapoptotic members of the BCL2 family in the outer membrane such as BCL2 and Bcl-XL [162-163].

As in tumor cells p53 is a key element in stress-induced apoptosis, alterations of the p53 pathway influence the sensitivity of tumors to apoptosis. Tumors that are deficient in Trp53 (the gene that encodes p53 in mice) in immunocompromised mice and cell lineages from transgenic mice that express mutant *Trp53* showed a poor response to γ-irradiation or chemotherapy [164]. Specific mutations in *TP53* (the gene that encodes p53 in humans) have been linked to primary resistance to *doxorubicin* treatment and early relapse in patients with *breast cancer* [165]. In cancer cell lines, the specific disruption of the *TP53* gene conferred resistance to 5-FU, but greater sensitivity to *adriamycin* or radiation *in vitro* [166]. Adenoviral transfer of the wild-type *Trp53* gene into tumor cells with mutated p53 induced apoptosis and suppressed tumor growth in nude mice [167-171]. Mutations of *CDKN2A* (also named Ink4a/Arf locus encodes two tumor suppressor genes, $p16^{Ink4a}$ and $p19^{Arf}$) are almost as widespread in tumors as are *TP53* mutations [172]. Lymphomas from *Trp53*-knockout mice and from *Cdkn2a*-knockout mice are highly invasive, display apoptotic defects and are markedly resistant to chemotherapy *in vitro* and *in vivo* [173].

In addition to the fact that most human cancers have either mutations in p53 or defects in the pathway, p53-null mice are highly prone to developing cancers. Moreover, crosses of most tumor-prone strains to p53-null mice result in increased tumorigenesis that is clearly correlated with, at least in some cases, loss of the apoptotic function of p53 [174]. Therefore, inactivation of the apoptosis pathway allows cancer cells to develop (Figure 4). So important is this inactivation of apoptosis to cancer development that evasion of apoptosis is considered to be one of the six fundamental hallmarks of cancer. A corollary of this is that if the apoptotic pathway inactivated in tumor development is the same as, or overlaps, that leading to cell death by DNA damage, most cancers would be expected to be resistance to apoptosis in response to DNA damage [175] ; most tumors lose the ability to die by apoptosis. For example, apoptosis has a relatively modest role in the tumor response to radiation. The anti-tumor effect of radiation is realized through mitotic catastrophe or in senescence-like irreversible growth arrest [176].

5. Further Mechanisms

Among the main forces limiting cell death to its appropriate and physiological level are survival signals derived from the activation of the PI3K/AKT/PTEN, the NF-κB and MAPK pathways. These signaling cascades are organized in intricate networks, and a detailed description of their regulation and outcome is outside the scope of this review. Below is a summary of how these pathways may contribute to tumor resistance.

5.1. Altered several signaling pathways

PI3K/AKT/PTEN-This pathway is typically engaged in response to multiple mitogens (including oncogenes such as Ras) that bind to receptor kinases at the plasma membrane and lead to the activation of the phosphoinositide 3-kinase (PI3K) [177]. Once activated, PI3K converts the lipid PIP_2 into PIP_3. PIP_3 activates the protein kinase B/AKT which, in turn, targets multiple factors involved in cell proliferation, migration and survival. Regarding survival functions, AKT promotes the transcription of Bcl-X_L, and the inactivation of the proapoptotic protein BAD, caspase- 9, and the transcription factor FKHRL1 (an inducer of a number of proapoptotic factors). In addition, AKT can activate NF-κB and potentiate its survival functions, illustrating the intricate crosstalk between both pathways [177].

Such a prosurvival force is too dangerous to be unchecked [178] , and thus mammalian cells have developed intrinsic mechanisms to regulate AKT activity. At the center of these protective mechanisms is the tumor suppressor PTEN, a phosphatase that targets PIP_3 and prevents the activation of AKT [179]. Therefore, PTEN counteracts survival signals and promotes apoptosis. Moreover, feedback loops have been recently found between PTEN and the p53 tumor suppressor pathway [180-181]. One-third of primary melanomas and about 50% of metastatic melanoma cell lines showed reduced expression of PTEN as a result of allelic deletion, mutation or transcriptional silencing [182-183] , suggesting that inactivation of PTEN is a late, but frequent, event on melanomagenesis [184-187].

The NF-κB pathway-NF-κB is a transcription factor considered 'at the crossroads of life and death' by its function as a modulator of inflammation, angiogenesis, cell cycle, differentiation, adhesion, migration and survival [188]. In the context of cell death control, NF-*k*B modulates the expression of a plethora of survival factors that interfere with mitochondrial and death receptor-mediated apoptosis [188] (Figure 2). In melanoma cells, the NF-*k*B pathway can be altered by upregulation of the NF-κB subunits p50 and Rel A [189-190] and downregulation of the NF-κB inhibitor IκB [191-192]. Consequently, downstream NF-κB targets like c-*myc*, cyclin D1, the antiapoptotic factor TRAF2, the invasion-associated proteins Mel-CAM or the proangiogenic chemokine GRO are also frequently upregulated in melanoma [178]. Depending on the stimulus and the cellular context, NF-κB can activate pro-apoptotic genes, such as those encoding CD95, CD95L, and TRAIL receptors, and anti-apoptotic genes, such as those encoding IAPs and BCL-X_L. Genes encoding NF-κB or IκB proteins are amplified or translocated in human cancer [193]. In Hodgkin's disease cells, constitutive activity of NF-κB has been observed [194].

The MAP kinases pathways-The MAP kinases are a superfamily of proteins that transmit signaling cascades from extracellular stimuli into cells ; examples of MAP kinases include extracellular signal-regulated kinases (ERKs), SAPK, stress-activated protein kinase (also known c-jun N-terminal protein kinases, JNKs), and p38 MAP kinases. Like NF-κB, MAP kinases participate in a wide variety of cellular processes, including immunoregulation, inflammation, cell growth, cell differentiation, and cell death [144]. SAPK can regulate the activity of AP-1 transcription factors. Usually, activation of ERKs in response to death stimuli is believed to have an antiapoptotic effect. In support of this conclusion were findings that TRAIL induced rapid ERK1/2 activation in a group of melanoma cell lines, and the inhibition of that activation sensitized TRAIL-resistant melanoma cells to TRAIL-induced apoptosis, suggesting that ERK1/2 activation can itself protect against TRAIL-induced cell death in these TRAIL-resistant cell lines [146]. However, TRAIL also induced rapid ERK1/2 activation in TRAIL-sensitive melanoma cell lines, indicating that ERK1/2 activation by itself is not sufficient to protect against TRAIL-induced cell death in these TRAIL-sensitive cell lines. Zhang et al. hypothesize that TRAIL treatment involves different MAP kinases, different cell environments, and different cytokines, all of which interact to tip the balance in favor of cell survival or cell death [11] . In TNF-resistant human breast carcinoma cells several mechanisms contribute to protect cells to TNF-α-induced cell death including abnormal cleavage of cytosolic phospholipase A2, alteration the cellular redox state and the loss of p53 wild-type function. However, the effect elicited by TNF-α on cell death was NF-kappaB- and SAPK/JNK-independent [195]. Probably, the implication of MAP Kinase pathway engaged in tumor resistance depends on the stress stimulus, the cell type, the tumor environment and many other factors.

5.2. Chemoresistance Mechanisms

Cancer treatment by chemotherapy and γ-irradiation kills target cells primarily by inducing apoptosis. Therefore, key elements of resistance to apoptosis influences resistance to chemotherapy and γ-irradiation. Traditionally, chemoresistance was attributed to a failure of drug-target interactions either due to a reduction of the effective concentration of the drug, via enhanced drug efflux pumps [196], or to detoxification enzymes or the drug's target(s) itself(s) [197-199]. The two transporters that are commonly found to confer chemoresistance in cancer are the *MDR1* gene products P-glycoprotein and MRP (multidrug resistance-associated protein) [200]. P-glycoprotein protects cells not only from chemotherapy-induced apoptosis, but also from other caspase-dependent death stimuli such as CD95L, TNF and UV-radiation. However, it does not confer resistance to the perforin/granzymes pathway [196, 201-202]. MRP is often associated with an ATP-dependent decrease in cellular drug accumulation which is attributed to the overexpression of certain ATP-binding cassette (ABC) transporter proteins. ABC proteins that confer drug resistance include (but are not limited to) P-glycoprotein (gene symbol *ABCB1*), the multidrug resistance protein 1 (MRP1, gene symbol *ABCC1*), MRP2 (gene symbol *ABCC2*), and the breast cancer resistance protein (BCRP, gene symbol *ABCG2*). In addition to their role in drug resistance, there is substantial evidence that these efflux pumps have overlapping functions in tissue defense. Collectively, these proteins are capable of transporting a vast and chemically diverse array of toxicants including bulky lipophilic cationic, anionic, and neutrally charged drugs and toxins as well as conjugated organic anions that encompass dietary and environmental carcinogens, pesticides, metals, metalloids, and lipid peroxidation products. P-glycoprotein, MRP1, MRP2, and BCRP/ABCG2 are expressed in tissues important for absorption (e.g., lung and gut) and metabolism and elimination (liver and kidney). In addition, these transporters have an important role in maintaining the barrier function of sanctuary site tissues (e.g., blood–brain barrier, blood–cerebral spinal fluid barrier, blood–testis barrier and the maternal–fetal barrier or placenta). Thus, these ABC transporters are increasingly recognized for their ability to modulate the absorption, distribution, metabolism, excretion, and toxicity of xenobiotics [203].

How these mechanisms contribute to drug resistance in melanoma is a matter of controversy [133, 204]. For example, although some reports indicate an upregulation of drug pumps such as P-glycoprotein and the multidrug resistance-associated protein MPR-1 upon treatment, others fail to do so [133, 204]. Similar contradictory results have been reported detoxification factors such as the glutathione/glutathione S transferase, associated with the inactivation of alkylating drugs. For drugs like etoposide, it has been argued that tumor cells may avoid DNA damage by actually downregulating its target, topoisomerase II [205-206]. Specifically, topoisomerase II has been found to be downregulated or mutated in melanoma cells [207-208], although its levels may not necessarily correlate with drug sensitivity [209]. Another possible mechanism to counteract the deleterious effects of DNA-damaging drugs could be a hyperactivation of DNA repair mechanisms, either by upregulating mismatch repair genes or by potentiating enzymes that remove DNA-alkylation damage. Again, reports in the melanoma literature come in different flavors. MSH2 and other mismatch repair genes can be found upregulated [210] or downregulated [211]. Therefore, the melanoma field is in great need of standardized pharmacological studies that unequivocally determine how chemotherapeutic cells are incorporated into tumor and normal melanocytes. However, the fact that chemotherapeutic drugs do activate classical DNA damage sensors in melanoma, for example, related to the p53 pathway [21, 212-214] and that restoring apoptotic defects increases drug sensitivity, indicates that melanoma cells actually *sense* the drugs but have developed clever escape alternatives to prevent or compensate for their action.

The extracellular matrix might also contribute to drug resistance *in vivo* [215]. *Small-cell lung cancer* is surrounded by an extensive stroma of extracellular matrix, and adhesion of the cancer cells to the extracellular matrix suppresses chemotherapy-induced apoptosis through integrin signalling. Furthermore, in myeloma, constitutive activation of STAT3 signalling upregulates BCL-X$_L$ and so confers resistance to apoptosis [216].

III. Prion Protein

The prion hypothesis (proteinaceous infectious particules) that spongiform encephalopathies are caused and transmitted by a misfolded prion protein, a protease-resistant protein referred to as PrP^{Sc}, has been accepted for some time, but has not yet been proved. Recent data lead us a step closer to proving the infectivity of prion protein [217-220]. These authors report for the first time that synthetic mammalian prions cause disease when they are transferred to transgenic mice [217]. Two recent papers using yeast system provide also the strongest evidence yet to support the protein-only hypothesis for the transmission of prion diseases. Similar to the events that occur in prion diseases, Sup35 can be converted into an 'infectious' form that can propagate itself and form aggregates — known as amyloids.

Cells in which this has occurred are known as [*PSI*⁺] cells, and can be distinguished from their normal counterparts, [*psi*⁻] cells, by alterations in their colour under certain conditions. The authors used *Escherichia coli* to overexpress a region of Sup35 that is sufficient to stimulate amyloid formation and purified aggregates of this protein. Expression in a bacterial system ensured the absence of any virus from the yeast cells that might be responsible for infectivity. They then used novel methods to deliver the aggregates into [*psi*⁻] cells and showed that this resulted in conversion to the [*PSI*⁺] state. Protease treatment greatly decreased the infectivity of the aggregates, whereas nuclease treatment had no effect, providing the strongest evidence so far that prion proteins, in the absence of genetic material, are sufficient for infectivity. In addition, it confirms that distinct 'strains' of prion arise in the absence of genetic alterations owing to differences in protein conformation [218-219]. This conformational conversion and subsequent pathologies absolutely require the presence of PrPc since the absence of endogenous PrPc totally precludes the PrP^{Sc}-mediated infectivity and neurotoxicity [220]. We believe that the issue of identifying PrPc function is important because understanding these functions have an impact on the pathophysiological manifestation of prion disease and cancer.

Structure and Location of PrPc

The gene *PRNP* that encodes prion protein is located on human chromosome 20pter-p12, approximately 20 kbp upstream of *PRND* gene which encodes a biochemically and structurally similar protein (Doppel) to the prion protein. *PRNP* gene spans 20 kbp and composed of 2 exons. Mutations in the repeat region as well as elsewhere in *PRNP* gene have been associated with Creutzfeldt-Jakob disease, fatal familial insomnia, Gerstmann-Straussler disease, Huntington disease-like 1, and kuru. Two transcript variants (2479 nucleotides) encoding the same protein have been found for this gene (Sanger Institute- Ensembl Protein Report, 2005). This gene code for the cellular prion protein (PrPc) consists of 253 amino acids (Figure 5), an ubiquitous protein of 32-35 kDa expressed by all known mammals predominantly in the brain, lymphocytes and stroma cells of lymphoid organs [6].

Trafficking of Cellular Prion Proteins

In cells, PrPc is post-translationally processed (posttranslational GPI anchoring, disulfide bonding, glycosylation) and transported along the secretory pathway to the plasma membrane, where it is attached to the cell surface by a glycosylphosphatidylinositol anchor. The PrPc can adopt multiple membrane topologies, including a fully translocated form two transmembrane forms (^{Ntm}PrP and ^{Ctm}PrP), and a cytosolic form [221]. At the plasma membrane PrPc can be constitutively internalised. The route and mechanism of internalization of PrPc are controversial [222]. Seminal studies by Harris and colleagues showed that ectopic expression of chicken PrPc into mouse N2a neuroblastoma cells was internalized via coated pits in a process dependent upon its N-terminal domain [223]. However, the concept that the GPI-anchor determines PrPc trafficking has been challenged.

Indeed, Nunziante et al.[224] have shown that mouse PrPc in which the N-terminal domain has been substituted by the non-Cu^{2+}-binding *Xenopus* homologue is endocytosed on N2a cells. PrPc endocytosis can also be increased by intercellular Cu^{2+} [223]. In neuronal cells, a major pathway for internalization of PrPc appears to use clathrin-mediated endocytosis. PrPc seems to coexist in the cell surface in raft and non raft components of the membrane (lipid rafts are defined as region of membranes resistant to cold detergent extraction), and it has been suggested that PrPc may leave rafts to be internalizes by coated-pits [225]. In contrast, subsequent studies [226-227], also using N2a cells, have argued that mammalian PrPc is internalized by one of the non-coated pit mechanisms characterized for raft-associated proteins [228-229]. Peters *et al.* using immunoelectron microscopy analysis in CHO cells have shown that PrPc uses an atypical endocytic pathway to reach the lysosomes. This pathway does not involve clathrin-coated vesicles, but contains caveolin-1, a protein that is characteristic of caveolae [230]. Particularly interesting was the demonstration that PrPc interacts functionally with Fyn, in a caveolin-1 manner in 1C11 cells, leading to activation of this tyrosine kinase activity [231]. More recently, Sunyach and collaborators have shown that experimental conditions aimed at blocking endocytosis prevent PrPc internalization and concomitantly abolish PrPc-mediated p53-dependent caspase-3 activation in human cells. Thus, it can be concluded that p53-dependent caspase-3 activation triggered by PrPc is directly dependent upon its endocytosis in human cells lines [232].

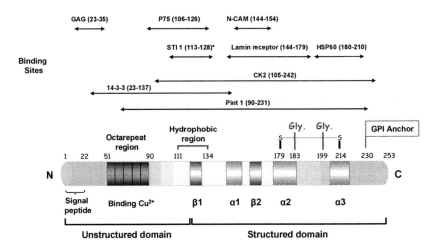

PrPc

Figure 5. Representation of the binding domains of PrPc ligands on the human PrPc molecule. The human PrPc molecule contains a signal peptide (1-22), five octapeptide repeats (51-91), highly conserved hydrophobic domain (106-126), three peptide sequences responsable for α-helix structure (α1-α2-α3), two peptide sequences responsable for β-helix structure, and a signal sequence for GPI anchor (231-254). Double arrows indicated the binding site for each PrPc-binding molecule, which is also represented by amino acid numbers in parentheses and correspond to the mouse PrPc amino acid sequence and to the bovine sequence [242]. Highly conserved octapeptide repeats of PrPc are similar to the BCL2 homology domain 2 (BH2) of BCL2 family proteins. Recently, the 14-3-3-binding domain was located in the N-terminal half of PrPc spanning amino acid residues 23-137[258].

It is still unclear whether caveolae-mediated endocytosis is involved in PrPc trafficking in epithelial or/and tumor cells. However, the data raise the possibility that transit through this unusual pathway might be involved in the implication of PrPc in tumor resistance to cell death induced by death receptor ligand.

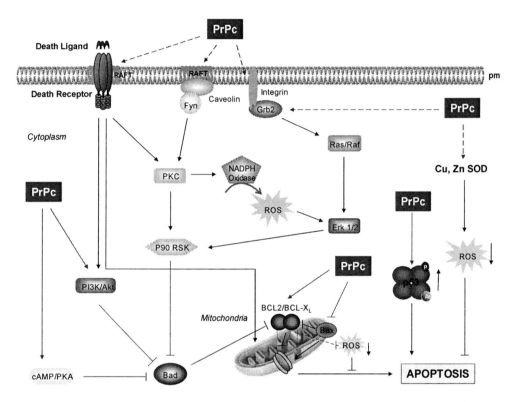

Figure 6. Signaling through PrPc. PrPc by interaction with various partners is involved in apoptosis. PrPc would inhibit the apoptotic process through multiple mechanisms, such as death receptor ligand pathway through interaction with lipid rafts and the endocytosis event, mitochondria pathway through its anti-BAX function and cellular redox homeostasis through decrease of ROS production. Dotted lines indicate the hypothetic involvement of PrPc.

Signal Transduction of Cellular Prion Proteins

Recent studies have advanced the hypothesis that PrPc may a role in signal transduction (Figure 6). Antibody-mediated PrP dimerization elicits rapid phosphorylation of extracellular regulated kinases 1 and 2 (ERK1/2) in neuron-like cell line and BW5147 lymphoid cells [231]. Particularly interesting was the demonstration that PrPc interacts functionally with Fyn, in a caveolin-1 manner in 1C11 differentiated neuronal cells leading to activation of this tyrosine kinase activity [231-233]. Here it is argued that since ERK1/2 are fully controlled through NADPH oxidase-dependent reactive oxygen species production, PrPc basically functions in maintaining the cellular redox homeostasis [231-233]. In another study, a PrPc binding peptide activated both the ERK- and cAMP-dependent protein kinase A pathway in retinal explants from neonatal mice [9]. Finally, mouse primary cerebellar granule neurons seeded either onto PrPc-coated dishes or onto Chinese hamster ovary cells overexpressing PrPc at their cell surface have increased neurite outgrowth and neuronal survival [234]. Neuronal survival involves the activation of the phosphatidyl-inositol-3-kinase/Akt and the mitogen-activated protein kinase/ERK kinases pathways. PrPc also binds to laminin and this interaction promotes neuritogenesis by a mechanism that could involve the mitogen-activated protein kinase/ERK pathway [234-,35]. Laminin is a major adhesive molecule, and the interaction with PrPc may promote neuronal adhesion. Intracellular endocytosed PrPc may interact with proteins involved in signaling pathways including Grb2, an adaptor protein involved in neuronal survival [236]. As with other GPI-anchor proteins, PrPc can transfer from cell to cell and may transduce signals to a large number of cells [237]. Furthermore, recent findings in neuronal and non-neuronal cell models indicate prion protein association with secreted exosomes. Exosomes are membrane vesicles released into the extracellular environment upon exocytic fusion of multivesicular endosomes with the cell surface. Exosome secretion can be used by cells to eject molecules targeted to intraluminal vesicles of multivesicular bodies, but particular cell types may exploit exosomes as intercellular communication devices for transfer of proteins and lipids among cells [238].

Taken together, these studies indicate that endogenously expressed PrPc can change several intracellular signaling pathways specially involving in cell adhesion, cell trafficking, and transmembrane signaling those to determine cellular survival.

PrPc Protects Human Neurons Against Cell Death

Several intriguing lines of evidence have emerged recently indicating that PrPc may function to protect human neurons from various kinds of internal or environmental stress (for review see [12]). The original of a possible PrPc participation in programmed cell death came from the identification of a significant similarity with the BCL2 homology domain 2 (BH2) and its ability to bind to BCL2 in yeast double hybrid approach [239-240]. Particularly, interesting studies from Kuwahara laboratory indicated that PrPc could protect human neurons from serum withdrawal [241] and BAX-induced apoptosis [10]. The octapeptide repeat domain that displays some similarity with the BH2 domain of BCL2 is essential for PrP's neuroprotective function against BAX. Furthermore, familial PrPc mutations D178N and T183A associated with the human prion diseases fatal familial insomnia and familial atypical spongiform encephalopathy partially or completely abolish PrPc anti-BAX function [12]. Therefore, the authors propose that the normal structure of the PrPc is likely important for the anti-BAX function. While PrPc needs to achieve some form of maturity (glycosylation and transport) in for its anti-BAX function, the GPI anchor of PrPc is not required for this function [10]. Furthermore, PrPc expressed uniquely in the cytosol is also able to inhibit BAX [12]. Recently the same laboratory show that PrPc is very specific for BAX and cannot prevent BAK -, tBID-, staurosporine- caspase- or thapsigargin-mediated cell death. PrPc does not colocalize with BAX in normal or apoptotic primary neurons and cannot prevent BAX-mediated cytochrome *c* release in a mitochondrial cell-free system. PrPc could protects against BAX-mediated cell death by preventing the BAX proapoptotic conformational change that occurs initially in BAX activation [11]. The authors propose that both BCL2 and

PrPc maintain BAX in an inactive state and confer neuroprotection. The conversion of PrP into PrP^{Sc} would inactivate antiapoptotic PrP, leaving BCL2 as the only major BAX inhibitor. At this stage the balance between BAX and BCL2 is equilibrated, but neurons would be more sensitive to apoptotic insults . Any further event leading to decreased expression of BCL2 as observed in the aging CNS and loss of PrPc neuroprotective function would increase neurotoxicity [12]. However, expression of PrPc can prevents the release of cytochrome c induced by serum deprivation in hippocampal cell lines HW8-2 and HpL3-4 [242]. Li and collaborators used mouse PrPc containing a modified signal peptide and yeast system to analyse the implication of PrPc in BAX-induced cell death in yeast [243]. The authors found that this PrPc potently suppressed the death of yeast cells expressing mammalian BAX. In contrast, cytosolic PrP-(23–231) failed to rescue growth of BAX-expressing yeast, indicating that protective activity requires targeting of PrPc to the secretory pathway. In opposite to the data of Leblanc's group in neuron cells; the octapeptide repeat domain is not essential for PrPc's neuroprotective function against BAX in yeast system, while a charged region encompassing residues 23–31 is essential for this function. Although *S. cerevisiae* does not contain endogenous *BCL2* family members or caspases, the initial events underlying BAX activity in yeast and mammalian cells are similar, including translocation of the protein to mitochondria, release of cytochrome c, and alterations in mitochondrial function [244]. How does PrPc protect yeast from BAX-induced cell death? In contrast of PrPc's neuroprotective function against BAX which it become from directly association between PrPc with BAX and inactivation of BAX, in yeast the protective effect of PrPc seems to be related to its interacts with endogenous yeast proteins that lie downstream of BAX in a cellular stress or toxicity pathway.

Another studies show that the PrPc binds to a heat-shock-related protein, stress-inducible protein 1 (*STI1*) (Figure 5), and that the interaction between these two proteins at the cell surface can rescue cultured retinal cells from apoptosis induced by treatment with anisomycin. A functional role for this interaction is supported by the finding of both proteins on neuronal cell surfaces in the central nervous system. The same research group shows that binding of PrPc by a peptide that recognizes the STI1-binding site (known as PrR) also protects neurons against apoptosis *in vitro*. It goes further to demonstrate that this protection depends on an increase in the levels of cyclic AMP that activates protein kinase A. the first question is how the binding of PrPc at the outer membrane activates adenylyl cyclase, which is normally regulated by G proteins on the inner membrane. We can find a beginning of explanation in the data of Mouillet-Richard *et al* [245]. The authors propose recently that the antibody-mediated ligation of PrPc affects the potency or dynamics of G-protein activation by agonist-bound serotonergic receptors. The PrPc-dependent modulation of 5-HT receptor couplings is restricted to 1C115-HT cells expressing a complete serotonergic phenotype. It critically involves a PrPc-caveolin platform implemented on the neurites of 1C115-HT cells during differentiation. In addition, PrPc-null mice are more susceptible to neuronal loss after experimental brain injury [246]. Examination the phenotype of PrPc-null mice show increased levels of nuclear factor NF-κB and Mn superoxide dismutase, COX-IV, decreased levels of Cu/Zn superoxide dismutase activity, decreased p53, and altered melatonin levels. Additionally, neurons cultured from these animals display abnormalities related to increased susceptibility to oxidative stress [247]. Finally, expression of wild-type PrPc completely abrogates the neurodegenerative phenotype of mice expressing the PrP paralogue, Doppel, or N-terminally truncated forms of PrP (Δ32–121 and Δ32–134) [12, 248-249]. Contradictory results have been reported on phenotype of PrP-null mice. Some studies found no significant abnormalities whereas others reported impairments in neuronal functioning, loss of cerebellar Purkinje cells and defects in sleep patterns and circadian activity. The divergences among PrP-null mice arise from differences in the methods used to suppress protein expression, and more specifically whether the expression of an adjacent PrP homologue protein, *Doppel*, has been artificially induced in the brain [250]; for more detail see [12].

PrPc and Tumor Resistance

In order to define genetic determinants of tumor cell resistance to the cytotoxic action of TNF, we have applied cDNA microarrays to a human breast carcinoma TNF-sensitive MCF7 cell line and its established TNF-resistant clone. A great number of genes involved in the PI3K/Akt signalling pathway were differentially expressed. Unexpected, endogenous PrPc was found overexpressed at both mRNA (17 fold) and protein levels (10 fold) in TNF-resistant derivative cells as compared to TNF-sensitive MCF7 cell line. The confocal scanning fluorescence analysis showed that PrPc was highly expressed in the Golgi apparatus of TNF-resistant MCF7 cell line (Figure 7A, left). In addition, TNF-resistant MCF7 cells is sensitive to phosphoinositol phospholipase C (PIPLC) treatment confirming that PrPc form is overexpressed at the surface of this tumor cells (Figure 7A, right).

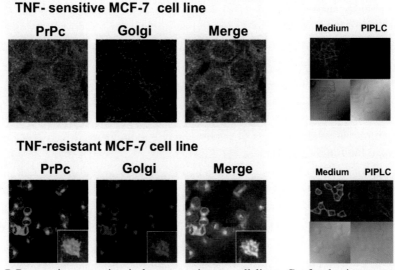

Figure 7A. PrPc protein expression in breast carcinoma cell lines. Confocal microscopy analysis for PrPc localization in breast carcinoma cell line. Left, TNF-sensitive MCF7 and TNF-resistant cell lines were immunostained with either with mouse Pri 308 anti-human PrPc monoclonal antibody (gift from Dr Grassi, CEA, France) or the rabbit RM130 Golgi-specific antibody (gift from Dr Bornons, Institut curie, France). Right: Cells were treated (PIPLC) or not (Medium) with 1 unit/ml PIPLC for 1 h at 37° C, followed by immunofluorescence staining with Pri 308 antibody. Nuclei were counterstained with Topro-3. The confocal scanning fluorescence micrographs are representative for the vast majority of the cells analyzed.

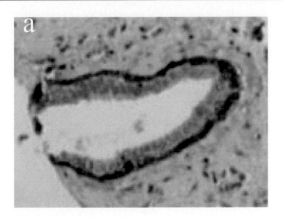

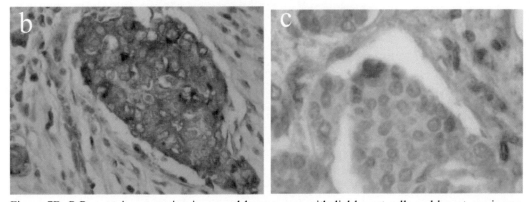

Figure 7B. PrPc protein expression in normal human myoepithelial breast cells and breast carcinoma. Immunohistochemistry analysis for PrPc. Tumor samples were obtained from 100 patients at Institue Gustave Roussy and were immunostained with SAF 69 anti-human PrPc antibody (gift from Dr Grassi, CEA, France). a, normal human breast tissue showing myoepithelial cells positive and luminal negative cells for PrPc. b, tumor tissue showing cancer cells and stromal cells positive for PrPc. c, tumor tissue showing cancer cells negative and stromal cells positive for PrPc. Photomicrographs are representative for the vast majority of the samples analyzed.

Immunohistochemical staining for PrPc in human breast carcinoma (one hundred patients) showed that PrPc was highly expressed in the tumor cells of 30 % of breast carcinoma and in the myoepithelial cells of all of normal tissue while it was absent in normal luminal cells (Figure 7B).

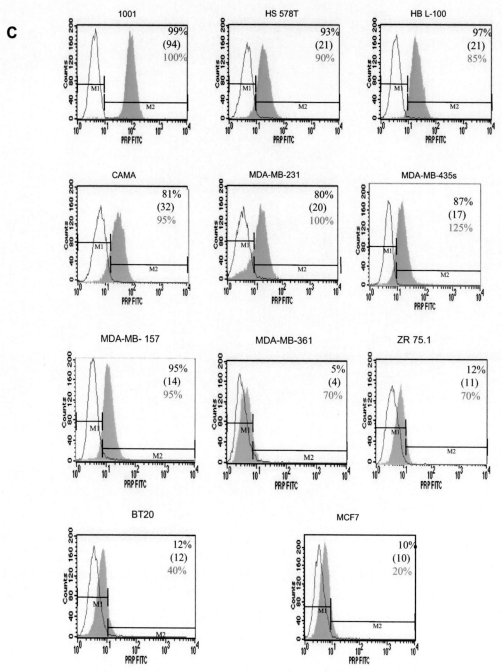

Figure 7C. PrPc protein expression in breast carcinoma cell lines. PrPc protein expression as determined by indirect immunofluorescence analysis using Pri 308 anti-human PrPc monoclonal antibody (green curve). Isotypic IgG1 control was included (open curve). The level of cell surface expression is indicated by the shift of the green curve to the right from the open control curve. Black numbers indicate percentages of positive cells. Numbers in parentheses correspond to mean fluorescence intensity. Cells were also treated or not with TNF (100 ng/ml) for 72 hours and the MTT assays were performed in replicate of three samples. Red numbers indicate percentages of MTT activity.

Examine of a panel of human breast carcinoma cell lines for their sensitivity to TNF and their PrPc expression show a good correlation between susceptibility to the cytotoxic action of TNF and expression for PrPc (Figure 7 C). Furthermore, ectopic expression of human PrPc converted TNF sensitive MCF7 cells into TNF resistant, by a mechanism involving alteration of cytochrome c release from mitochondria and nuclear condensation. These data show for the first time that ectopic expression of PrPc protects breast cancer cell line from TNF mediated cell death, by interfering with mitochondrial pathway and also its involvement in tumor resistance. More recently, Du *et al.* using the adriamycin-senstivie gastric carcinoma cell line SGC7901/ADR and its derivative resistance clone reported that PrPc is also involved in multidrug resistance. Overexpression of PrPc conferred resistance of both P-glycoprotein (P-gp)-related and P-gp-nonrelated drugs on SGC7901. PrPc knock down expression partially reverse multidrug-resistant phenotype of SGC7901/ADR. PrPc significantly upregulated the expression of the classical MDR-related molecule P-gp but not multidrug resistance associated protein and glutathione S-transferase pi. PrPc could also suppress adriamycin-induced apoptosis and alter the expression of BCL2 and BAX [15].

In contrast, we did not observed that PrPc could also suppress adriamycin-induced apoptosis in breast human breast carcinoma cell line (manuscript in preparation). The molecular target of PrPc is unknown. It is quite possible that the antiapoptotic function of PrPc observed in breast adenocarcinoma cells after TNF treatment occurs through its anti-BAX function. We can also imagine their involvement in early events in TNFR family signalling, particularly in lipid rafts. More recently, Sunyach and collaborators show that experimental conditions aimed at blocking endocytosis prevent PrPc internalization and concomitantly abolish PrPc-mediated p53-dependent caspase- 3 activation in human cells. Thus, they propose that p53-dependent caspase- 3 activation triggered by PrPc is directly dependent upon its endocytosis in human cells lines [232]. PrPc like 14-3-3 chapron protein and p53 seems to be involved in multiple cellular processes.

The Protective Function of PrPc: Controversy Story

To our knowledge, normal PrPc has never been shown to be toxic upon overexpression in tumor sample or transfected cells. Therefore, PrPc is not toxic by itself and tumor cell endogenously, transiently or stably overexpressing PrPc can be routinely obtained in laboratories (Figure 8). However, in human embryonic kidney 293 cell line, rabbit epithelial Rov9 cell line, and murine cortical TSM1 cell line, PrPc overexpression increases the susceptibility of these cells to the apoptotic inducer, staurosporine. The authors showed that this cellular response was due to activation of caspase- 3 via transcriptional and post-transcriptional control of the proapoptotic oncogene *p53* [251-252]. These results are in contradiction with reports indicating that the absence of PrPc sensitizes cells to apoptotic stimuli, and also with the fact that activation of ERKs in response to death stimuli is believed to have an antiapoptotic effect. The major cause to explain this discrepancy is that the effect of PrPc may depend on cell type and death stimuli. Recent reports show that BCL2 conformational change occurs during the onset of apoptosis, although it is suggested that this conformational change may be integral property for its antiapoptotic function [253].

Interaction of BCL2 with nuclear orphan receptor may convert the BCL2 from a protector to a killer molecule via conformational change [254]. We propose (Figure 9) that in the cell with BCL2 and PrPc protector both PrPc and BCL2 maintain BAX in an inactive state and confer cell protection. Interaction of BCL2 with PrPc or other factor convert BCL2 from a protector to a killer molecule via conformational change. Among others, BCL2 and BCL-X$_L$ can be cleaved by endogenous caspases to give a potent proapoptotic carboxy-terminal fragment. As a consequence, overexpression of these proteins would accelerate cell death after caspase activation [255]. This hypothesis is in agreement with other reports indicating that the transfection of the BCL2 gene in MCF-7 did not reduce TNF sensitivity [256]. While BCL2 was overexpressed in TNF-sensitive MCF7 cell line, it is downregulated in TNF-resistant MCF7 cell lin [167]. Adenovirus-mediated transfer of wild type p53 gene sensitizes TNF-resistant MCF7 cell line to the cytotoxic action of TNF without affecting the expression of BCL2 level [167]. PrPc like BCL2 may be converted into a lethal protein. PrPc has a caspase-like site at amino acid 145, and it may generate a potent proapoptotic PrPc fragment as seen with the 145 STOP codon mutation that is associated with a vascular form of prion disease [257]. Lastly, it is possible that unknown factors convert PrPc from a protector to a killer molecule via conformational change that induces cell death.

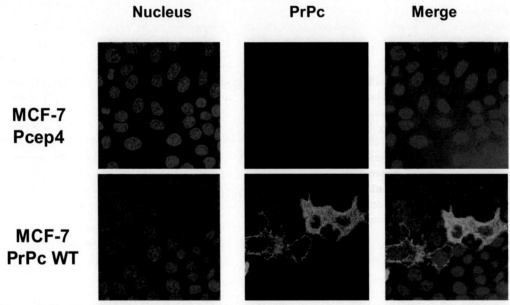

Figure 8. Overexpression of PrPc is not toxic for human breast carcinoma cell-line. The PrPc was expressed by the PrPc Wild type (WT) Pcep4 recombinant vector which encodes the human *PRNP* gene. MCF7 cell line was transfected for 24 hours, followed by immunofluorescence staining using Pri308 monoclonal antibody. Nuclei were counterstained with Topro-3. The confocal scanning fluorescence micrographs are representative for the vast majority of the cells analyzed.

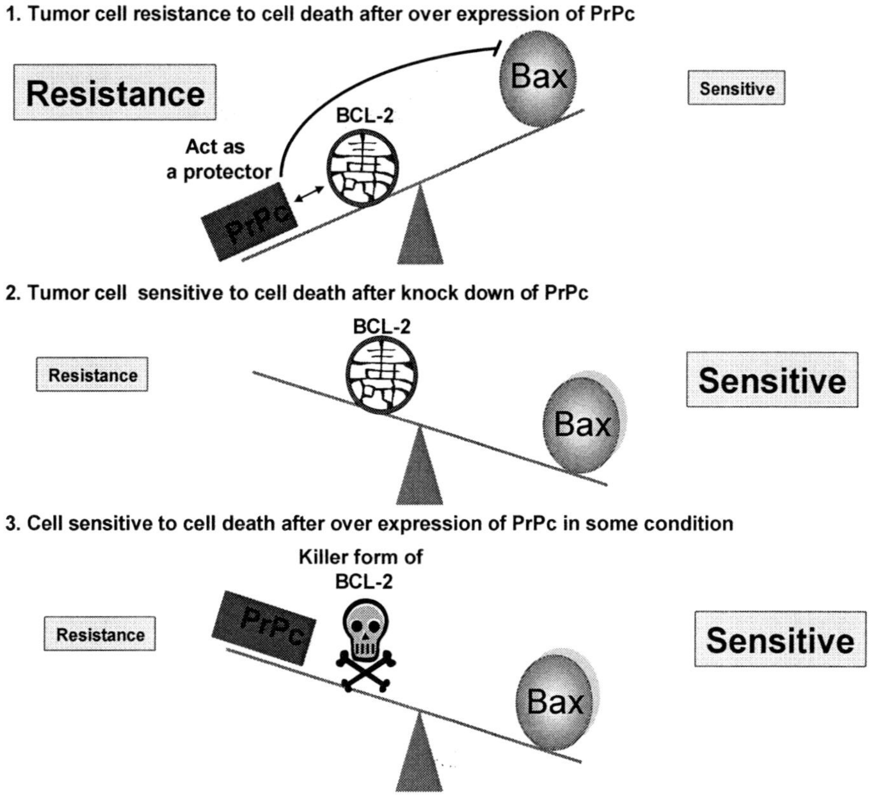

Figure 9. PrPc takes part in the balance between resistance and sensitivity to cell death. 1. Like Roucou et al.[13] we propose that both Bcl-2 and overexpression of PrPc maintain Bax in an inactive form and protect cancer cell from cell death. In some conditions, including Bax expression, serum withdrawal, TNF and anti-cancer drug PrPc overexpression confers resistance to cell death. 2. Cells deficient of PrPc are sensitive to cell death because BCL2 alone can not maintain Bax in an inactive form and protect cancer cell from cell death. 3. In some cells and/or under some death stimuli (for example staursporine), in the presence of PrPc and interaction with some partenaires (for example nuclear orphan recptor), Bcl-2 conformational change, convert the BCL2 protector form to killer form. Cells become sensitives to cell death. In the cell's deficience for PrPc, Bcl-2 conformational change does not happen and cells continue to resist to apoptosis.

Acknowledgments

We apologize to those investigators whose work was not cited or discussed because of space limitations. We thank the members of our laboratories, past and present, for their support and advice. This work was supported by grants from INSERM, Association pour la Recherche sur le Cancer (grant 3520 to MD). Franck Melin and Ahmed Hamaï are recipients of fellowship from Ligue Nationale Contre le Cancer and canceropole Ile de France, respectively.

References

[1]   Hanahan, D. and R.A. Weinberg, The hallmarks of cancer. *Cell,* 2000. 100: p. 57-70.

[2]   Green, D.R. and G.I. Evan, A matter of life and death. *Cancer Cell,* 2002. 1: p. 19-30.

[3]   Johnstone, R.W., A.A. Ruefli, and S.W. Lowe, Apoptosis: a link between cancer genetics and chemotherapy. *Cell,* 2002. 108: p. 153-164.

[4]   Prusiner, S.B., Early evidence that a protease-resistant protein is an active component of the infectious prion. *Cell,* 2004. 116(2 Suppl): p. S109, 1 p following S113.

[5]   Prusiner, S.B., Prions. *Proc Natl Acad Sci U S A,* 1998. 95(23): p. 13363-83.

[6]   Aguzzi, A. and M. Polymenidou, Mammalian prion biology: one century of evolving concepts. *Cell,* 2004. 116(2): p. 313-27.

[7]   Pauly, P.C. and D.A. Harris, Copper stimulates endocytosis of the prion protein. *J Biol Chem,* 1998. 273(50): p. 33107-10.

[8]   Mange, A., et al., PrP-dependent cell adhesion in N2a neuroblastoma cells. *FEBS Lett,* 2002. 514(2-3): p. 159-62.

[9]   Chiarini, L.B., et al., Cellular prion protein transduces neuroprotective signals. *Embo J,* 2002. 21(13): p. 3317-26.

[10]   Bounhar, Y., et al., Prion protein protects human neurons against Bax-mediated apoptosis. *J Biol Chem,* 2001. 276(42): p. 39145-9.

[11]   Roucou, X., et al., Cellular prion protein inhibits proapoptotic Bax conformational change in human neurons and in breast carcinoma MCF-7 cells. *Cell Death Differ,* 2005. 12(7): p. 783-95.

[12]   Roucou, X. and A.C. LeBlanc, Cellular prion protein neuroprotective function: implications in prion diseases. *J Mol Med,* 2005. 83(1): p. 3-11.

[13]   Roucou, X., M. Gains, and A.C. LeBlanc, Neuroprotective functions of prion protein. *J Neurosci Res,* 2004. 75(2): p. 153-61.

[14]   Roucou, X., et al., Cytosolic prion protein is not toxic and protects against Bax-mediated cell death in human primary neurons. *J Biol Chem,* 2003. 278(42): p. 40877-81.

[15]   Du, J., et al., Overexpression and significance of prion protein in gastric cancer and multidrug-resistant gastric carcinoma cell line SGC7901/ADR. *Int J Cancer,* 2005. 113(2): p. 213-20.

[16]   Diarra-Mehrpour, M., et al., Prion protein prevents human breast carcinoma cell line from tumor necrosis factor alpha-induced cell death. *Cancer Res,* 2004. 64(2): p. 719-27.

[17]   Senator, A., et al., Prion protein protects against DNA damage induced by paraquat in cultured cells. *Free Radic Biol Med,* 2004. 37(8): p. 1224-30.

[18]   Walczak, H. and P.H. Krammer, The CD95 (APO-1/Fas) and the TRAIL (APO-2L) Apoptosis Systems. *Exp. Cell Res.,* 2000. 256(1): p. 58-66.

[19]   Muppidi, J.R., J. Tschopp, and R.M. Siegel, Secondary Complexes and Lipid Rafts in TNF Receptor Family Signal Transduction. *Immunity,* 2004. 21(4): p. 461-465.

[20]   Walczak, H., Tumoricidal activity of tumor necrosis factor-related apoptosis- inducing ligand in vivo. *Nature Med.,* 1999. 5: p. 157-163.

[21] Krammer, P.H., CD95(APO-1/Fas)-mediated apoptosis: live and let die. *Adv. Immunol.*, 1999. 71: p. 163-210.

[22] Ashkenazi, A. and V.M. Dixit, Apoptosis control by death and decoy receptors. *Curr Opin Cell Biol*, 1999. 11(2): p. 255-60.

[23] Debatin, K., and and P. Krammer, Death receptors in chemotherapy and cancer. *Oncogene*, 2004. 23(16): p. 2950-66.

[24] Kischkel, F.C., Cytotoxicity-dependent APO-1 (Fas/CD95)-associated proteins form a death-inducing signaling complex (DISC) with the receptor. 1995. 14: p. 5579-5588.

[25] Green, D.R. and T.A. Ferguson, The role of Fas ligand in immune privilege. *Nat Rev Mol Cell Biol*, 2001. 2(12): p. 917-24.

[26] Igney, F.H.a. and P.H. Krammer, Death and anti-death: tumor resistance to apoptosis. *Nature Reviews Cancer Nat Rev Cancer*, 2002. 2(4): p. 277-288.

[27] Roy, S. and D.W. Nicholson, Programmed cell-death regulation: basic mechanisms and therapeutic opportunities. American Association for Cancer Research Special Conference: Lake Tahoe, CA, USA, 27 February-2 March 2000. *Mol Med Today*, 2000. 6(7): p. 264-6.

[28] Eggert A, Grotzer MA , and Z. TJ, Resistance to tumor necrosis factor-related apoptosis-inducing ligand (TRAIL)-induced apoptosis in neuroblastoma cells correlates with a loss of caspase-8 expression. *Cancer Res*, 2001. 61: p. 1314-1319.

[29] Daniel, P.T., et al., The kiss of death: promises and failures of death receptors and ligands in cancer therapy. *Leukemia*, 2001. 15(7): p. 1022-32.

[30] Cowling, V. and J. Downward, Caspase-6 is the direct activator of caspase-8 in the cytochrome *c*-induced apoptosis pathway: absolute requirement for removal of caspase-6 prodomain. *Cell Death Differ*, 2002. 9(10): p. 1046-56.

[31] Zinkel, S.S., A role for proapoptotic BID in the DNA-damage response. *Cell*, 2005. 122: p. 579-591.

[32] Kamer, I., Proapoptotic BID is an ATM effector in the DNA-damage response. *Cell*, 2005. 122: p. 593-603.

[33] Wiley, S.R., et al., Identification and characterization of a new member of the TNF family that induces apoptosis. *Immunity*, 1995. 3(6): p. 673-82.

[34] Danial, N.N. and S.J. Korsmeyer, *Cell*, 2004. 116: p. 205-219.

[35] LeBlanc, H.N. and A. Ashkenazi, Apo2L/TRAIL and its death and decoy receptors. *Cell Death Differ*, 2003. 10(1): p. 66-75.

[36] Van Geelen, C.M.M., E.G.E. de Vries, and S. de Jong, Lessons from TRAIL-resistance mechanisms in colorectal cancer cells: paving the road to patient-tailored therapy. *Drug Resistance Updates*, 2004. 7(6): p. 345-358.

[37] Chaudhary, P.M., et al., Death receptor 5, a new member of the TNFR family, and DR4 induce FADD-dependent apoptosis and activate the NF-kappaB pathway. *Immunity*, 1997. 7(6): p. 821-30.

[38] McDonnell, T.J., BCL-2-immunoglobulin transgenic mice demonstrate extended B cell survival and follicular lymphoproliferation. *Cell*, 1989. 57: p. 79-88.

[39] Pan, G., et al., An antagonist decoy receptor and a death domain-containing receptor for TRAIL. *Science*, 1997. 277(5327): p. 815-8.

[40] MacFarlane, M., et al., Identification and molecular cloning of two novel receptors for the cytotoxic ligand TRAIL. *J Biol Chem*, 1997. 272(41): p. 25417-20.

[41] Walczak, H., et al., TRAIL-R2: a novel apoptosis-mediating receptor for TRAIL. *Embo J*, 1997. 16(17): p. 5386-97.

[42] Wu, G.S., KILLER/DR5 is a DNA damage-inducible p53-regulated death receptor gene. *Nature Genet.*, 1997. 17: p. 141-143.

[43] Debatin, K.M. and P.H. Krammer, Death receptors in chemotherapy and cancer. *Oncogene,* 2004. 23(16): p. 2950-66.

[44] Degli-Esposti, M.A., et al., The novel receptor TRAIL-R4 induces NF-kappaB and protects against TRAIL-mediated apoptosis, yet retains an incomplete death domain. *Immunity*, 1997. 7(6): p. 813-20.

[45] Degli-Esposti, M.A., et al., Cloning and characterization of TRAIL-R3, a novel member of the emerging TRAIL receptor family. *J Exp Med*, 1997. 186(7): p. 1165-70.

[46] Emery, J.G., et al., Osteoprotegerin is a receptor for the cytotoxic ligand TRAIL. *J Biol Chem*, 1998. 273(23): p. 14363-7.

[47] Kischkel FC, et al., Apo2L/TRAIL-dependent recruitment of endogenous FADD and caspase-8 to death receptors 4 and 5. *Immunity*, 2000. 12: p. 611-20.

[48] Micheau, O. and J. Tschopp, Induction of TNF receptor I-mediated apoptosis via two sequential signaling complexes. *Cell*, 2003. 114(2): p. 181-90.

[49] Schneider, P., et al., TRAIL receptors 1 (DR4) and 2 (DR5) signal FADD-dependent apoptosis and activate NF-kappaB. *Immunity*, 1997. 7(6): p. 831-6.

[50] Ozoren, N. and W.S. El-Deiry, Cell surface Death Receptor signaling in normal and cancer cells. *Semin Cancer Biol*, 2003. 13(2): p. 135-47.

[51] Wajant, H., Death receptors. *Essays Biochem*, 2003. 39: p. 53-71.

[52] Barnhart, B.C. and M.E. Peter, The TNF receptor 1: a split personality complex. *Cell*, 2003. 114(2): p. 148-50.

[53] Schutze, S., et al., Inhibition of receptor internalization by monodansylcadaverine selectively blocks p55 tumor necrosis factor receptor death domain signaling. *J Biol Chem*, 1999. 274(15): p. 10203-12.

[54] Legler, D.F., et al., Recruitment of TNF receptor 1 to lipid rafts is essential for TNFalpha-mediated NF-kappaB activation. *Immunity*, 2003. 18(5): p. 655-64.

[55] Schneider-Brachert, W., et al., Compartmentalization of TNF receptor 1 signaling: internalized TNF receptosomes as death signaling vesicles. *Immunity*, 2004. 21(3): p. 415-28.

[56] Zamzami, N. and G. Kroemer, The mitochondrion in apoptosis: how Pandora's box opens. *Nature Rev. Mol. Cell Biol.*, 2001. 2: p. 67-71.

[57] Martinou, J.C. and D.R. Green, Breaking the mitochondrial barrier. *Nature Rev. Mol. Cell Biol.*, 2001. 2: p. 63-67.

[58] Huang, D.C. and A. Strasser, BH3-only proteins-essential initiators of apoptotic cell death. *Cell*, 2000. 103: p. 839-842.

[59] Du, C., et al., Smac, a mitochondrial protein that promotes cytochrome *c*-dependent caspase activation by eliminating IAP inhibition. *Cell,* 2000. 102(1): p. 33-42.

[60] Kroemer, G. and J.C. Reed, Mitochondrial control of cell death. *Nat Med*, 2000. 6(5): p. 513-9.

[61] Li, L.Y., X. Luo, and X. Wang, Endonuclease G is an apoptotic DNase when released from mitochondria. *Nature*, 2001. 412(6842): p. 95-9.

[62] van Loo, G., et al., The role of mitochondrial factors in apoptosis: a Russian roulette with more than one bullet. *Cell Death Differ*, 2002. 9(10): p. 1031-42.

[63] Fulda, S., et al., Smac agonists sensitize for Apo2L/TRAIL- or anticancer drug-induced apoptosis and induce regression of malignant glioma in vivo. *Nat Med*, 2002. 8(8): p. 808-15.

[64] Li, J., et al., Human ovarian cancer nd cisplatin resistance: possible role of inhibitor of apoptosis proteins. *Endocrinology*, 2001. 142(1): p. 370-80.

[65] Ferri, K.F. and G. Kroemer, Control of apoptotic DNA degradation. *Nat Cell Biol*, 2000. 2(4): p. E63-4.

[66] Kroemer, G. and S.J. Martin, Caspase-independent cell death. 2005. 11(7): p. 725-730.

[67] Nakagawa, T., et al., Cyclophilin D-dependent mitochondrial permeability transition regulates some necrotic but not apoptotic cell death. *Nature*, 2005. 434(7033): p. 652-8.

[68] Annis, M.G., et al., Bax forms multispanning monomers that oligomerize to permeabilize membranes during apoptosis. *Embo J*, 2005. 24(12): p. 2096-103.

[69] Kandasamy, K., et al., Involvement of proapoptotic molecules Bax and Bak in tumor necrosis factor-related apoptosis-inducing ligand (TRAIL)-induced mitochondrial disruption and apoptosis: differential regulation of cytochrome *c* and Smac/DIABLO release. *Cancer Res*, 2003. 63(7): p. 1712-21.

[70] De Giorgi, F., et al., The permeability transition pore signals apoptosis by directing Bax translocation and multimerization. *Faseb J*, 2002. 16(6): p. 607-9.

[71] Desagher, S., et al., Bid-induced conformational change of Bax is responsible for mitochondrial cytochrome *c* release during apoptosis. *J Cell Biol*, 1999. 144(5): p. 891-901.

[72] Kuwana, T., et al., BH3 domains of BH3-only proteins differentially regulate Bax-mediated mitochondrial membrane permeabilization both directly and indirectly. *Mol Cell*, 2005. 17(4): p. 525-35.

[73] Korsmeyer, S.J., et al., Pro-apoptotic cascade activates BID, which oligomerizes BAK or BAX into pores that result in the release of cytochrome *c*. *Cell Death Differ*, 2000. 7(12): p. 1166-73.

[74] Shin, M.S., Mutations of tumor necrosis factor-related apoptosis-inducing ligand receptor 1 (TRAIL-R1) and receptor 2 (TRAIL-R2) genes in metastatic breast cancers. *Cancer Res.*, 2001. 61: p. 4942-4946.

[75] Pai, S.I., Rare loss-of-function mutation of a death receptor gene in head and neck cancer. *Cancer Res*, 1998. 58: p. 3513-3518.

[76] Fisher, M.J., Nucleotide substitution in the ectodomain of trail receptor DR4 is associated with lung cancer and head and neck cancer. *Clin. Cancer Res.*, 2001. 7: p. 1688-1697.

[77] Lee, S.H., Alterations of the DR5/TRAIL receptor 2 gene in non-small cell lung cancers. *Cancer Res.*, 1999. 59: p. 5683-5686.

[78] Mullauer, L., et al., Mutations in apoptosis genes: a pathogenetic factor for human disease. *Mutat Res*, 2001. 488(3): p. 211-31.

[79] Zhang, L. and B. Fang, Mechanisms of resistance to TRAIL-induced apoptosis in cancer. *Cancer Gene Ther*, 2005. 12(3): p. 228-37.

[80] Held, J. and K. Schulze-Osthoff, Potential and caveats of TRAIL in cancer therapy. *Drug Resist Updat*, 2001. 4(4): p. 243-52.

[81] Marsters, S.A., et al., Activation of apoptosis by Apo-2 ligand is independent of FADD but blocked by CrmA. *Curr Biol*, 1996. 6(6): p. 750-2.

[82] Wu, D., H.D. Wallen, and G. Nunez, Interaction and regulation of subcellular localization of CED-4 by CED-9. *Science*, 1997. 275(5303): p. 1126-9.

[83] Petak, I., et al., Hypermethylation of the gene promoter and enhancer region can regulate Fas expression and sensitivity in colon carcinoma. *Cell Death Differ*, 2003. 10(2): p. 211-7.

[84] Hopkins-Donaldson, S., et al., Silencing of death receptor and caspase-8 expression in small cell lung carcinoma cell lines and tumors by DNA methylation. *Cell Death Differ*, 2003. 10(3): p. 356-64.

[85] Cheng, J., Protection from Fas-mediated apoptosis by a soluble form of the Fas molecule. *Science*, 1994. 263: p. 1759-1762.

[86] Midis, G.P., Y. Shen, and L.B. Owen-Schaub, Elevated soluble Fas (sFas) levels in nonhematopoietic human malignancy. *Cancer Res.*, 1996. 56: p. 3870-3874.

[87] Ugurel, S., et al., Increased soluble CD95 (sFas/CD95) serum level correlates with poor prognosis in melanoma patients. *Clin. Cancer Res.*, 2001. 7: p. 1282-1286.

[88] Gerharz, C.D., Resistance to CD95 (APO-1/Fas)-mediated apoptosis in human renal cell carcinomas: an important factor for evasion from negative growth control. *Lab. Invest.*, 1999. 79: p. 1521-1534.

[89] Pitti, R.M., Genomic amplification of a decoy receptor for Fas ligand in lung and colon cancer. *Nature*, 1998. 396: p. 699-703.

[90] Yu, K.Y., A newly identified member of tumor necrosis factor receptor superfamily (TR6) suppresses LIGHT-mediated apoptosis. *J. Biol. Chem.*, 1999. 274: p. 13733-13736.

[91] Roth, W., Soluble decoy receptor 3 is expressed by malignant gliomas and suppresses CD95 ligand-induced apoptosis and chemotaxis. *Cancer Res.*, 2001. 61: p. 2759-2765.

[92] Strand, S., Lymphocyte apoptosis induced by CD95 (APO-1/Fas) ligand-expressing tumor cells - a mechanism of immune evasion? *Nature Med.*, 1996. 2: p. 1361-1366.

[93] Moller, P., Expression of APO-1 (CD95), a member of the NGF/TNF receptor superfamily, in normal and neoplastic colon epithelium. *Int. J. Cancer*, 1994. 57: p. 371-377.

[94] Leithauser, F., Constitutive and induced expression of APO-1, a new member of the nerve growth factor/tumor necrosis factor receptor superfamily, in normal and neoplastic *cells. Lab. Invest.*, 1993. 69: p. 415-429.

[95] Volkmann, M., Loss of CD95 expression is linked to most but not all p53 mutants in European hepatocellular carcinoma. *J. Mol. Med.*, 2001. 79: p. 594-600.

[96] Peli, J., Oncogenic Ras inhibits Fas ligand-mediated apoptosis by downregulating the expression of Fas. *EMBO J.*, 1999. 18: p. 1824-1831.

[97] Maeda, T., Fas gene mutation in the progression of adult T cell leukemia. *J. Exp. Med.*, 1999. 189: p. 1063-1071.

[98] Landowski, T.H., et al., Mutations in the Fas antigen in patients with multiple myeloma. *Blood*, 1997. 90: p. 4266-4270.

[99] Cascino, I., et al., Fas/Apo-1 (CD95) receptor lacking the intracytoplasmic signaling domain protects tumor cells from Fas-mediated apoptosis. *J. Immunol.*, 1996. 156: p. 13-17.

[100] Straus, S.E., The development of lymphomas in families with autoimmune lymphoproliferative syndrome with germline Fas mutations and defective lymphocyte apoptosis. *Blood*, 2001. 98: p. 194-200.

[101] Griffith, T.S., et al., Intracellular regulation of TRAIL-induced apoptosis in human melanoma cells. *J Immunol*, 1998. 161(6): p. 2833-40.

[102] Hao, C., et al., Induction and intracellular regulation of tumor necrosis factor-related apoptosis-inducing ligand (TRAIL) mediated apotosis in human malignant glioma cells. *Cancer Res*, 2001. 61(3): p. 1162-70.

[103] Sprick, M.R., FADD/MORT1 and caspase-8 are recruited to TRAIL receptors 1 and 2 and are essential for apoptosis mediated by TRAIL receptor 2. *Immunity*, 2000. 12: p. 599-609.

[104] Kischkel, F.C., et al., Apo2L/TRAIL-dependent recruitment of endogenous FADD and caspase-8 to death receptors 4 and 5. *Immunity*, 2000. 12(6): p. 611-20.

[105] Truneh, A., et al., Temperature-sensitive differential affinity of TRAIL for its receptors. DR5 is the highest affinity receptor. *J Biol Chem*, 2000. 275(30): p. 23319-25.

[106] Wajant, H., et al., Differential activation of TRAIL-R1 and -2 by soluble and membrane TRAIL allows selective surface antigen-directed activation of TRAIL-R2 by a soluble TRAIL derivative. *Oncogene*, 2001. 20(30): p. 4101-6.

[107] Muhlenbeck, F., et al., The tumor necrosis factor-related apoptosis-inducing ligand receptors TRAIL-R1 and TRAIL-R2 have distinct cross-linking requirements for initiation of apoptosis and are non-redundant in JNK activation. *J Biol Chem*, 2000. 275(41): p. 32208-13.

[108] Zhang, X.D., et al., Differential localization and regulation of death and decoy receptors for TNF-related apoptosis-inducing ligand (TRAIL) in human melanoma cells. *J Immunol*, 2000. 164(8): p. 3961-70.

[109] Zhang, X.D., et al., Mechanisms of resistance of normal cells to TRAIL induced apoptosis vary between different cell types. *FEBS Lett*, 2000. 482(3): p. 193-9.

[110] Jin, Z., et al., Deficient tumor necrosis factor-related apoptosis-inducing ligand (TRAIL) death receptor transport to the cell surface in human colon cancer cells selected for resistance to TRAIL-induced apoptosis. *J Biol Chem*, 2004. 279(34): p. 35829-39.

[111] Delmas, D., et al., Redistribution of CD95, DR4 and DR5 in rafts accounts for the synergistic toxicity of resveratrol and death receptor ligands in colon carcinoma cells. *Oncogene*, 2004. 23(55): p. 8979-86.

[112] Krueger, A., et al., FLICE-inhibitory proteins: regulators of death receptor-mediated apoptosis. *Mol. Cell. Biol.*, 2001. 21: p. 8247-8254.

[113] French, L.E. and J. Tschopp, Defective death receptor signaling as a cause of tumor immune escape. *Semin Cancer Biol*, 2002. 12(1): p. 51-5.

[114] Longley, D.B., et al., c-FLIP inhibits chemotherapy-induced colorectal cancer cell death. *Oncogene,* 2005.

[115] Fulda, S., E. Meyer, and K.M. Debatin, Metabolic inhibitors sensitize for CD95 (APO-1/Fas)-induced apoptosis by down-regulating Fas-associated death domain-like interleukin 1-converting enzyme inhibitory protein expression. *Cancer Res,* 2000. 60(14): p. 3947-56.

[116] Thome, M., Viral FLICE-inhibitory proteins (FLIPs) prevent apoptosis induced by death receptors. *Nature,* 1997. 386: p. 517-521.

[117] Hu, S., et al., I-FLICE, a novel inhibitor of tumor necrosis factor receptor-1- and CD-95-induced apoptosis. *J. Biol. Chem.,* 1997. 272: p. 17255-17257.

[118] Bertin, J., Death effector domain-containing herpesvirus and poxvirus proteins inhibit both Fas- and TNFR1-induced apoptosis. *Proc. Natl Acad. Sci. USA,* 1997. 94: p. 1172-1176.

[119] Sturzl, M., Expression of K13/v-FLIP gene of human herpesvirus 8 and apoptosis in Kaposi's sarcoma spindle cells. J. Natl Cancer Inst., 1999. 91: p. 1725-1733.

[120] Vaux, D.L. and J. Silke, Iaps, Rings and ubiquitylation. Nature Reviews Molecular Cell Biology *Nat Rev Mol Cell Biol,* 2005. 6(4): p. 287-297.

[121] Altieri, D.C., The molecular basis and potential role of survivin in cancer diagnosis and therapy. *Trends Mol. Med.,* 2001. 7: p. 542-547.

[122] Altieri, D.C., Survivin, versatile modulation of cell division and apoptosis in cancer. *Oncogene,* 2003. 22: p. 8581-8589.

[123] Adida, C., et al., Anti-apoptosis gene, survivin, and prognosis of neuroblastoma. *Lancet,* 1998. 351: p. 882-883.

[124] Datta, R., et al., XIAP regulates DNA damage-induced apoptosis downstream of caspase-9 cleavage. *J Biol Chem,* 2000. 275(41): p. 31733-8.

[125] Tamm, I., et al., Peptides targeting caspase inhibitors. *J Biol Chem,* 2003. 278(16): p. 14401-5.

[126] Xu, Y., et al., A mutation found in the promoter region of the human survivin gene is correlated to overexpression of survivin in cancer cells. DNA Cell Biol, 2004. 23(7): p. 419-29.

[127] Chawla-Sarkar, M., et al., Downregulation of Bcl-2, FLIP or IAPs (XIAP and survivin) by siRNAs sensitizes resistant melanoma cells to Apo2L/TRAIL-induced apoptosis. *Cell Death Differ,* 2004. 11(8): p. 915-23.

[128] Griffith, T.S., et al., Induction and regulation of tumor necrosis factor-related apoptosis-inducing ligand/Apo-2 ligand-mediated apoptosis in renal cell carcinoma. *Cancer Res,* 2002. 62(11): p. 3093-9.

[129] Cummins, J.M., et al., X-linked inhibitor of apoptosis protein (XIAP) is a nonredundant modulator of tumor necrosis factor-related apoptosis-inducing ligand (TRAIL)-mediated apoptosis in human cancer cells. *Cancer Res,* 2004. 64(9): p. 3006-8.

[130] Liston, P., Identification of XAF1 as an antagonist of XIAP anti-caspase activity. *Nature Cell Biol.,* 2001. 3: p. 128-133.

[131] Reed, J.C., The Survivin saga goes in vivo. *J. Clin. Invest.,* 2001. 108: p. 965-969.

[132] Ambrosini, G., C. Adida, and D.C. Altieri, A novel anti-apoptosis gene, survivin, expressed in cancer and lymphoma. *Nature Med.,* 1997. 3: p. 917-921.

[133] Grossman, D., Transgenic expression of survivin in keratinocytes counteracts UVB-induced apoptosis and cooperates with loss of p53. *J. Clin. Invest.*, 2001. 108: p. 991-999.

[134] Mesri, M., et al., Cancer gene therapy using a survivin mutant adenovirus. *J. Clin. Invest*, 2001. 108: p. 981-990.

[135] Dierlamm, J., The apoptosis inhibitor gene API2 and a novel 18q gene, MLT, are recurrently rearranged in the t(11;18)(q21;q21) associated with mucosa-associated lymphoid tissue lymphomas. *Blood*, 1999. 93: p. 3601-3609.

[136] Vucic, D., et al., ML-IAP, a novel inhibitor of apoptosis that is preferentially expressed in human melanomas. *Curr. Biol.*, 2000. 10: p. 1359-1366.

[137] Medema, J.P., Blockade of the granzyme B/perforin pathway through overexpression of the serine protease inhibitor PI-9/SPI-6 constitutes a mechanism for immune escape by tumors. *Proc. Natl Acad. Sci. USA*, 2001. 98: p. 11515-11520.

[138] Miyashita, T. and J.C. Reed, BCL-2 gene transfer increases relative resistance of S49.1 and WEHI7.2 lymphoid cells to cell death and DNA fragmentation induced by glucocorticoids and multiple chemotherapeutic drugs. *Cancer Res.*, 1992. 52: p. 5407-5411.

[139] Schmitt, C.A., C.T. Rosenthal, and S.W. Lowe, Genetic analysis of chemoresistance in primary murine lymphomas. *Nature Med.*, 2000. 6: p. 1029-1035.

[140] Findley, H.W., et al., Expression and regulation of BCL-2, BCL-XL, and BAX correlate with p53 status and sensitivity to apoptosis in childhood acute lymphoblastic leukemia. *Blood*, 1997. 89: p. 2986-2993.

[141] Coustan-Smith, E., Clinical relevance of BCL-2 overexpression in childhood acute lymphoblastic leukemia. *Blood*, 1996. 87: p. 1140-1146.

[142] Campos, L., High expression of BCL-2 protein in acute myeloid leukemia cells is associated with poor response to chemotherapy. *Blood*, 1993. 81: p. 3091-3096.

[143] Hermine, O., Prognostic significance of BCL-2 protein expression in aggressive non-Hodgkin's lymphoma. Groupe d'Etude des Lymphomes de l'Adulte (GELA). *Blood*, 1996. 87: p. 265-272.

[144] Tarodi, B., T. Subramanian, and G. Chinnadurai, Epstein-Barr virus BHRF1 protein protects against cell death induced by DNA-damaging agents and heterologous viral infection. *Virology*, 1994. 201: p. 404-407.

[145] Henderson, S., Epstein-Barr virus-coded BHRF1 protein, a viral homologue of Bcl-2, protects human B cells from programmed cell death. *Proc. Natl Acad. Sci. USA*, 1993. 90: p. 8479-8483.

[146] Sarid, R., et al., Kaposi's sarcoma-associated herpesvirus encodes a functional bcl-2 homologue. *Nature Med.*, 1997. 3: p. 293-298.

[147] Boise, L.H., BCL-X, a BCL-2-related gene that functions as a dominant regulator of apoptotic cell death. *Cell*, 1993. 74: p. 597-608.

[148] Dole, M.G., BCL-XL is expressed in neuroblastoma cells and modulates chemotherapy-induced apoptosis. *Cancer Res.*, 1995. 55: p. 2576-2582.

[149] Nagane, M., et al., Drug resistance of human glioblastoma cells conferred by a tumor-specific mutant epidermal growth factor receptor through modulation of BCL-XL and caspase-3-like proteases. *Proc. Natl Acad. Sci. USA*, 1998. 95: p. 5724-5729.

[150] Minn, A.J., et al., Expression of BCL-XL can confer a multidrug resistance phenotype. *Blood*, 1995. 86: p. 1903-1910.

[151] Zhou, P., et al., MCL-1, a BCL-2 family member, delays the death of hematopoietic cells under a variety of apoptosis-inducing conditions. *Blood*, 1997. 89: p. 630-643.

[152] Teitz, T., Caspase-8 is deleted or silenced preferentially in childhood neuroblastomas with amplification of MYCN. *Nature Med.*, 2000. 6: p. 529-535.

[153] Rampino, N., Somatic frameshift mutations in the BAX gene in colon cancers of the microsatellite mutator phenotype. *Science*, 1997. 275: p. 967-969.

[154] Meijerink, J.P., Hematopoietic malignancies demonstrate loss-of-function mutations of BAX. *Blood*, 1998. 91: p. 2991-2997.

[155] Molenaar, J.J., Microsatellite instability and frameshift mutations in BAX and transforming growth factor-[beta] RII genes are very uncommon in acute lymphoblastic leukemia in vivo but not in cell lines. *Blood*, 1998. 92: p. 230-233.

[156] Krajewski, S., Reduced expression of proapoptotic gene BAX is associated with poor response rates to combination chemotherapy and shorter survival in women with metastatic breast adenocarcinoma. *Cancer Res.*, 1995. 55: p. 4471-4478.

[157] Soengas, M.S., Apaf-1 and caspase-9 in p53-dependent apoptosis and tumor inhibition. *Science*, 1999. 284: p. 156-159.

[158] Sarasin, A., An overview of the mechanisms of mutagenesis and carcinogenesis. *Mutat. Res.*, 2003. 544: p. 99-106.

[159] Levine, A.J., p53, the cellular gatekeeper for growth and division. *Cell,* 1997. 88: p. 323-331.

[160] Vousden, K.H. and X. Lu, Live or let die: the cell's response to p53. *Nature Rev. Cancer*, 2002. 2: p. 594-604.

[161] Berube, C., et al., Apoptosis caused by p53-induced protein with death domain (PIDD) depends on the death adapter protein RAIDD. *Proc Natl Acad Sci U S A*, 2005. 102(40): p. 14314-20.

[162] Erster, S. and U.M. Moll, Stress-induced p53 runs a transcription-independent death program. *Biochem Biophys Res Commun*, 2005. 331(3): p. 843-50.

[163] Chipuk, J.E., et al., PUMA couples the nuclear and cytoplasmic proapoptotic function of p53. *Science*, 2005. 309(5741): p. 1732-5.

[164] Lee, J.M. and A. Bernstein, p53 mutations increase resistance to ionizing radiation. *Proc. Natl Acad. Sci. USA*, 1993. 90: p. 5742-5746.

[165] Aas, T., Specific P53 mutations are associated with de novo resistance to doxorubicin in breast cancer patients. *Nature Med.*, 1996. 2: p. 811-814.

[166] Bunz, F., Disruption of p53 in human cancer cells alters the responses to therapeutic agents. *J. Clin. Invest.*, 1999. 104: p. 263-269.

[167] Ameyar, M., et al., Adenovirus-mediated transfer of wild-type p53 gene sensitizes TNF resistant MCF7 derivatives to the cytotoxic effect of this cytokine: relationship with c-myc and Rb. *Oncogene,* 1999. 18(39): p. 5464-72.

[168] Ameyar-Zazoua, M., et al., Wild-type p53 induced sensitization of mutant p53 TNF-resistant cells: role of caspase-8 and mitochondria. *Cancer Gene Ther*, 2002. 9(3): p. 219-27.

[169] Thiery, J., et al., Role of p53 in the sensitization of tumor cells to apoptotic cell death. *Mol Immunol*, 2002. 38(12-13): p. 977-80.

[170] Asgari, K., Inhibition of the growth of pre-established subcutaneous tumor nodules of human prostate cancer cells by single injection of the recombinant adenovirus p53 expression vector. *Int. J. Cancer*, 1997. 71: p. 377-382.

[171] Wang, C.Y., et al., Control of inducible chemoresistance: enhanced anti-tumor therapy through increased apoptosis by inhibition of NF-[kappa]B. *Nature Med.*, 1999. 5: p. 412-417.

[172] Sherr, C.J., The INK4A/ARF network in tumor suppression. *Nature Rev. Mol. Cell Biol.*, 2001. 2: p. 731-737.

[173] Schmitt, C.A., et al., INK4A/ARF mutations accelerate lymphomagenesis and promote chemoresistance by disabling p53. *Genes Dev.*, 1999. 13: p. 2670-2677.

[174] Attardi, L.D., The role of p53-mediated apoptosis as a crucial anti-tumor response to genomic instability: lessons from mouse models. *Mutat. Res.*, 2005. 569: p. 145-157.

[175] Brown, J.M. and L.D. Attardi, The role of apoptosis in cancer development and treatment response. *Nature Reviews Cancer Nat Rev Cancer*, 2005. 5(3): p. 231-237.

[176] Gudkov, A.V. and E.A. Komarova, The role of p53 in determining sensitivity to radiotherapy. *Nature Rev. Cancer*, 2003. 3: p. 117-129.

[177] Cantley, L.C., The phosphoinositide 3-kinase pathway. *Science*, 2002. 296(5573): p. 1655-7.

[178] Baldwin, A.S., Control of oncogenesis and cancer therapy resistance by the transcription factor NF-kappaB. *J Clin Invest*, 2001. 107(3): p. 241-6.

[179] Maehama, T., G.S. Taylor, and J.E. Dixon, PTEN and myotubularin: novel phosphoinositide phosphatases. *Annu Rev Biochem*, 2001. 70: p. 247-79.

[180] Mayo, L.D., et al., PTEN protects p53 from Mdm2 and sensitizes cancer cells to chemotherapy. *J Biol Chem*, 2002. 277(7): p. 5484-9.

[181] Stambolic, V., et al., Regulation of PTEN transcription by p53. *Mol Cell*, 2001. 8(2): p. 317-25.

[182] Birck, A., et al., Mutation and allelic loss of the PTEN/MMAC1 gene in primary and metastatic melanoma biopsies. *J Invest Dermatol*, 2000. 114(2): p. 277-80.

[183] Zhou, X.P., et al., Epigenetic PTEN silencing in malignant melanomas without PTEN mutation. *Am J Pathol*, 2000. 157(4): p. 1123-8.

[184] Whiteman, D.C., et al., Nuclear PTEN expression and clinicopathologic features in a population-based series of primary cutaneous melanoma. *Int J Cancer*, 2002. 99(1): p. 63-7.

[185] Poetsch, M., T. Dittberner, and C. Woenckhaus, PTEN/MMAC1 in malignant melanoma and its importance for tumor progression. *Cancer Genet Cytogenet*, 2001. 125(1): p. 21-6.

[186] Celebi, J.T., et al., Identification of PTEN mutations in metastatic melanoma specimens. *J Med Genet*, 2000. 37(9): p. 653-7.

[187] Guldberg, P., et al., Disruption of the MMAC1/PTEN gene by deletion or mutation is a frequent event in malignant melanoma. *Cancer Res*, 1997. 57(17): p. 3660-3.

[188] Karin, M. and A. Lin, NF-kappaB at the crossroads of life and death. *Nat Immunol*, 2002. 3(3): p. 221-7.

[189] Meyskens, F.L., Jr., et al., Activation of nuclear factor-kappa B in human metastatic melanomacells and the effect of oxidative stress. *Clin Cancer Res*, 1999. 5(5): p. 1197-202.

[190] McNulty, S.E., N.B. Tohidian, and F.L. Meyskens, Jr., RelA, p50 and inhibitor of kappa B alpha are elevated in human metastatic melanoma cells and respond aberrantly to ultraviolet light B. *Pigment Cell Res*, 2001. 14(6): p. 456-65.

[191] Yang, J. and A. Richmond, Constitutive IkappaB kinase activity correlates with nuclear factor-kappaB activation in human melanoma cells. *Cancer Res*, 2001. 61(12): p. 4901-9.

[192] Dhawan, P. and A. Richmond, A novel NF-kappa B-inducing kinase-MAPK signaling pathway upregulates NF-kappa B activity in melanoma cells. *J Biol Chem*, 2002. 277(10): p. 7920-8.

[193] Rayet, B. and C. Gelinas, Aberrant REL/NF-[kappa]B genes and activity in human cancer. *Oncogene*, 1999. 18: p. 6938-6947.

[194] Wood, K.M., M. Roff, and R.T. Hay, Defective I[kappa]B[alpha] in Hodgkin cell lines with constitutively active NF-[kappa]B. *Oncogene*, 1998. 16: p. 2131-2139.

[195] Bentires-Alj, M., et al., Stable inhibition of nuclear factor kappaB in cancer cells does not increase sensitivity to cytotoxic drugs. *Cancer Res*, 1999. 59(4): p. 811-5.

[196] Smyth, M.J., et al., The drug efflux protein, P-glycoprotein, additionally protects drug-resistant tumor cells from multiple forms of caspase-dependent apoptosis. *Proc. Natl Acad. Sci. USA*, 1998. 95: p. 7024-7029.

[197] Gottesman, M.M. and I. Pastan, Biochemistry of multidrug resistance mediated by the multidrug transporter. *Annu Rev Biochem*, 1993. 62: p. 385-427.

[198] Soengas, M. and S. Lowe, Apoptosis and melanoma chemoresistance. *Oncogene*, 2003. 22: p. 3138-3151.

[199] Zhang, K., P. Mack, and K.P. Wong, Glutathione-related mechanisms in cellular resistance to anticancer drugs. *Int J Oncol*, 1998. 12(4): p. 871-82.

[200] Cole, S.P., Overexpression of a transporter gene in a multidrug-resistant human lung cancer cell line. *Science*, 1992. 258: p. 1650-1654.

[201] Ellerby, H.M., Anti-cancer activity of targeted pro-apoptotic peptides. *Nature Med.*, 1999. 5: p. 1032-1038.

[202] Johnstone, R.W., E. Cretney, and M.J. Smyth, P-glycoprotein protects leukemia cells against caspase-dependent, but not caspase-independent, cell death. *Blood*, 1999. 93: p. 1075-1085.

[203] Leslie, E.M., R.G. Deeley, and S.P.C. Cole, Multidrug resistance proteins: role of P-glycoprotein, MRP1, MRP2, and BCRP (ABCG2) in tissue defense. *Toxicology and Applied Pharmacology*, 2005. 204(3): p. 216-237.

[204] Helmbach, H., et al., Drug-resistance in human melanoma. *Int J Cancer*, 2001. 93(5): p. 617-22.

[205] Withoff, S., et al., Human DNA topoisomerase II: biochemistry and role in chemotherapy resistance (review). *Anticancer Res*, 1996. 16(4A): p. 1867-80.

[206] Larsen, A.K. and A. Skladanowski, Cellular resistance to topoisomerase-targeted drugs: from drug uptake to cell death. *Biochim Biophys Acta*, 1998. 1400(1-3): p. 257-74.

[207] Campain, J.A., et al., Acquisition of multiple copies of a mutant topoisomerase IIalpha allele by chromosome 17 aneuploidy is associated with etoposide resistance in human melanoma cell lines. *Somat Cell Mol Genet*, 1995. 21(6): p. 451-71.

[208] Lage, H., et al., Modulation of DNA topoisomerase II activity and expression in melanoma cells with acquired drug resistance. *Br J Cancer*, 2000. 82(2): p. 488-91.

[209] Satherley, K., et al., Relationship between expression of topoisomerase II isoforms and chemosensitivity in choroidal melanoma. *J Pathol*, 2000. 192(2): p. 174-81.

[210] Rass, K., et al., DNA mismatch repair enzyme hMSH2 in malignant melanoma: increased immunoreactivity as compared to acquired melanocytic nevi and strong mRNA expression in melanoma cell lines. *Histochem J*, 2001. 33(8): p. 459-67.

[211] Korabiowska, M., et al., Comparative study of the expression of DNA mismatch repair genes, the adenomatous polyposis coli gene and growth arrest DNA damage genes in melanoma recurrences and metastases. *Melanoma Res*, 2000. 10(6): p. 537-44.

[212] Rieber, M. and M. Strasberg Rieber, Induction of p53 without increase in p21WAF1 in betulinic acid-mediated cell death is preferential for human metastatic melanoma. *DNA Cell Biol*, 1998. 17(5): p. 399-406.

[213] Rieber, M. and M. Strasberg-Rieber, Induction of p53 and melanoma cell death is reciprocal with down-regulation of E2F, cyclin D1 and pRB. *Int J Cancer*, 1998. 76(5): p. 757-60.

[214] Soengas, M.S., Inactivation of the apoptosis effector Apaf-1 in malignant melanoma. *Nature*, 2001. 409: p. 207-211.

[215] Sethi, T., Extracellular matrix proteins protect small cell lung cancer cells against apoptosis: a mechanism for small cell lung cancer growth and drug resistance in vivo. *Nature Med.*, 1999. 5: p. 662-668.

[216] Catlett-Falcone, R., Constitutive activation of STAT3 signaling confers resistance to apoptosis in human U266 myeloma cells. *Immunity*, 1999. 10: p. 105-115.

[217] Legname, G., Synthetic mammalian prions. *Science,* 2004. 305: p. 673-676.

[218] King, C.-Y. and R. Diaz-Avalos, Protein-only transmission of three yeast prion strains. *Nature*, 2004. 428: p. 319-323.

[219] Tanaka, M., et al., Conformational variations in an infectious protein determine prion strain differences. *Nature,* 2004. 428: p. 323-327.

[220] Bueler, H., Mice devoid of PrP are resistant to scrapie. Cell, 1993. 73: p. 1339-1347.

[221] Stewart, G.S., et al., MDC1 is a mediator of the mammalian DNA damage checkpoint. *Nature*, 2003. 421: p. 961-966.

[222] Prado, M.A., et al., PrPc on the road: trafficking of the cellular prion protein. *J Neurochem*, 2004. 88(4): p. 769-81.

[223] Harris, D.A., Trafficking, turnover and membrane topology of PrP. *Br Med Bull*, 2003. 66: p. 71-85.

[224] Nunziante, M., S. Gilch, and H.M. Schatzl, Essential role of the prion protein N terminus in subcellular trafficking and half-life of cellular prion protein. *J Biol Chem*, 2003. 278(6): p. 3726-34.

[225] Sunyach, C., et al., The mechanism of internalization of glycosylphosphatidylinositol-anchored prion protein. *Embo J,* 2003. 22(14): p. 3591-601.

[226] Marella, M., et al., Filipin prevents pathological prion protein accumulation by reducing endocytosis and inducing cellular PrP release. *J Biol Chem*, 2002. 277(28): p. 25457-64.

[227] Kaneko, K., et al., COOH-terminal sequence of the cellular prion protein directs subcellular trafficking and controls conversion into the scrapie isoform. *Proc Natl Acad Sci U S A*, 1997. 94(6): p. 2333-8.

[228] Sabharanjak, S., et al., GPI-anchored proteins are delivered to recycling endosomes via a distinct cdc42-regulated, clathrin-independent pinocytic pathway. *Dev Cell*, 2002. 2(4): p. 411-23.

[229] Johannes, L. and C. Lamaze, Clathrin-dependent or not: is it still the question? *Traffic*, 2002. 3(7): p. 443-51.

[230] Weigelt, B., et al., Gene expression profiles of primary breast tumors maintained in distant metastases. *Proc Natl Acad Sci U S A*, 2003. 100(26): p. 15901-5.

[231] Mouillet-Richard, S., Signal transduction through prion protein. *Science*, 2000. 289: p. 1925-1928.

[232] Sunyach, C. and F. Checler, Combined pharmacological, mutational and cell biology approaches indicate that p53-dependent caspaseandnbsp;3 activation triggered by cellular prion is dependent on its endocytosis. *Journal of Neurochemistry*, 2005. 92(6): p. 1399-1407.

[233] Schneider, B., et al., NADPH oxidase and extracellular regulated kinases 1/2 are targets of prion protein signaling in neuronal and nonneuronal cells. *Proc Natl Acad Sci U S A*, 2003. 100(23): p. 13326-31.

[234] Chen, S., et al., Prion protein as trans-interacting partner for neurons is involved in neurite outgrowth and neuronal survival. *Mol. Cell. Neurosci.*, 2003. 22: p. 227-233.

[235] Graner, E., Cellular prion protein binds laminin and mediates neuritogenesis. *Mol. Brain Res.*, 2000. 76: p. 85-92.

[236] Spielhaupter, C. and H.M. Schatzl, PrPC directly interacts with proteins involved in signaling pathways. *J Biol Chem*, 2001. 276(48): p. 44604-12.

[237] Liu, T., et al., Intercellular transfer of the cellular prion protein. *J Biol Chem*, 2002. 277(49): p. 47671-8.

[238] Porto-Carreiro, I., et al., Prions and exosomes: from PrPc trafficking to PrPsc propagation. *Blood Cells Mol Dis*, 2005. 35(2): p. 143-8.

[239] Kurschner, C. and J.I. Morgan, Analysis of interaction sites in homo- and heteromeric complexes containing Bcl-2 family members and the cellular prion protein. *Brain Res Mol Brain Res*, 1996. 37(1-2): p. 249-58.

[240] Kurschner, C. and J.I. Morgan, The cellular prion protein (PrP) selectively binds to Bcl-2 in the yeast two-hybrid system. *Brain Res Mol Brain Res*, 1995. 30(1): p. 165-8.

[241] Kuwahara, C., Prions prevent neuronal cell-line death. *Nature*, 1999. 400: p. 225-226.

[242] Kim, B., et al., The cellular prion protein (PrP (C)) prevents apoptotic neuronal cell death and mitochondrial dysfunction induced by serum deprivation. *Brain Res Mol Brain Res*, 2004. 124: p. 40-50.

[243] Li, A. and D.A. Harris, Mammalian prion protein suppresses Bax-induced cell death in yeast. *J Biol Chem*, 2005. 280(17): p. 17430-4.

[244] Zha, H., et al., Structure-function comparisons of the proapoptotic protein Bax in yeast and mammalian cells. *Mol Cell Biol*, 1996. 16(11): p. 6494-508.

[245] Mouillet-Richard, S., et al., Modulation of serotonergic receptor signaling and cross-talk by prion protein. *J Biol Chem*, 2005. 280(6): p. 4592-601.

[246] Hoshino, S., et al., Prions prevent brain damage after experimental brain injury: a preliminary report. *Acta Neurochir Suppl*, 2003. 86: p. 297-9.

[247] Brown, D.R., R.S. Nicholas, and L. Canevari, Lack of prion protein expression results in a neuronal phenotype sensitive to stress. *J Neurosci Res*, 2002. 67(2): p. 211-24.

[248] Shmerling, D., et al., Expression of amino-terminally truncated PrP in the mouse leading to ataxia and specific cerebellar lesions. *Cell*, 1998. 93(2): p. 203-14.

[249] Rossi, D., Onset of ataxia and Purkinje cell loss in PrP null mice inversely correlated with Dpl level in brain. *EMBO J.*, 2001. 20: p. 694-702.

[250] Behrens, A. and A. Aguzzi, Small is not beautiful: antagonizing functions for the prion protein PrP(C) and its homologue Dpl. *Trends Neurosci*, 2002. 25(3): p. 150-4.

[251] Paitel, E., et al., Primary cultured neurons devoid of cellular prion display lower responsiveness to staurosporine through the control of p53 at both transcriptional and post-transcriptional levels. *J Biol Chem*, 2004. 279(1): p. 612-8.

[252] Paitel E, Fahraeus R, and Checler F, Cellular prion protein sensitizes neurons to apoptotic stimuli through Mdm2-regulated and p53-dependent caspase 3-like activation. *J Biol Chem*, 2003. 278: p. 10061-10066.

[253] Kim, P.K., et al., During apoptosis bcl-2 changes membrane topology at both the endoplasmic reticulum and mitochondria. *Mol. Cell*, 2004. 14: p. 523-529.

[254] Lin, B. and e. al., Conversion of Bcl-2 from protector to killer by interaction with nuclear orphan receptor Nur77/TR3. . *Cell* 2004. 116: p. 527-540.

[255] Clem, R., et al., Modulation of cell death by Bcl-XL through caspase interaction. *Proc Natl Acad Sci USA* 1998. 95: p. 554-559.

[256] Vanhaesebroeck, B., et al., Effect of bcl-2 proto-oncogene expression on cellular sensitivity to tumor necrosis factor-mediated cytotoxicity. *Oncogene*, 1993. 8(4): p. 1075-81.

[257] Ghetti B, et al., Vascular variant of prion protein cerebral amyloidosis with tau-positive neurofibrillary tangles: the phenotype of the stop codon 145 mutation in PRNP. *Proc Natl Acad Sci USA* 1996. 93: p. 744-748.

[258] Satoh, J., et al., The 14-3-3 protein detectable in the cerebrospinal fluid of patients with prion-unrelated neurological diseases is expressed constitutively in neurons and glial cells in culture. *Eur. Neurol.*, 1999. 41: p. 216-225.

[259] Weller, M., et al., Protooncogene bcl-2 genetransfer abrogates Fas/APO-1 antibody-mediated apoptosis of human malignant glioma cells and confers resistance to chemotherapeutic drugs and therapeutic irradiation. *J. Clin. Invest.*, 1995. 95: p. 2633-2643.

[260] Kaufmann, S.H., Elevated expression of the apoptotic regulator MCL-1 at the time of leukemic relapse. *Blood*, 1998. 91: p. 991-1000.

[261] Irmler, M., Inhibition of death receptor signals by cellular FLIP. *Nature*, 1997. 388: p. 190-195.

[262] Medema, J.P., Cleavage of FLICE (caspase-8) by granzyme B during cytotoxic T lymphocyte-induced apoptosis. 1997. 27: p. 3492-3498.

[263] Mueller, C.M. and D.W. Scott, Distinct molecular mechanisms of Fas resistance in murine B lymphoma cells. *J. Immunol.*, 2000. 165: p. 1854-1862.

[264] Tepper, C.G. and M.F. Seldin, Modulation of caspase-8 and FLICE-inhibitory protein expression as a potential mechanism of Epstein-Barr virus tumorigenesis in Burkitt's lymphoma. *Blood,* 1999. 94: p. 1727-1737.

[265] Kataoka, T., FLIP prevents apoptosis induced by death receptors but not by perforin/granzyme B, chemotherapeutic drugs, and [gamma] irradiation. J. *Immunol.*, 1998. 161: p. 3936-3942.

[266] Medema, J.P., et al., Immune escape of tumors in vivo by expression of cellular FLICE- inhibitory protein. *J. Exp. Med.*, 1999. 190: p. 1033-1038.

[267] Djerbi, M., The inhibitor of death receptor signaling, FLICE-inhibitory protein defines a new class of tumor progression factors. *J. Exp. Med.*, 1999. 190: p. 1025-1032.

[268] Okada, H. and T.W. Mak, Pathways of apoptotic and non- apoptotic death in tumor cells. *Nature Reviews Cancer Nat Rev Cancer*, 2004. 4(8): p. 592-603.

[269] Amundson, S.A., An informatics approach identifying markers of chemosensitivity in human cancer cell lines. *Cancer Res.*, 2000. 60: p. 6101-6110.

[270] Barrie, M.B., et al., Antiviral cytokines induce hepatic expression of the granzyme B inhibitors, proteinase inhibitor 9 and serine proteinase inhibitor 6. *J Immunol*, 2004. 172(10): p. 6453-9.

[271] Bird, C.H., et al., Selective regulation of apoptosis: the cytotoxic lymphocyte serpin proteinase inhibitor 9 protects against granzyme B-mediated apoptosis without perturbing the Fas cell death pathway. *Mol Cell Biol*, 1998. 18(11): p. 6387-98.

[272] Du, J.P., et al., [The overexpression of prion protein in drug resistant gastric cancer cell line SGC7901/ADR and its significance]. *Zhonghua Yi Xue Za Zhi*, 2003. 83(4): p. 328-32.

[273] Yin, C., et al., Bax suppresses tumorigenesis and stimulates apoptosis in vivo. *Nature,* 1997. 385: p. 637-640.

[274] Bargou, R.C., Overexpression of the death-promoting gene Bax-[alpha] which is downregulated in breast cancer restores sensitivity to different apoptotic stimuli and reduces tumor growth in SCID mice. *J. Clin. Invest.*, 1996. 97: p. 2651-2659.

[275] Ionov, Y., et al., Mutational inactivation of the proapoptotic gene BAX confers selective advantage during tumor clonal evolution. *Proc. Natl Acad. Sci. USA*, 2000. 97: p. 10872-10877.

[276] Ryan, K.M., A.C. Phillips, and K.H. Vousden, Regulation and function of the p53 tumor suppressor protein. *Curr. Opin. Cell Biol.*, 2001. 13: p. 332-337.

[277] Okada, H., Survivin loss in thymocytes triggers p53-mediated growth arrest and p53-independent cell death. *J. Exp. Med.*, 2004. 199: p. 399-410.

[278] Okada, H. and T.W. Mak, Pathways of apoptotic and non-apoptotic death in tumor cells. Nature Rev. *Cancer*, 2004. 4: p. 592-603.

[279] Shatrov, V.A., et al., Adenovirus-mediated wild-type-p53-gene expressionsensitizes TNF-resistant tumor cells to TNF-induced cytotoxicity by altering the cellular redox state. *Int J Cancer*, 2000. 85(1): p. 93-7.

[280] Lowe, S.W., p53 status and the efficacy of cancer therapy in vivo. *Science*, 1994. 266: p. 807-810.

[281] Liu, Z.-J., X. Lu, and S. Zhong, ASPP--Apoptotic specific regulator of p53. *Biochimica et Biophysica Acta (BBA) - Reviews on Cancer*, 2005. 1756(1): p. 77-80.

In: Cell Apoptosis Research Trends
Editor: Charles V. Zhang, pp. 93-109

ISBN: 1-60021-424-X
© 2007 Nova Science Publishers, Inc

Chapter III

Pancratistatin a Novel Highly Selective Anti-Cancer Agent that Induces Apoptosis by the Activation of Membrane-Fas-Receptor Associated Caspase-3

Carly Griffin[†], James McNulty[‡], Caroline Hamm[*]
and Siyaram Pandey[†3]

[†] Department of Chemistry and Biochemistry, University of Windsor, Windsor, Ontario
[‡] Department of Chemistry, McMaster University, Hamilton, Ontario
[*] Windsor Regional Cancer Center, Windsor, Ontario

ABSTRACT

Pancratistatin is a natural compound extracted from the Hawaiian spider lily (*Hymenocalis littoralis*) known to have anti-cancer capabilities. Previous studies with Pancratistatin have demonstrated that this compound rapidly and efficiently induces apoptosis (programmed cell death) in various types of cancer cell lines, including breast, colon, prostate, neuroblastoma, melanoma and leukemia. Most importantly, when Pancratistatin was tested for toxicity on peripheral white blood cells from healthy volunteers, there was little or no demonstrable effect on their viability and nuclear morphology, indicating the relative specificity of this compound for cancer cells. Our previous results have shown that upon treatment with Pancratistatin, phosphatidyl-serine flips to the outer leaflet of the plasma membrane, caspase-3 is rapidly activated, and there is no occurrence of massive DNA damage at early time-points in cancerous cells. To better understand the selectivity of Pancratistatin, our recent work focuses on deducing

[3] Correspondence: Siyaram Pandey, Ph.D., Associate Professor, Department of Chemistry and Biochemistry, University of Windsor, 401 Sunset Ave., Windsor, ON, N9B 3P4 Canada, Email: spandey@uwindsor.ca, Phone: 519-253-3000 Ex: 3701, Fax: 519-973-7098

the mechanism of action of this compound particularly in the acute human T-cell leukemia (Jurkat) cell line. This model of cancer was used in conjunction with normal, nucleated blood cells, obtained from healthy volunteers, as well as with samples from relapsed acute leukemia patients, prior to chemotherapeutic treatment at the Windsor Regional Cancer Centre. Our results suggest that Pancratistatin caused caspase-3 activation in association with the Fas-receptor within membranous lipid rafts, upstream of caspase-8 activation in Jurkat cells. Early events were followed by the loss of mitochondrial membrane potential, DNA degradation and subsequent cell death. Incubation of Jurkat cells with a caspase-3 inhibitor prior to treatment with Pancratistatin delayed these events dramatically, which implied that caspase-3 activation is an essential step in Pancratistatin-induced apoptosis. Interestingly, in contrast with these results, Pancratistatin caused none of these reactions in normal nucleated blood cells under identical treatment conditions. Taken together, our results indicate that Pancratistatin is a novel anti-cancer agent that specifically induces apoptosis in human leukemia cells through Fas-receptor associated caspase-3 activation, while leaving healthy white blood cells unscathed.

Key words: cancer, apoptosis, caspases, leukemia, chemotherapy

INTRODUCTION

Cancer

Despite aggressive research efforts, cancer remains unconquered and is the second most frequent cause of death, killing more than half a million Americans in 2003. A few cancers account for most of this mortality, with cancers of the lung, colon, breast and prostate accounting for more than 50% of all cancer related deaths [8]. Most cancers have a multi-factorial etiology including obvious environmental factors in cancers like lung cancer. At the molecular level cancer is known to be due to the cumulative dysfunction of multiple regulatory mechanisms in pathways that control proliferation and apoptosis [6]. Mutations responsible for the majority of cancers occur in regulatory genes that result in either inactivated tumor suppressor genes, or activated proto-oncogenes [14].

This chapter reports an exciting new development with a novel cytotoxic molecule that seems to function by exploiting the differences between normal and malignant cells. We have used hematopoietic malignancy as our model. Leukemias are cancers of circulating white cells and are primarily categorized as lymphoid, myeloid or monocytic leukemia, depending on the lineage of the cell involved. While it is clear that as a liquid tumor they differ from cancers arising in solid tissues, in certain fundamental ways they share many of the traits of all malignant cells while providing significant advantages for *ex-vivo* experimentation.

Apoptosis

Apoptosis, alternatively referred to as programmed cell death, is a physiological process required for proper development and tissue homeostasis [3]. Cysteine proteases known as caspases are key players of apoptosis; they cleave other proteins after an aspartic acid residue in an apoptotic cascade. Executioner caspases, such as caspase-3, -7 and -9, become active by cleavage from other upstream initiator caspases, such as caspase-8 and -10 [2]. Initiator caspases can be activated by both the intrinsic and extrinsic pathways of apoptosis. Intrinsic apoptosis involves generation of reactive oxygen species (ROS), disruption of the mitochondrial membrane and potential, activation of pro-apoptotic proteins like Bax and Bid, and release of cytochrome c into the cytoplasm. Cytochrome c forms the apoptosome with apoptotic protease activating factor-1 (Apaf-1) and pro-caspase-9, activating caspase-9, which then activates caspase-3 resulting in DNA fragmentation and apoptosis [10, 21].

The extrinsic pathway of apoptosis involves oligomerization of death receptor proteins on the plasma membrane upon stimulation with their respective ligand. Upon Fas ligand binding in Type I cells, three Fas receptor proteins come together to recruit Fas-associated Death Domain (FADD) and Death Effector Domain (DED) to the cytosolic portion of the receptors. FADD then recruits c-FLIP (FADD-like interleukin-1B converting inhibitory enzyme) and the DED region aligns with that of pro-caspase-8 or -10, which becomes active and can subsequently activate executioner caspases. Assembly of these proteins is referred to as the death-inducing signaling complex (DISC), and is required for extrinsic apoptosis [12]. In Type II cells, formation of the DISC is somewhat impaired following stimulation by Fas ligand; however, enough active caspase-8 is produced to allow cleavage and activation of cytosolic BID (Bcl_2-interacting domain) [12]. Activation of BID triggers the activation of a mitochondrial-dependent pathway of apoptosis in which caspase-9 becomes active and in-turn cleaves and activates caspase-3. In both cases, caspase-3 activation results in DNA fragmentation by caspase-activated DNases (CAD), ultimately leading to apoptosis [12].

It has recently been reported by Aouad *et al.*, 2004 that the Fas ligand/receptor interaction in Jurkat cells occurs in the vicinity of complex lipid rafts in the plasma membrane that house executioner proteins, including capsase-3. This finding contradicts the traditional finding that caspase-3 activation occurs downstream of caspase-8 activation in Fas-mediated apoptosis.

CURRENT THERAPEUTIC OPTIONS AND STRATEGIES

The modern therapeutic approach to cancer is multimodal comprising of three main categories; surgical excision, radiotherapy and cytotoxic chemotherapy. Most classical chemotherapy treatments, such as anti-metabolites, DNA repair enzyme inhibitors, DNA synthesis inhibitors, DNA damaging agents act through targets that are present in both healthy and cancerous cells [24]. As a consequence, agents that target the cell cycle machinery are not only effective against malignant cells but also against cells that undergo mitosis at a high rate, such as skin cells and immune system cells which is the basis of the observed toxicity like increased susceptibility to infections and alopecia. [25]. Radiation therapy, like chemotherapy, causes massive DNA damage to ultimately induce apoptosis of both cancerous and healthy neighboring cells [15].

To circumvent these disadvantages, an emerging trend in the development of novel cancer therapeutics has been to target molecules that may be unique to or selectively enriched in cancer cells [7]. Various biochemical components involved in apoptosis are being targeted, such as TRAIL receptors, the anti-apoptotic protein Bcl$_2$, and proteasomal proteins [4, 11, 23].

Chronic myeloid leukemia (CML) is a good example of the successful treatment of a cancer with a drug that selectively affects only cancerous cells while leaving normal myeloid cells unaffected. CML is caused by a specific chromosomal translocation t(9;12) resulting in the *Philadelphia chromosome* [5]. The result of this translocation is the juxtaposition of the C-abl tyrosine kinase with the bcr gene producing a chimera termed bcr-abl. [22]. CML is currently treated by a revolutionary designer drug, Imatinib (Gleevec^{TM}), which blocks the tyrosine kinase activity of the fusion protein bcr-abl [22].

Drugs that target the cell death machinery, such as CD95 membrane receptors common in leukemia cells, exploit the finding that cancerous cells tend to over-express these proteins. Laboratories including ours have found that targeting CD95 in Jurkat cells activates caspase-3 at the membrane level, causing rapid induction of apoptosis through a non-traditional mechanism [1, 9].

Some of the natural compounds extracted from plants (e.g. Paclitaxel) have shown promising results as anti-cancer therapy. Pancratistatin is a natural alkaloid extracted from the Hawaiian spider lily (*Hyemocallis littoralis*) and has been shown to have anti-neoplastic effects. This compound has a molecular weight of 325 Daltons and a complex poly-oxygenated phenanthridone skeleton (figure 1) [17]. Recent results from our laboratory have indicated that Pancratistatin (PST) is able to specifically induce apoptosis in both Jurkat cells (a human T-cell leukemia cell line) and blasts from acute leukemia patients collected prior to chemotherapeutic interference [9, 16].

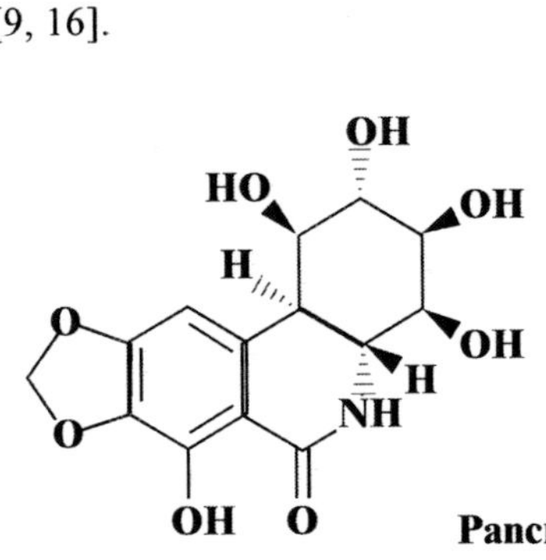

Molecular Weight = 325.28
Exact Mass = 325
Molecular Formula = C14H15NO8
Molecular Composition= C   51.70%   H   4.65%   N   4.31%   O   39.35%

Figure 1. The chemical structure of Pancratistatin.

RESEARCH OBJECTIVE

The objectives of the work described here were to further investigate the efficacy of Pancratistatin in inducing apoptosis in blast cells obtained from acute leukemia patients, and to compare the effect of Pancratistatin on normal blood cells from age-matched healthy volunteers. Furthermore, we investigated the possible mechanism of action of Pancratistatin-induced apoptosis in Jurkat cells.

MATERIALS AND METHODS

Cell Culture

A human T-cell Leukemia cell line (Jurkat cells) was purchased from ATCC, Manassas, VA. These cells were grown and maintained in an incubator set at 37°C, 5% CO_2 and 95% humidity. These cells were cultured in RPMI-1640 media supplemented with 10% fetal bovine serum (FBS) and 20µg/mL gentamycin (Life Technologies, Mississauga, ON, Canada).

Human nucleated blood cells were purified from whole blood obtained from either healthy male and female volunteers, or acute leukemia patients at the Windsor Regional Cancer Centre as approved by the University of Windsor ethical committee, REB#04-060 and the W.R.C.C. ethical committee, REB#04-044. Seven milliliters of whole blood was collected in a BD Vacutainer™ CPT Tube (Cell Preparation Tube) obtained from Becton Dickinson (Franklin Lakes, NJ). With both sample types, whole blood was spun down in a table-top low-speed centrifuge at 2900 rpm for 30 min at 25°C. The top layer containing mononuclear cells, platelets and plasma was collected; these cells were maintained in the same incubator as the Jurkat cells (37°C, 5% CO_2 and 95% humidity).

For the induction of apoptosis by treatment with Pancratistatin, Jurkat cells were grown to a cellular density of 0.5 million cells/ml then treated for varying periods of time. Pancratistatin was isolated from *Hymenocallis littoralis* following published procedures (99% pure) [20]. Jurkat cells were directly treated with 1µM concentrations of Pancratistatin. The normal nucleated blood cells and leukemia-containing cells were treated in a similar manner after being purified from whole blood.

Cellular Viability Assay

To examine apoptotic morphological changes, Jurkat cells and lymphocytes were grown treated and stained with 10µM final concentration of cell permeable Hoechst 33342 (Molecular Probes, Eugene, OR) and incubated for 10 min at 25°C. The cells were then examined under a fluorescent microscope (Leica DM IRB, Germany) where both phase-contrast and fluorescent images were captured. Cells in five fields at 10X magnification were used to count apoptotic versus live cells (where brightly stained cells with condensed nuclei were considered apoptotic). These results were then calculated and tabulated as a percentage of apoptotic cells using Microsoft® Excel 6.0 software. The pictures at higher magnification were compiled using Adobe® Photoshop 7.0 software.

Annexin-V Binding Assay

After cell treatment, the annexin-V binding assay was conducted using a commercial kit and the manufacturer's protocol (Molecular Probes, Eugene, OR). After treatment, cells were washed in phosphate-buffered saline (PBS) and re-suspended in annexin-V binding buffer (10mM HEPES/NaOH pH 7.5, 140mM NaCl, 2.5mM $CaCl_2$), containing 1:50 annexin-V Alexa Fluor® 488 conjugate for 15 min at 25°C. Cells were then examined under a fluorescent microscope (Leica DM IRB, Germany) and the images were captured and stored. All pictures were processed using Adobe® Photoshop 7.0 software.

Caspase-3 Activity Assay

The caspase-3 assay was performed using a previously published method [18]. To determine caspase-3 activity, total protein from Jurkat or lymphocyte (leukemia or normal) cell lysates were incubated with the fluorogenic substrate (DEVD-AFC), which is a tetra peptide sequence that corresponds to the substrate cleavage site. This assay was carried out according to the manufacturer's protocol (Enzyme System Products, USA). The fluorescence was measured by excitation at 400nm and emission at 505nm using the Spectra Max Gemini XS (Molecular Devices, Sunnyvale, CA). Caspase-3 activity was calculated per microgram of protein, and protein concentration was determined with BioRad protein assay reagent (BioRad, Mississauga, ON, Canada) using bovine serum albumin as a standard. Microsoft® Excel 6.0 software was used for data representation and statistical analysis.

Sub-Cellular Fractionation of Jurkat Cells

After treatment, the cells were collected by centrifugation and washed twice in PBS, then lysed in hypotonic buffer (10mM Tris-HCl pH 7.2, 5mM KCl, 1mM $MgCl_2$, 1mM EGTA, 1% Triton-X-100). Following 10 min incubation at 4°C, the cell suspension was transferred to a glass cell homogenizer (Kontes Glass Co., Vineland, NJ) and the cell membranes were mechanically disrupted by at least 30 strokes. Estimation of protein concentration in each total cell lysate was performed after 5 min at 4°C with the BioRad protein assay reagent (BioRad, Mississauga, ON, Canada) using bovine serum albumin as a standard. The nuclear pellet (and non-lysed whole cells) was collected by centrifugation at 3000 rpm for 5 min at 4°C. The supernatant was then collected and spun down at 10000 rpm at 4°C for 5 min; the resulting pellet containing the mitochondrial fraction. The supernatant was collected and spun down at 55000 rpm at 4°C for 1 hr, the final pellet being the mixed membrane fraction and the supernatant the S100 cytosolic fraction. Each fraction obtained from the Jurkat cells was tested for caspase-3 activity as previously described. In addition, a batch of untreated Jurkat cells was fractionated as above, and then incubated *in-vivo* with 1μM Pancratistatin for 15 min at 37°C. These fractions were also tested for caspase-3 activity.

Immuno-Precipitation / SDS-PAGE / Western Blotting

After treatment, cell lysis and protein estimation were performed as described above. First, a mixture of 500μL RIPA buffer (20mM Tris-OH pH 7.5, 150mM NaCl, 10mM KCl, 1% Triton-X-100), 10μL protein-G-sepharose and 2μL primary antibody (α-Fas Ab Human (Rabbit) polyclonal, Oncogene Research Products, San Diego, CA; or α-caspase-3 Ab Human (Rabbit) polyclonal, Sigma-Aldrich, St. Louis, MO) was mixed gently for 1 hr at 25°C. The sepharose-antibody complex was collected by centrifugation at 4000 rpm for 1 min at 25°C. To the pellet, 300μg protein from the cell lysate was added to a total volume of 500μL RIPA buffer, and nutated at 25°C for 1 hr. The protein-antibody-sepharose complex was then centrifuged at 4000 rpm for 1 min, re-suspend in RIPA buffer (20mM Tris-OH pH 7.5, 450mM NaCl, 10mM KCl, 1% Triton-X-100) and nutated for 10 min at 25°C; the wash process was repeated four times. After the last wash, 10μL SDS-PAGE loading buffer was added to approximately 50μL protein-complex in RIPA buffer and boiled at 95°C for 4 min. The complex was then resolved on a 12% agarose gel at 30mA and 250V, and then transferred to nitrocellulose paper at 80V and 250mA. The nitrocellulose paper was blocked with a 5% milk solution in TBST (20mM Tris-HCl pH 7.2, 205mM NaCl, 0.2% Tween). The nitrocellulose paper was then incubated with primary antibody (either 1:1000 dilution for α-Fas Ab, or 1:2000 dilution for α-caspase-3 Ab) in 2% milk solution overnight at 4°C. The blot was rinsed twice and washed once for 15 min, then twice for 5 min each at 25°C with TBST. The nitrocellulose was then incubated with secondary antibody (1:5000 dilution α-Rabbit (Goat) peroxidase, Sigma-Aldrich, St. Louis, MO) in 2% milk solution for 1 hr at 25°C. The blot was rinsed twice, and then washed three times for 5 min each with TBST. The blot was then incubated with Chemi-Glow[TM] soluble peroxide buffer for 5 min; the band was captured on the Alpha Innotech photo-documentation system (Alpha Innotech, Merced St. San Leandro, CA).

Caspase-3 Inhibition Assay

Caspase-3 activity was inhibited prior to treatment by incubating Jurkat or lymphoma cells with the fluorogenic substrate Z-DEVD-fmk (EMD Biosciences Inc., San Diego, CA). This O-methylated tetra peptide sequence corresponds to the substrate cleavage site of caspase-3, where the fluoromethyl ketone group at the C-terminus reacts and gets cross-linked in the active site thus blocking its activity. Cells were treated with a final concentration of 30μM Z-DEVD-fmk for 30 min, followed by treatment with 1μM Pancratistatin. Cells were then collected by centrifugation, washed twice with PBS and incubated with a final concentration of 10μM cell permeable Hoechst 33342 dye for 10 min at 25°C. Cells were then examined under a fluorescent microscope (Leica DM IRB, Germany) and the images were captured and stored.

In-Vivo Studies with Murine Leukemia in Balb/C Mice

In preliminary studies, balb/c mice were injected with one million T27a cells (murine leukemia, ATCC, Manassas, VA) in the intra-peritoneal cavity, in compliance with the University of Windsor REB#05-011. Half of the animals were treated with PST (10μM final concentration); the other half received placebo (PBS). After several treatments, the mice were sacrificed and cells were removed from the intra-peritoneal cavity of all mice for examination of nuclear morphology. The cells were incubated with a final concentration of 10μM cell permeable Hoechst 33342 dye for 10 min at 25°C and then examined under a fluorescent microscope (Leica DM IRB, Germany); the images were captured and stored.

RESULTS

Ex-Vivo Treatment of Acute Leukemia Samples

Pancratistatin has been shown in induce apoptosis in a number of established human and rat cancer cell lines grown in culture [21]. As the next important step, we needed to examine the effect of this compound on primary human cancers. For this purpose we chose to test PST on human leukemia cells *ex-vivo*. In collaboration with the Windsor Regional Cancer Center (Windsor, ON, CA), blood samples were obtained from acute leukemia patients and treated with PST for 24 hours. As a control, nucleated blood cells were collected from age-matched healthy volunteers and treated with PST under similar conditions. Patients included in the study were either newly diagnosed or had a relapse, but had not yet undergone chemotherapy treatments. Five patient samples were tested. Figures 2 – 5 represent our results. The results obtained from these experiments indicated that Pancratistatin efficiently and selectively induced apoptosis in blasts cells of acute leukemia patients.

Effect of PST in Inducing Apoptosis in Human T-Cell Leukemia (Jurkat) Cells

For the majority of our cellular studies, the stable Jurkat cell line was used to study the effects of PST treatment at various concentrations and durations. We have shown that PST is highly selective for inducing apoptosis in Jurkat cells (Figure 6), as well as in neuroblastoma (SHSY-5Y), rat hepatoma (5123t) and breast cancer cells (MCF-7 and HTB-126), while having no effect on normal counterpart cells [16, 19]. Previous experiments with PST revealed that its target is non-genomic since it does not cause DNA double-stranded breaks or DNA fragmentation prior to caspase activation [9]. Preliminary studies into the mechanism of action of PST in Jurkat cells at early treatment time points showed caspase-3 activation and intact mitochondrial membrane potential, with an increase in ROS production a later event [9].

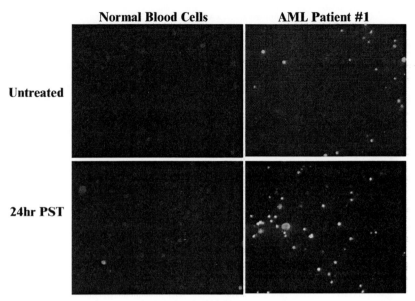

Figure 2. Effect of PST treatment on AML cells obtained from Patient #1 and a healthy volunteer: AML cells were treated in parallel with an aged-matched healthy cell volunteer. Hoechst pictures were taken on a fluorescent microscope at 20X objective; Hoechst brightly stains condensed nuclei, a characteristic marker of apoptosis.

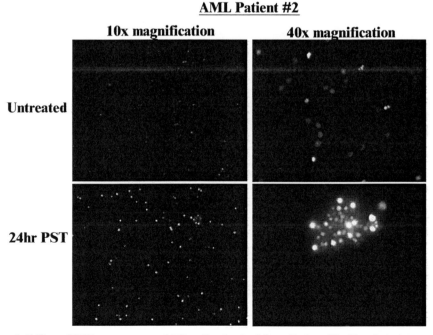

Figure 3. Effect of PST treatment on AML cells obtained from Patient #2. Hoechst pictures were taken on a fluorescent microscope at both 10X and 40X objective to better observe bright, condensed nuclei, following Hoechst incubation, characteristic to apoptosis.

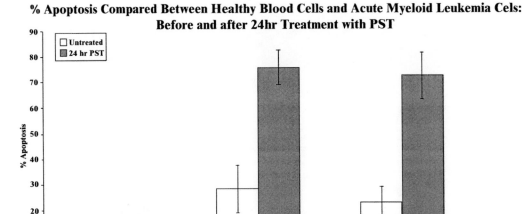

% Apoptosis Compared Between Healthy Blood Cells and Acute Myeloid Leukemia Cels: Before and after 24hr Treatment with PST

Figure 4. Degree of apoptosis: a comparison between AML patients and healthy age-matched volunteers before and after 24hr 1μM PST treatment. Hoechst pictures were taken and bright, condensed cells were considered apoptotic. Number of apoptotic cells was calculated as a percentage of total cell number over several fields.

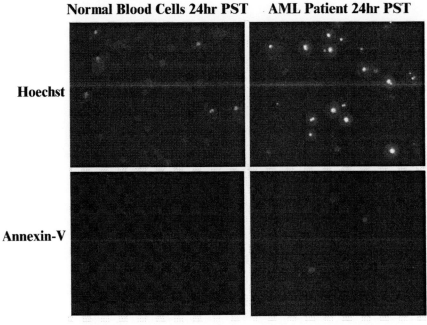

Figure 5. AML cells treated with 1μM PST show flipping of phosphatidyl-serine to the outer leaflet of the plasma membrane. Apoptosis is characterized by brightly stained, condensed nuclei by Hoechst dye, as well as positive staining (green rings) by Annexin-V dye, which binds to phosphatidyl-serine once flipped to the outer membrane. Phosphatidyl-serine flipping is also a characteristic feature of apoptosis.

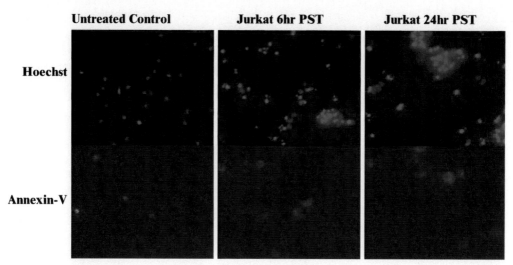

Figure 6. Human T-cell leukemia (Jurkat) cells undergo apoptosis at 6hr and 24hr treatments with 1μM PST. Hoechst and Annexin-V dye were used to characterize apoptosis in Jurkat cells following treatment with PST. Condensed and brightly stained nuclei following Hoechst staining is indicative of apoptosis; a bright green ring around a cell indicates phosphatidyl-serine plasma membrane flipping, a characteristic feature of apoptosis.

Activation of Membrane-Bound Caspase-3 by PST

At early treatment time points with PST, Jurkat cells were tested for caspase-3 activation. Caspase activation is a necessary step in the apoptotic pathway, and determination of whether it occurs early or late in the apoptotic process is imperative in understanding the pathway activated upon treatment with PST. Later activation of caspase-3 may indicate mitochondrial permeability occurs first, leading to a cascade of events ending in caspase-3 activation prior to DNA fragmentation. However, early activation of caspase-3 suggests that the extrinsic pathway of apoptosis is being followed, possibly due to stimulation of cell-membrane death receptor proteins. Figure 7 shows caspase-3 activation occurring at early time points in whole cell lysate from Jurkat cells treated with PST. After only 1hr, 1μM PST causes a 1.5 fold increase over baseline caspase-3 activity in untreated (control) cells. To further analyze the process of caspase-3 activation in Jurkat cells following PST treatment, the cells were fractionated after a 1hr treatment with 1μM PST; the fractions collected were the nuclear and mitochondrial pellet, the cytosolic S100 fraction and the mixed membrane pellet. Figure 8 shows the highest level of active caspase-3 was detected in the mixed membrane fraction.

The mixed membrane fraction contains cell-death machinery proteins, such as Fas-receptors known to be over-expressed in Jurkat cells [13], which may interact with lipid-raft proteins like caspase-3, initiating an apoptosis signaling response [1]. To test this hypothesis in Jurkat cells, the control portion of the mixed membrane fraction was incubated with 1μM PST for 15 minutes, and caspase-3 activity was re-measured. Caspase-3 activity increased approximately two-fold compared to control (figure 9), indicating that PST either activates caspase-3 directly or indirectly through a plasma membrane protein. PST was added at 1μM to a control whole cell lysate and incubated for 15 minutes; however, levels of activated caspase-3 did not change (data not shown). Therefore, PST might be activating caspase-3 indirectly through an undetermined protein associated with it in the membrane.

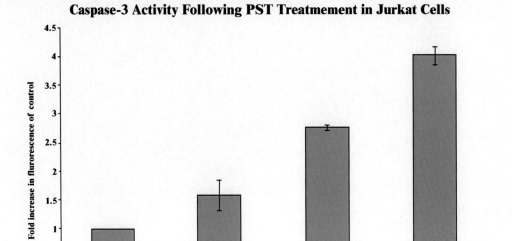

Figure 7. PST causes early caspase-3 activation in Jurkat cells. Whole cell lysate was collected (in the absence of protease inhibitors) after specified treatment times and tested for caspase-3 activity with DEVD-afc peptide; DEVD-afc read at emission 400nm, excitation 505nm.

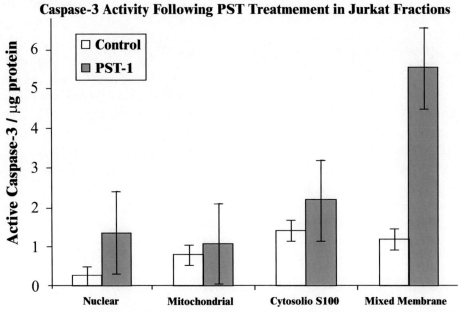

Figure 8. Caspase-3 activity is the highest in the mixed membrane fraction of Jurkat cells treated with 1µM PST for 1hr. Cellular fractions were collected from both control and 1hr PST treated Jurkat cells. The mixed membrane fraction displayed increased caspase-3 activity compared to control; DEVD-afc read at emission 400nm, excitation 505nm.

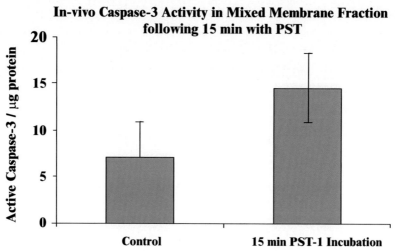

Figure 9. Activation of caspase-3 in control mixed membrane fraction following 15 min incubation with PST. When the control portion of the mixed membrane fraction was treated with PST, caspase-3 activation occurred very rapidly; DEVD-afc read at emission 400nm, excitation 505.

Active Caspase-3 Associated with Fas-Receptor as Seen Through Immuno-Precipitation and Western Blotting

As reported by Aouad *et al.*, 2004, caspase-3 is present in lipid rafts of untreated Jurkat cells and upon stimulation by Fas ligand, caspase-3 and caspase-8 is further recruited to the DISC, where active caspase-3 is required for activation of caspase-8. In order to investigate if PST activates membrane bound caspase-3 associated with Fas receptor we carried out immuno-precipitation experiments. Total protein extracts from the control and PST treated Jurkat cells were immuno-precipitated separately with both anti-capase-3 antibody and anti-Fas receptor antibody and analyzed by western blots using anti-caspase-3 antibodies in order to observe any interaction between these two proteins. Our results show that the cleaved, active form of caspase-3 (17kDa) is pulled down with anti-Fas antibody at treatment times of 3 and 6 hours (figure 10a). Figure 10b shows that both pro-caspase-3 (32kDa) and its active form were brought down by anti-caspase-3 immuno-precipitation, in amounts increasing with treatment duration. These results suggest the possibility of interaction between PST, Fas receptor protein and caspase-3 probably within the plasma membrane. High expression of Fas receptors or a conformational change or the presence of caspase-3 in the plasma membrane and its association with Fas in the blast cells could be the features responsible for the selective targeting of these cells for apoptosis induction by Pancratistatin.

Caspase-3 Inhibition Halts Apoptosis in Jurkat Cells Treated with PST

The results described above indicate a strong connection between PST-induced apoptosis and the activation of caspase-3. In order to determine whether caspase-3 activation is a required step necessary for apoptosis to occur, we carried the following experiment. Jurkat cells were incubated with the caspase-3 inhibitor molecule z-DEVD-fmk for 30 minutes prior

to treatment with PST for 24 hours. In the usual case PST is able to induce apoptosis in over
90% of Jurkat cells at this time-point [9]. However, when caspase-3 activity is inhibited,
nuclear morphology as seen by Hoechst staining indicates healthy, smooth nuclei; the
characteristic nuclear shrinking, membrane blebbing and formation of apoptotic bodies is not
observed (figure 11). This result indicated that activation of caspase-3 is a requirement of
PST-induced apoptosis, and that an alternate mechanism is not activated.

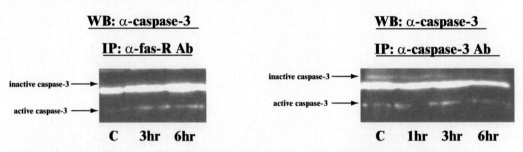

Figure 10. Caspase-3 interacts with Fas receptor in Jurkat cells following 1hr PST treatment.
Immunoprecipition with anti-fas antibody or anti-caspase-3 antibody, followed by western blot with
anti-caspase-3 antibody demonstrates an interaction between Fas receptor protein and active caspase-3;
active caspase-3 (17kDa) was observed only in the anti-Fas Ab immuno-precipitate, whereas both the
inactive (32kDa) and active subunits were visible in the anti-caspase-3 immuno-precipitate.

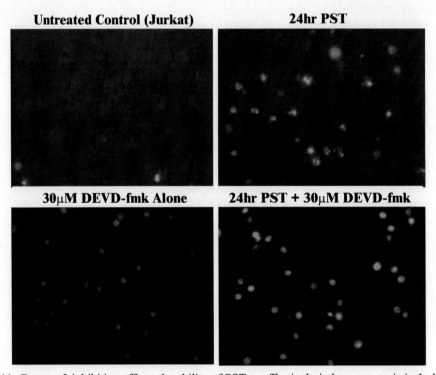

Figure 11. Caspase-3 inhibition affects the ability of PST to effectively induce apoptosis in Jurkat cells.
Jurkat cells were incubated with DEVD-fmk prior to a 24hr treatment with PST. Nuclear morphology
as seen by Hoechst staining indicates that cells pretreated with caspase-3 inhibitor have healthy, smooth
intact nuclei, as apposed to the brightly stained, condensed nuclei observed with PST treatment alone.

PST is Effective Against Leukemia *In-Vivo*: Preliminary Results

PST has been shown effective and specific for inducing apoptosis *in-vitro*, with little to no effect on healthy nucleated blood cells treated under the same conditions [9]. Preliminary *in-vivo* studies are currently underway to investigate the toxicity and efficacy of PST using balb/c mice injected intra-peritoneally with murine leukemia (T27a cells) as a model. Preliminary results indicate that PST has some effect on leukemia cells *in-vivo* (figure 12). In addition, over several treatments, PST did not cause adverse effects (weight loss, skin changes) in un-injected healthy balb/c mice (data not shown).

Preliminary In-vivo Results: Leukemia Treated with PST

Saline Treated T27A Injected Mouse   **PST Treated T27A Injected Mouse**

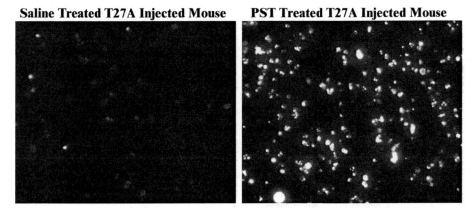

Figure 12. PST treatment of murine leukemia *in-vivo* results in apoptosis of these cells compared to placebo. Leukemia cells were removed from the intra-peritoneal cavity of balb/c mice treated with either PST or placebo. Cells were stained with Hoechst; brightly stained, condensed nuclei indicate apoptotic morphology.

CONCLUSION

In conclusion our results have clearly indicated that Pancratistatin could be a very effective and selective inducer of apoptosis in acute leukemia cells. Our preliminary finding that Pancratistatin did not induce apoptosis in nucleated blood cells obtained from age-matched healthy volunteers, suggests that this compound will possibly be well tolerated *in-vivo*. Furthermore, we have shown that Pancratistatin induces apoptosis by indirectly activating caspase-3 by targeting the Fas receptor.

ACKNOWLEDGEMENTS

We are grateful to Dr. T. S. Sridhar, M.D., Ph.D. for critically reviewing the manuscript and providing valuable suggestions. We acknowledge the occasional technical assistance provided by Natasha Sharda and Divya Sood. Research funding for this work was provided by the Lotte and John Hecht Foundation, Vancouver, B.C.; the Knights of Columbus, Council #9671, Windsor, ON; and Windsor Regional Cancer Foundation, Windsor, ON.

REFERENCES

[1] Aouad, SM; Cohen, LY; Sharif-Askari, E; Haddad, EK; Alam, A; Sekaly, RP. Caspase-3 is a component of Fas death-inducing signaling complex in lipid rafts and its activity is required for complete caspase-8 activation during Fas-mediated cell death. *J Immunol*, 2004, 172(4), 2316-23.

[2] Boatright, KM; Salvesen, GS. Mechanisms of caspase activation. *Curr Opin Cell Biol.* 2003, 15(6), 725-31.

[3] Cho, SG; Choi, EJ. Apoptotic signaling pathways: caspases and stress-activated protein kinases. *J Biochem Mol Biol.* 2002, 35(1), 24-7. *Review.*

[4] Ciardiello, F; Tortora, G. Inhibition of Bcl-2 as cancer therapy. *Annual Oncology.* 2001, 13, 501-502.

[5] Copland, M; Jorgensen, HG; Holyoake, TL. Evolving molecular therapy for chronic myeloid leukaemia--are we on target? *Hematology.* 2005, 10(5), 349-359. *Review.*

[6] Green, DR; Evan, GI. A matter of life and death. *Cancer Cell.* 2002, 1, 19-28. *Review.*

[7] Hu, W; Kavanagh, J. Anticancer therapy targeting the apoptotic pathway. *The LANCET Oncology.* 2003, 4, 721-729.

[8] Jemal, A; Murray, T; Samuels, A; Ghafoor, A; Ward, E; Thun, MJ. Cancer Statistics, 2003. *CA Cancer J Clin* 2003, 53, 5-26.

[9] Kekre, N; Griffin, C; McNulty, J; Pandey, S. Pancratistatin causes early activation of caspase-3 and the flipping of phosphatidyl serine followed by rapid apoptosis specifically in human lymphoma cells. *Cancer Chem Pharmacol.* 2005, 56(1), 29-38.

[10] Kim, R. Recent Advances in Understanding the Cell Death Pathways Activated by Anticancer Therapy. *Cancer.* 2005, 103(8), 1551-1560.

[11] LeBlanc, R; Catley, LP; Hideshima, T; Lentzsch, S; Mitsiades, CS; Mitsiades, N; Neuberg, D; Goloubeva, O; Pien, CS; Adams, J; Gupta, D; Richardson, PG; Munshi, NC; Anderson, KC. Proteasome inhibitor PS-341 inhibits human myeloma cell growth in vivo and prolongs survival in a murine model. *Cancer Res.* 2002, 62(17), 4996-5000.

[12] Legembre, P; Daburon, S; Moreau, P; Ichas, F; de Giorgi, F; Moreau, JF; Taupin, JL. Amplification of Fas-mediated apoptosis in type II cells via microdomain recruitment. *Mol Cell Biol.* 2005, 25(15), 6811-6820.

[13] Legembre, P ; Daburon, S ; Moreau, P ; Moreau, JF ; Taupin, JL. Modulation of Fas-mediated apoptosis by lipid rafts in T lymphocytes. *J Immunol.* 2006, 176(2), 716-720.

[14] Loeb, LA; Loeb, KR; Anderson, JP. Multiple mutations in cancer. *Proc. Natl. Acad. Sci. USA.* 2003, 100, 776–781.

[15] Martin, NMB. DNA Repair Inhibition and Cancer Therapy. *J Photochem. Photobio. B, Biology.* 2001, 63, 162-170.

[16] McLachlan, A; Kekre, N; McNulty, J; Pandey, S. Pancratistatin: a natural anti-cancer compound that targets mitochondria specifically in cancer cells to induce apoptosis. *Apoptosis*. 2005, 10(3), 619-630.

[17] McNulty, J; Larichev, V; Pandey, S. A synthesis of 3-deoxydihydrolycoricidine: Refinement of a structurally minimum Pancratistatin parmacophore. *Bioorg. and Med. Chem. Lett.* 2005, 15, 5315-5318.

[18] Naderi, J; Hung, M; Pandey, S. Oxidative stress-induced apoptosis in dividing fibroblasts involves activation of p38 MAP kinase and over-expression of Bax: Resistance of quiescent cells to oxidative stress. *Apoptosis*. 2003, 8, 91-100.

[19] Pandey, S; Kekre, N; Naderi, J; McNulty, J. Induction of Apoptotic Cell Death Specifically in Rat and Human Cancer Cells by Pancratistatin. *Artificial Cells, Blood Substitutes, and Biotechnology*. 2005, 33, 279-295.

[20] Pettit, GR; Pettit, GR III; Backhaus, RA; Boyd, MR; Meerow, AW. Antineoplastic Agents 256. Cell Growth Inhibitory Isocarbostyrils from *Hymenocallis. J. Nat. Prod.* 1993, 56, 1682.

[21] Sato, T; Machida, T; Takahashi, S; Iyama, S; Sato, Y; Kuribayashi, K; Takada, K; Oku, T; Kawano, Y; Okamoto, T; Takimoto, R; Matsunaga, T; Takayama, T; Takahashi, M; Kato, J; Niitsu, Y. Fas-mediated apoptosome formation is dependent on reactive oxygen species derived from mitochondrial permeability transition in Jurkat cells. *J. Immunol.* 2004, 173(1), 285-296.

[22] Sharifi, N; Steinman, RA. Targeted chemotherapy: chronic myelogenous leukemia as a model. *J. Molec. Med.* 2002, 80, 219-232.

[23] Srivastava, RK. TRAIL/Apo-21: mechanism and clinical application in cancer. *Neoplasia*. 2001, 3, 535-546.

[24] Thatte, U; Bagadey, S; Dahanukar, S. Modulation of programmed cell death by medicinal plants. *Cell. Molec. Bio.* 2000, 46(1), 199-214.

[25] Travis, LB. Therapy-associated solid tumors. *Acta. Oncol.* 2002, 41(4), 323-333. *Review*.

In: Cell Apoptosis Research Trends
Editor: Charles V. Zhang, pp. 111-149
ISBN: 1-60021-424-X
© 2007 Nova Science Publishers, Inc

Chapter IV

Neuropeptides as Modulators of Apoptosis in Prostate Cancer

Jose Vilches and Mercedes Salido

Departament of Histology. School of Medicine. University of Cadiz.
Dr Marañon 3. 11002 Cadiz. Spain.

Dedicated to Prof. Gómez Sanchez, Emeritus Professor of Histology

ABSTRACT

Apoptosis is the result of an evolutionarily conserved program with common mechanisms for all euchariotic cells, in the same way that cell cycle does, thus creating a network of regulatory pathways that ensure the right development of cell populations included as essential parts of multicellular organisms. Cell death program is activated more easily and quickly in different cell types, and even in the same cells depending on their evolutionary stage. A single cell can also be sensitive to several types of death signals.

Physiological importance of apoptosis is enhanced by the fact that we are facing a phenomenon which, when disregulated, can be harmful for the organism, being in the core of a number of pathologies.Thus, the most relevant biological interest of apoptosis is the possibility of its modulation and so, the identification of inductive factors and their mechanisms of action appear as the most relevant challenges in apoptosis research.

Our research interest centers on neuroendocrine differentiation in prostatic carcinoma and its role on apoptosis modulation in the disease, with the development and applications of novel methodologies.

Neuroendocrine cells appear phisiologically inserted between epithelial cells, may be acting as trophyc stimuli in normal prostate. Of uncertain origin, the neuroendocrine cells are positive for neural markers and secrete neuropeptides with paracrine action on neighbouring cells. The presence of neuroendocrine differentiation -understood as an exaggerated ratio of neuroendocrine cells to normal or neoplastic epithelial cells- in the carcinoma is associated to worse prognostic, higher tumor progression and an androgen-independent status of the tumor, that becomes unresponsive to hormonal therapy.

Our research is based on three main goals. First: demonstrating the capability of prostate carcinoma epithelial cells in androgen sensitive and androgen insensitive stages to die by apoptosis after induction. That is, a preserved cell death machinery in these cells. Second, the role of representative neuropeptides in modulating the induction of cell death on the cells. Third, the study of intracellular alterations in both processes, i.e. ionic changes, mitochondrial permeability transition alterations. Novel methodologies are applied to develop working protocols for light and electron microscopy, electron probe analysis or confocal microscopy. As conclusions, we have demonstrated a role of neuropeptides bombesin and calcitonin on modulating etoposide-induced cell death, in vitro, on androgen sensitive and insensitive prostate cancer cells together with the fact that, in the presence of neuropeptides, apoptosis modulation is accompanied by reversion of ionic changes of apoptosis and of mitochondrial membrane potential alterations.

INTRODUCTION

Prostate cancer is a frequently diagnosed condition in the aging population and the incidence of this disease is expected to increase in the years to come, with an average age at diagnosis of 70 years, and 80% of the cases being diagnosed in men over 65. Due to the detection at an earlier age, there has also been a rapid increase in men diagnosed under the age of 65. Mortality from prostate cancer has increased at a slower rate but overall has doubled in the last 50 years.

Growth of a cancer is determined by the relationship between the rate of cell proliferation and the rate of cell death. Only when the rate of cell proliferation is greater than cell death does the tumor growth continue. If the rate of cell proliferation is lower than the rate of cell death, then regression of the cancer ocurrs.

Metastatic prostatic cancer, like the normal prostates from which they arise, are sensitive to androgenic stimulation of their growth, due to the presence of androgen-dependent prostatic cancer cells. These cells are androgen-dependent since androgens stimulate their daily rate of cell proliferation, while inhibiting their rate of death. In contrast, following androgen ablation, androgen-dependent prostate cancer cells stop proliferating and activate a cellular suicide pathway termed programmed cell death or apoptosis.

The programmed cell death induced in the prostate by androgen ablation is cell type specific: only the prostatic glandular epithelial cells and not the basal epithelial cells or stromal cells are androgen-dependent and then undergo programmed cell death following castration. [Bonkhoff 1993, Oberhammer 1993, Berchem 1995, Bonkhoff, 1995, Tang 1998, Baretton 1999]. The androgen deprivation results in the elimination of the androgen-dependent prostatic cancer cells from the patient. Due to this elimination, 80-90% of all men with metastatic prostate cancer treated with androgen ablation therapy have an initial positive response. All of these patients relapse eventually to a state unresponsive to further antiandrogen therapy, no matter how completely given. This is due to the presence of androgen independent prostatic cancer cells within such metastatic patients with a rate of proliferation exceeding their rate of cell death even after complete androgen blockade is performed.

Attempts to use nonandrogen ablative chemotherapeutic agents to adjust the kinetics parameters of androgen-independent prostate cancer cells so that their rate of cell death exceeds their rate of proliferation leads to the novel concept of antisurvival factor therapy for prostate cancer as a component of anticancer treatments [Sciarra 2006]. These agents have been targeted at inducing DNA damage directly or indirectly via inhibition of DNA metabolism or repair and are critically dependent on an adequate rate of proliferation to be citotoxic. In vitro studies have demonstrated that, when these cells are rapidly proliferating, they are highly sensitive to the induction of programmed cell death via exposure to the same antiproliferative chemotherapeutic agents which are of limited value in vivo, due to major differences in the rate of proliferation occurring in the two states. Likewise, for chemotherapeutic agents to be effective, not only must the cancer cells have a critical rate of proliferation but also a critical sensitivity to induction of cell death. Anyway, the proliferative rate for androgen independent prostatic cancer cells is very low, explaining why antiproliferative chemotherapy is of limited value against metastatic prostatic cells. Based on this, what is needed is some type of cytotoxic therapy which induces the death of androgen-independent as well as androgen-dependent prostate cancer cells without requiring the cells to proliferate. Recently, the use of somatostatin analogues has been proposed for inhibiting the release of various neuroendocrine thus acting as antiproliferative and proapoptotic factors. [Fairbairn 1993, Borner 1995, Wright 1996, Sciarra 2006]

An optimal approach is to activate the "programmed cell death " pathway within these cells leading to their suicide. Both androgen-dependent normal prostatic glandular cells and androgen-dependent prostatic cancer cells can be induced to undergo programmed cell death following androgen ablation and this process does not require the cells to be in the proliferative cell cycle. [Staack 2003] Androgen-independent tumor cells do not, due to a defect in the initiation step. However, and as we have previously demonstrated [Salido 1996, Salido 1999], they retain the basic machinery for undergoing apoptosis. (Figure 1)

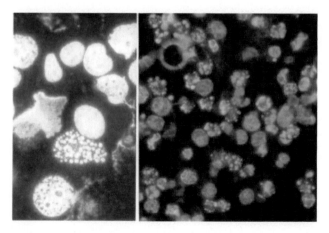

Figure 1. Apoptosis induced in Du-145 prostate cancer cell line after etoposide treatment. Nuclei present typical apoptotic bodies appear in different stages of the process. (DAPI, 25x, 100x).

Neuroendocrine differentiation in prostatic adenocarcinomas has received increasing attention in recent years as a result of possible implications on prognosis and therapy [Vashchenko and Abrahamsson 2005, Quek 2006, Sciarra 2006, Slovin 2006]

Prostatic adenocarcinoma reveals extensive and multilocal neuroendocrine features in aproximately 10% of the cases by using the neuroendocrine marker chromogranin A. These tumors tend to be more aggressive and resistant to hormonal therapy [Fixemer 2002]. Ultrastructural studies, biochemical analyses, and histochemical staining provide evidence for functionally diverse subtypes of NE cells within the prostate, that secrete a wide range of peptides known to stimulate cell growth in an autocrine and paracrine fashion [Clegg 2003]. Neuroendocrine secretory products and their interactions with epithelial prostate cells are currently under investigation in order to understand their significance for the pathogenesis of the prostate gland, prognosis and therapy [Wilson 2001, Quek 2006, Sciarra 2006, Slovin 2006].

Among the best-described neuroendocrine peptide growth factors in the prostate are members of the bombesin family and calcitonin. The mechanism by which bombesin stimulates growth of prostate cancer cells is still a matter of debate, and is probably complex. Bombesin appears to affect various signaling pathways, involving Ca^{2+} ions [Han 1997, Wasilenko 1997], cAMP [Jongsma 2000], and tyrosine kinases such as Src [Allard 2000]. It has also been proposed that bombesin can activate extracellular proteolytic activity and in that way contribute to metastatic spread [Festuccia 1998]. Recently, we have shown that bombesin and calcitonin inhibit etoposide-induced apoptosis in human androgen-independent prostatic cancer cell lines [Salido 2000, Salido 2002, Salido 2004, Vilches 2004], and these neuropeptides could thus disrupt the balance between cell death and cell growth in the tumor. This is an important point for the knowledge of the relationship between neuroendocrine differentiation and prognosis, because according to the kinetics of tumor growth, an increase in a neoplastic cell population is the result of imbalance between the two processes controlling tissue homeostasis: cell proliferation and cell death. Apoptosis, therefore, comprises a critical intrinsic cellular defense mechanism against tumorigenic growth which, when suppressed, may contribute to malignant development. Calcitonin, one of these neuropeptides, is reported to be associated with the growth of prostate cancer [Aprikian 1998, Chien 2001].

We have previously demonstrated that the neuropeptides bombesin and calcitonin inhibit etoposide-induced apoptosis both in androgen-dependent (LNCaP) and in androgen-independent (PC-3 and Du 145) prostate cancer cell lines [Salido 2000]. Furthermore, we have shown that apoptosis is associated with changes in elemental content of the cells, and that bombesin and calcitonin inhibit apoptosis-associated elemental changes [Salido 2002]. This finding strengthens the link between apoptosis and changes in the intracellular elemental content.

The development of in vitro models for an adequate approach to neuroendocrine differentiation of prostatic carcinoma and its implications on this disease is imperative. The Transmission and Scanning Electron Microscopy and X-ray microanalysis study of ionic changes that occur during etoposide-induced apoptosis in prostate cancer cell lines, as described below, shows how a combined treatment with etoposide and the neuropeptides bombesin and calcitonin inhibits etoposide-induced apoptosis in these cells, and also inhibits the major changes in intracellular ion concentrations in androgen independent and in androgen dependent prostate cancer cell lines.

In an effort to understand the role that neuroendocrine differentiation plays in the progression and metastatic spread of prostate carcinoma, the electron probe X-ray microanalysis is unique because it combines the ability to undertake chemical analysis with the high resolution of the electron microscope, and thus is capable of correlating chemical information with known ultraestructural details.[Salido 2001, Vilches 2004].Seven of the most relevant ions involved in cell viability and cell death can be simultaneously quantified under the same analytical conditions, as described below. Healthy cells have a characteristic elemental pattern of viable cells, with low Na and high K, Mg and P and we will thus try to identify the elemental alterations that appear in neoplastic cells and the role of neuropeptides in cultured neoplastic prostate cancer cells. Our research interest mainly focuses on those parameters that indicate cell viability (Na/K ratio) and those that reflect alterations in the nucleus to cytoplasm ratio, as a result of changes in cell volume and nuclear changes that are described as morphological hallmarks of apoptosis (P/S ratio, Cl, Na and K concentrations). (Figure 2)

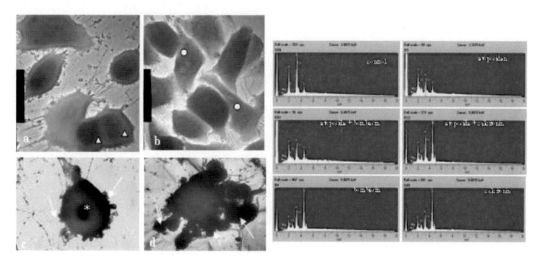

Figure 2. Du 145 cells as examined in the Transmission Electron Microscope, in STEM mode together with the spectra from quantitative analysis of the cells. Cells are grown on titanium grids covered with a formvar film. a) etoposide treated cells (triangle) show a ruffled membrane and condensed nuclei and become round prior to detachment from the substrate and neighbouring cells. As shown in b) for control cells, viable cells remain together and with well preserved morphology (o). As cells pass to later stages, b) nuclei appear more and more condensed (*), and blebs are more evident as shown in c). X-ray microanalysis reveals that etoposide treatment causes changes in the most relevant ions implicated in cell viability and cell death, with an increase in Na$^+$ and a decrease in K$^+$ and changes in Cl$^-$, P, and S that are blocked by neuropeptides, as described in the text.

Mitochondrial dysfunction in apoptosis is related with specific permeabilization of the outer mitochondrial membrane to large molecules including ions that are relevant in apoptotic process [Antonsson 2004]

The increasing number of studies enforcing the importance of mitochondria in apoptosis signalling, with an increase in complexity, result in a strong debate concerning the exact sequence of mitochondrial events.

In apoptosis mitochondria have two essential functions. First, provide energy, in the form of ATP, which is required for cells to die by the apoptosis pathway. Second, to release pro-apoptotic proteins normally sequestered in the intermembrane space into the cytosol where they trigger downstream apoptotic signaling pathways.

A recent study from our laboratory demonstrated the capability of etoposide in the induction of alterations in $\Delta\Psi$m, with subsequent release of IMs proteins [Vilches and Salido 2005] Concomitantly we have also described the increase in $[Ca^{++}]i$, which is regarded as inducer of mitochondrial permeability transition, in etoposide-induced apoptosis [Salido 2001] (Figure 3).

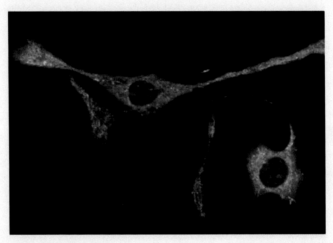

Figure 3. PC3 cells stained in our laboratory with JC-1 and examined in the confocal microscope. Upper left: green channel. The monomeric form of the dye emits green fluorescence in those mitochondria with low $\Delta\Psi_m$. Upper right: red channel. Red fluorescente emitted by J-aggregates present in mitochondria with preserved $\Delta\Psi_m$ Down: Overlay image for comparative stdy of green to red fluorescence.

The approach taken in the present study mimics the presence of neuroendocrine cells in prostatic carcinoma. We have investigated whether the neuropeptides bombesin and calcitonin inhibit the etoposide-induced changes in intracellular elemental concentrations and of mitochondrial membrane potential alterations in androgen-independent PC3. Neuroendocrine cells, thus, become an independent indicator of poor prognosis in patients with prostate carcinoma, independently of hormonal status of the epithelial cell tumors. Apoptosis, therefore, comprises a critical intrinsic cellular defense mechanism against tumorigenic growth which, when suppressed, may contribute to malignant development.

Clarify the interactions between epithelial and neuroendocrine cells in hormone-dependent microenvironments, such as prostate cancer, and a better comprehension of mechanisms of tumor progression are, therefore, of paramount importance.

PROTOCOL FOR CELL LINES PREPARATION

Cell Culture

Culture Protocols and Treatment Administration

Androgen independent cells PC-3, Du 145 (ATCC, Manassas USA) grown in Dulbecco Modified Essential Medium (DMEM) supplemented with 10% Fetal Bovine Serum (FBS), 4% penicillin-streptomycin and 0.4% gentamycin are keeped under standard conditions, in a water saturated atmosphere of 5% CO_2 until the experiment is started. LNCaP androgen dependent cells (ATCC, Manassas USA) are grown in RPMI 1640 under the same conditions. All experiments are started with unsynchronized exponentially growing cells, and the culture medium is changed to 5% FBS supplemented medium.

Androgen-Independent Cells

Cells are seeded in microplates at a density of 200.0000 cells/well under the conditions described above and 48 hours later exposed to different treatment protocols.

Androgen-dependent cells

A second protocol including hormonal deprivation is used. 200.0000 cells/well are seeded in microplates in RPMI-1640 under the conditions described above and 72 hours later changed to RPMI containing 5% Steroid Free Serum –SFS- and 48 hours later exposed to the different treatment protocols.

Treatment Protocols

Etoposide-induced apoptosis: Cells are exposed to etoposide -added from a 2mM stock solution in DMSO- doses 80 µM, (Du 145) and 150 µM (PC-3, LNCaP) for 48 h. Inhibition of etoposide-induced apoptosis: neuropeptides and androgen exposure. Cells are exposed to combined treatments with etoposide (as described above) and bombesin (1 nM), or calcitonin (500 pg/ml). In LNCaP a combined group with dihydrotestosterone -DHT- (1nM) is added. Control groups: A control group cultured in the standard medium during the experiment is used in all experiments. Positive controls are treated with bombesin (1 nM), calcitonin (500 pg/ml), and DHT(1nM), in LNCaP, for 48 h .

Growth Kinetics and Cell Viability

Determined by XTT viability assay (Boehringer Manheim)and trypan blue exclusion, with trypan blue in culture media (0.5%) After incubation of cells with trypan blue, non stained cells are regarded as viable cells, and blue cells are considered non viable, when observed in a hematocytometer.*Percentages of viable cells:* (Number of non stained cells/ total cell number) x 100.

XTT assay: cells are grown in a microtiter plate, 96 wells, flat bottom, in a final volume of 100 µl culture medium per well, in a humidified atmosphere (37°C and 5% CO2), during the assay. After 24 and 48 hours, 50 µl of the XTT labelling mixture is added to each well and cells are incubated for 4 hours in a humidified atmosphere. Absorbance is measured using an ELISA reader, at a wavelength of 450-500 nm.

CHARACTERIZATION OF APOPTOSIS WITH LIGHT AND FLUORESCENCE MICROSCOPY

Direct examination by phase contrast microscopy is performed once the experiment starts.

Microscopical Quantification of Apoptotic Cells

Slides Obtained From In Vitro Cell Cultures

1. Collect the supernatant, containing the floating apoptotic cells
2. Trypsinization of the rest of the monolayer, containing healthy cells.
3. Reconstitute the total population by adding together both fractions
4. Centrifugation at 1000 rpm for 5 minutes to get the pellet.
5. Wash cells twice in PBS and cytospIn by means of cytobuckets, at 1500 rpm for 5 minutes.
6. Air-dry samples and stain as desired for light (i.e. haematoxylin-eosin, TUNEL) and fluorescence microscopy (i.e. fluorescent Dapi)[4].

The percentage of apoptotic cells is defined as (number of apoptotic cells/ total cell number) x 100. At least 200 cells should be counted for each experiment.

CHARACTERIZATION OF APOPTOSIS WITH FLOW CYTOMETRY AND AGAROSE GEL ELECTROPHORESIS

Flow Cytometry

1. Detach cells by trypsinization
2. Centrifuge celll suspension (10^5 cels/ml) at 1000 rpm for 5 min
3. Wash three times in PBS.
4. Resuspend the pellet in 425 µl PBS and 25 µl propidium iodide
5. Add 50 µl of NP40 in 1% PBS , prior to cytometric analysis

---

[4] *Haematoxylin-eosin staining:* Air dried slides are fixed in 10% formaldehyde and stained in haematoxylin and counterstained with eosin. *Fluorescent DAPI:* Air dried slides are fixed in metanol at -20° for 20 min., air dried and stained with DAPI- at room temperature and in the dark for 20 minutes, and mounted with an antifading medium, O-phenylendiamine in glycerol, and preserved in the dark at -20° C until examination, at a fluorescence range between 300 and 400nm.*TUNEL:* Formalin-fixed cells are permeabilized after dehydration and rehydration of the specimens by treatment with 0.5 pepsin, washed in distilled water and TBS, and endogen peroxidase blocked for 30 min in blocking solution, prior to exposition to labeling solution for 30 min at 37°. Converter POD is added for 30 min at 37° and cells are washed in TBS prior to addition of DAB and counterstaining with haematoxilin.

Agarose Gel Electrophoresis. Analysis of DNA Fragmentation

1. Detach cells by trypsinization and centrifuge cell suspension (10^6 cels/ml) at 1000 rpm for 5 min
2. Wash in PBS
3. Resuspend in 50µl Tris Borate EDTA- pH:8-(Merck Darmstadt, Germany), and 2,4 µl Nonidet P40 (Sigma St Louis MO, USA).
4. Add 2 µl RNA-sa -1/100-, 1mg/ml- (Sigma, St Louis MO, USA) to each sample prior to incubation
5. Incubate at 37°C for 2 hours.
6. Add 10µl proteinase k (Boehringer Mannheim, Germany)and incubate at 37°C overnight.
7. Heat samples to 65°, and mix 20 µl agarose with each sample before loading them into the dry wells of a 2% agarose gel in TAE 1x (Merck Darmstadt, Germany)
8. Load mollecular weight marker with 4µl marker (Amresco, OH, USA), 8µl water and 0,25µl bromophenol blue (Merck, Darmstadt, Germany). in 10% agarose 1%. (Pronadisa, Madrid Spain).
9. Run the gels at 70 V until the marker dye had migrated 3-4 cm, and then at 15 V overnight.
10. Stain in ethidium bromide (Sigma St Louis MO, USA) and destain in water for DNA visualization.

TEM, SEM AND MORPHOLOGY OF ETOPOSIDE-INDUCED APOPTOSIS

Scanning Electron Microscopy

1. Grow cells directly on sterile glass coverslips.
2. Once the experiment is finished, fix the cells in 2.5 % glutaraldehyde in 0.1 M sodium cacodylate buffer,
3. Then postfix in 1% OsO4 in the same buffer,
4. Dehydrate in a graded acetone series
5. Critical point dry the samples.
6. Coat with gold prior to examination.

Transmission Electron Microscopy

1. Cells are directly grown on small Petri dishes.
2. Once the experiment is finished cells are fixed in 2.5 % glutaraldehyde in 0.1 M sodium cacodylate buffer,
3. Postfixation in 1% OsO4 in the same buffer,
4. Dehydrate in a graded acetone series

5. Resin-embedding

X-RAY MICROANALYSIS (STEM MODE)

Cell and tissue culture offer an interesting perspective to increase the number of problems to which x-ray microanalysis can be applied. Among the numerous applications of x-ray microanalysis in cell biology and cell pathology, we will focus our interest on the role of ions in programmed cell death in cancer cells [Smith 1998; Skeeper 1999; Gutierrez 1999; Fernandez-Segura 1999; Mason , 1999; Salido 2001, Salido 2002, Salido 2004, Vilches 2004], as cancer growth is the result of the imbalance between cell death and cell proliferation of the neoplastic population.

Whole Mounts, for Analysis in the Scanning Transmission Electron Microscope (STEM)

1. Grow the cells on titanium grids covered with a Formvar film and sterilized by ultra-violet light overnight.
2. Grids can be distributed in small Petri dishes, 3 or 4 in each, and then, a drop of about 10µl of the cell suspension carefully seeded on each of the grids.
3. Keep Petri dishes in the incubator for at least 30 min. to allow cells to attach to substrate, then fill with 2-3 ml of complete medium and keep in the incubator.
4. After the exposure period, rinse the grids with attached cells in cold distilled water (4°C), briefly and discard excess of water
5. Freeze the grids in liquid nitrogen-cooled liquid propane (-180°C), by quick immersion
6. Freeze-dry the samples vacuum overnight at -130°C in a K775X Emitech Turbo Freeze Drier (Emitech, Ashford, Kent, UK)
7. Bring the samples to room temperature under vacuum. Coat the freeze-dried specimens with a conductive carbon layer.

X-ray microanalysis is performed at 100 kV in the STEM mode of an electron microscope with an energy dispersive spectrometer system. Quantitative analysis is carried out based on the peak-to-continuum ratio after correction for extraneous background [Roomans, 1988] and by comparing the spectra from the cells with those of a standard, which consists of known concentrations of mineral salts in a 20% gelatin and 5% glycerol matrix, frozen, cryosectioned and freeze dried to resemble the specimen in its physical and chemical properties [Roomans, 1988]. Spectra are acquired for 100 seconds and only one spectrum is obtained from each cell.

X-RAY MICROANALYSIS (SEM MODE)

For analysis in the scanning electron microscope (SEM), the cells are grown on Millipore Millicell® filters [Fernández-Segura 1999; Salido 2001, Salido 2002], placed in each of the wells of a 24 well plate.

1. Place 10 µl of a 50,000 cells/ml suspension by pipetting onto the filter.
2. Store plates in the incubator for at least 30 min. to allow cells to attach to the substrate, and then fill the wells with complete medium.
3. After exposure, rinse the filters with the cells in distilled water at 4°C and blot excess fluid with a filter paper,
4. Freeze the cells immediately in liquid propane cooled by liquid nitrogen
5. Freeze-dry in vacuum overnight at -130°C in a K775X Emitech Turbo Freeze Drier (Emitech, Ashford, Kent, UK)
6. Bring the samples to room temperature under vacuum
7. Coat the dried filters with a conductive carbon layer to avoid charging in the electron microscope.

The cells on the filter are analyzed in a scanning electron microscope (SEM) with an energy-dispersive X-ray microanalysis system at 20 kV, spot mode. Quantitative analysis is performed by determining the ratio (P/B) of the characteristic intensity (peak, P) to the background intensity (B) in the same energy range as the peak and comparing this P/B ratio with that obtained by analysis of a standard[5] [Salido 2001, Roomans 1988]. Each spectrum is acquired for 200 seconds. Only one spectrum is acquired from each cell.

MORPHOLOGICAL MARKERS. NUCLEAR MORPHOLOGY AND CELL MEMBRANE ALTERATIONS: ANNEXIN V / SYTOX GREEN STAINING

Apoptosis is distinguished from necrosis, or accidental cell death, by characteristic morphological and biochemical changes, including compaction and fragmentation of the nuclear chromatin, shrinkage of the cytoplasm, and loss of membrane asymmetry.In normal viable cells, phosphatidyl-serine (PS) is located on the cytoplasmic surface of the cell

---

[5] The concentration of element x in the specimen (C_{xsp}) was obtained with the peak-to-local-background (P/B) ratio method, and was calculated according to the formula

$$C_{xsp} = \frac{(Px/Bx)_{sp}}{(Px/Bx)_{std}} \cdot \frac{G_{sp}}{G_{std}} \cdot C_{xstd}$$

where Cx is the concentration of element x in millimoles per kilogram (dry weight), (Px/Bx) is the peak to background ratio for the element x, the subscripts sp and std refer to specimen and standard, respectively, and the G value is the mean value of the atomic number squared (Z^2) and divided by the atomic weight (A) in the sample. Cellular elemental concentrations were obtained with reference to 20% dextran standards containing known amounts of inorganic salts. [Warley, 1990] membrane. However, in apoptotic cells, PS is translocated from the inner to the outer leaflet of the plasma membrane, thus exposing PS to the external cellular environment.

The human vascular anticoagulant, annexin V, is a 35-36 kD Ca $^{2+}$-dependent phospholipid-binding protein that has a high affinity for PS. Annexin V ® (Molecular Probes, Leiden The Netherlands) labelled with a fluorophore or biotin can identify apoptotic cells by binding to PS exposed on the outer leaflet

SYTOX® Green (Molecular Probes, Leiden The Netherlands)nucleic acid stain is a high-affinity nucleic acid stain that easily penetrates cells with compromised plasma membranes and yet will not cross the membranes of live cells. After brief incubation with SYTOX® Green nucleic acid stain, the nucleic acids of dead cells fluoresce bright green when excited with the 488 nm spectral line of the argon-ion laser, or any other 450-490 nm source. These properties, combined with its >500-fold fluorescence enhancement upon nucleic acid binding, make the SYTOX® Green stain a simple and quantitative single-step dead-cell indicator.

Protocol for Adherent Cells Cultured on BD Falcon ® Culture Slides

1. Prepare annexin-binding buffer: 10 mM HEPES, 140 mM NaCl, and 2.5 mM CaCl $_2$, pH 7.4.
2. Induce apoptosis in cells using the desired method. A negative control should be prepared by incubating cells in the absence of inducing agent.
3. Remove culture media
4. Wash the cells in cold (2°-8°)phosphate-buffered saline (PBS).
5. Prepare staining solution: 100 µl is enough for one slide. Prepare 20 µl of the annexin V conjugate and 80 µl of SYTOX Green stain for each sample
6. Discard PBS from slides
7. Add staining solution
8. Incubate the cells in darkness at room temperature (18° to 24° may be suitable) for 15 minutes.
9. Wash the cells with annexin-binding buffer.
10. Mount using appropriate antifading media prior to examination in a Leica TCS SL inverted confocal Microscope (Leica Microsystems, Barcelona,Spain)

MITOCHONDRIAL MEMBRANE POTENTIAL

In the Mito PT ^{TM} ® kit, provided by Immunohistochemistry Technologies (Bloomington MN,USA), JC1 is incorporated in a friendly user presentation for easy penetration in cells and healthy mitochondria. Once inside a healthy non-apoptotic cell, the lipophilic MitoPT ^{TM} reagent , bearing a delocalized positive charge, enters the negatively charged mitochondria where it aggregates and fluoresces red, the so called J-aggregates. When the mitochondrial $\Delta\psi$ collapses in apoptotic cells, the reagent no longer accumulates inside the mitochondria and, instead, it is distributed throughout the cell. When dispersed in this manner, it assumes a monomeric form which fluoresces green.

Mito PT Protocol

1. Culture cells for the assay, not exceeding a final number of 10^6 cells/ml
2. Prepare and warm to 37° assay buffer (supplied). Keep warm
3. Prepare 100 Mito PT stock
4. Prepare 1x Mito PT solution
5. Remove media from cultures
6. Add 1x Mito PT solution (0.5 ml is enough for one slide)
7. Incubate the cells at 37° for 15 min in a CO_2 incubator
8. Warm the 1x assay buffer to 37°
9. Carefully remove and discard staining media
10. Wash the monolayer cultures 1-2 ml
11. Discard wash
12. Add a drop of 1x assay buffer plus coverslip prior to examination in a Leica TCS SL inverted confocal microscope (Leica Microsystems, Barcelona,Spain)

APOPTOSIS, IONS AND NEUROENDOCRINE DIFFERENTIATION

Scanning electron microscopy and light and fluorescence microscopy of the cultured prostatic cancer cells show non apoptotic cells that are round with well-preserved cytoplasm and plasma membrane and apoptotic cells that pass through a series of morphologically identifiable stages in their pathway to death. In the initial phase of the apoptotic process, cells shrink due to a loss of cytoplasmic volume, become detached from their neighbours and from culture substrata and adopt a smooth contour. In a following phase, the plasma membrane ruffles and blebs are formed. In the third phase, progressive degeneration of residual nuclear and cytoplasmic structures can be observed. [Salido 1999]. (Figure 4) (Figure 5) (Figure 6)

Microprobe analytical techniques take advantage of the physical interactions that occurr when an electron beam impinges on a specimen, when examined in SEM and STEM. One such interaction is the production of x-rays with energy and wavelength characteristics indicative of the elemental composition of the specimen interacting with the electron beam. [Warley, 1997,Ingram, 1999] Use of in vitro systems and cell cultures may further increase the number of problems to which x-ray microanalysis can be applied. Among the numerous applications of x-ray microanalysis in cell biology and cell pathology, the role of ions in programmed cell death is of interest. [Bowen, 1988, Skepper 1999, Arrebola 2005,Arrebola 2006] (Figure 7)

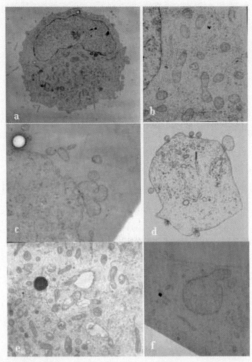

Figure 4. Prostate cancer cell line PC3 after etoposide treatment. Apoptotic cells pass through diferent stages with progressive condensation of nuclei. Control cells appear bigger, with round nuclei and well preserved nucleoli. Apoptotic cells show progressive hypercondensation of nuclear content and surface changes such as membrane blebs until the whole cell breaks into apoptotic bodies with or without nuclear fragments that are then phagocyted by neighboring cells: haematoxylin-eosin stained cells , x 40. x100).

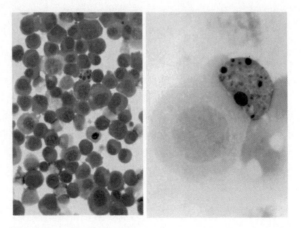

Figure 5. Transmission electron microscopy of prostate cancer cells. Control PC3 cells in a) without chromatin condensation and preserved nucleolus. The cytoplasm shows a number of functional organelles. x 8000. When the neuropeptide bombesin is added, b) the cytoplasm is full of granular endoplasmic reticulum with ribosomes and healthy mitochondria. After etoposide treatment, membrane starts blebbing as cells enter into apoptosis c) for Du 145 cells, x 7000, d) for PC3 cells, x 4000, as cytoplasmic organelles degenerate. Mitochondria are relatively well preserved with continuous membranes until the last stages of apoptosis e) in PC 3 cells after etoposide treatment, x 8000 and f) in Du 145 cells after etoposide treatment, x 8000.

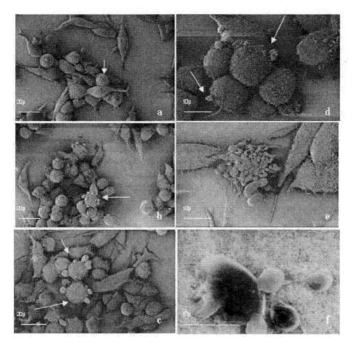

Figure 6. Scanning electron microscopy of LNCaP cells that shows different stages of apoptosis after treatment with etoposide. In non-treated cells only some androgen dependent cells become apoptotic under the deprivation protocol (arrow) and start rounding. After treatment, cells are progresively more altered as shown in b, c, d, and finally blow into apoptotic bodies covered with plasma membrane, e) and f).

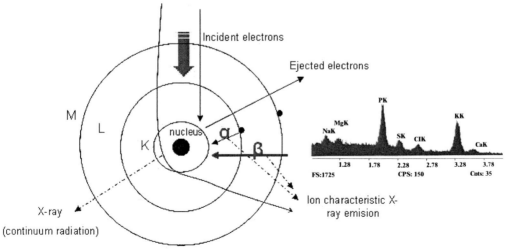

Figure 7. Collision between the incident electron beam and the orbiting electron in the specimen. The electron removed from inner shell causes the fall of an electron from the outer shells. The excess energy is emitted as an x-ray photon, characteristic of the element from which it is emitted. For analysis, the emitted x-ray are collected by a detector system and displayed as an x-ray spectrum.

Electron probe x-ray microanalysis has been used for the last 25 years by biologists to obtain information about the distribution of elements at the cell and tissue level. During this period, progress has mainly been made through the development of more adequate techniques for specimen preparation (mainly low temperature techniques) and quantitative analysis of physiologically important cellular ions can be carried out [Roomans 1990, Hongpaisan 1996, Roomans and von Euler 1996]. X-ray microanalysis can be used to study the distribution of elements in prostatic cells lines at the cellular and subcellular level in a variety of physiological and pathophysiological conditions.[Halgunset 1992, Salido 2001, Salido 2002, Salido 2004, Vilches 2004].

We have been interested in the cellular concentration and distribution of several ions in the androgen-independent prostate cancer cell lines PC-3 and Du 145 and the androgen-dependent prostate cancer cell line LNCaP to describe the different phases of etoposide-induced programmed cell death in these cells and their correlation with morphology. This may be useful for use in future studies and could contribute to a better understanding of the apoptotic process in prostatic epithelium. (Figure 8) (Figure 9) (Figure 10) (Figure 11) (Figure 12)

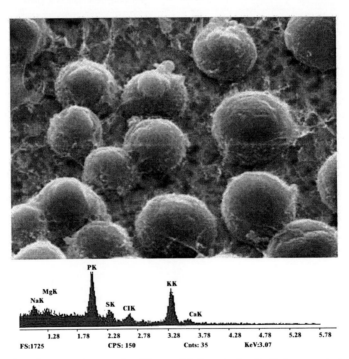

Figure 8. a.)Scanning electron micrograph of freeze dried control cells that appear to be round with well preserved cytoplasm and plasma membrane and display a smooth surface b.) x ray spectra of control cells.

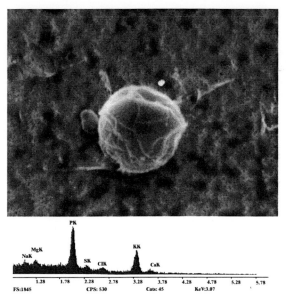

Figure 9. a.) Etoposide treated PC-3 cells. In the first stage of the process, cells shrank due to a loss of cytoplasmic volume, became detached from their neighbours and from culture substrata and adopted a smooth contour. b) x ray spectra corresponding to first stage of apoptosis in PC3 cells. Note that cells have lower Cl Kα and K Kα and higher P Kα peaks than control cells.

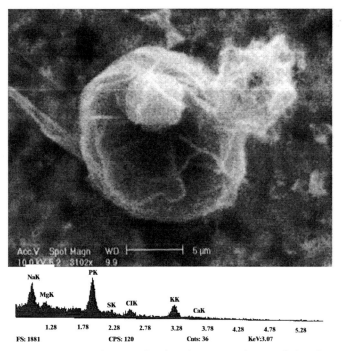

Figure 10. a.) In the second stage of apoptosis, the plasma membrane of shrunken etoposide treated cells ruffled and blebbed. b.) X-ray spectra in second stage of apoptosis. Na Kα and Mg Kα peaks progressively increase while Cl Kα and K Kα peaks decrease when compared to control cells andto previous stage of apoptosis.

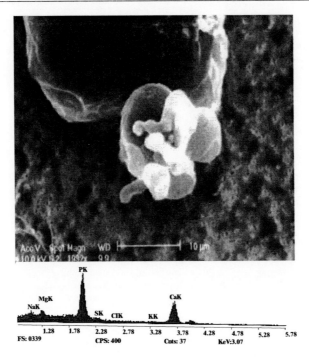

Figure 11. a.) In the third phase, progressive degeneration of residual nuclear and cytoplasmic structures was observed as cell disintegrate into apoptotic bodies while b.) an increase in P Kα and a substantial increase in Ca Kα appear in x-ray spectra.

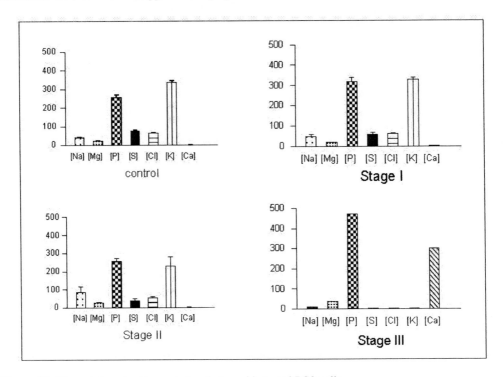

Figure 12. Elemental content in control and etoposide trated PC3 cells.

Non-apoptotic cells appear to have the characteristic elemental pattern of viable cells, with low Na and high K, Mg and P [Hongpaisan 1995, Grängsjö 1997, Fernandez-Segura 1999, Roomans 2001, Arrebola 2005 b, Arrebola 2006].

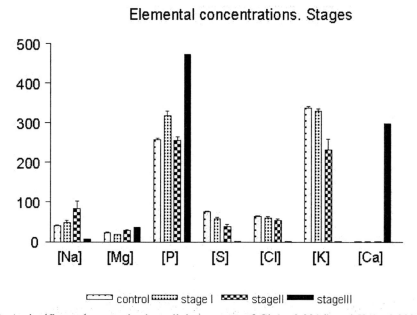

Figure 13. A significant decrease in the cellular content of Cl ($p<0.0016$) and K ($p<0.0001$) and an progressive increase in Mg ($p<0.01$) and Na ($p<0.0006$), was found in etoposide treated cells when compared to control cells.

Chloride has also been implicated in apoptosis [Fujita 1997, Szabo 1998, Nilius 2001, Lang 2005]. Treatment of prostate cancer cell lines with etoposide causes a significant decrease in the cellular content of Cl in the PC3 and Du 145 cells. LNCaP cells show an increase in the cellular Cl concentration, which can also be expressed as an increased Cl/K ratio. Activation of Cl channels is requires for stimulation of apoptosis, which is blunted or even disrupted by Cl channel inhibitors. [Lang 2005] This may appear to contradict the notion that chloride channel activation occurs at the onset of apoptosis [Maeno, 2000]. However, if the increase in the cellular Na/K ratio is to be interpreted as an effect of energy deficiency, the cell would be unable to maintain its low intracellular chloride concentration, and chloride ions would flow into the cell along the electrochemical gradient. (Figure 13)

Both in androgen-independent prostatic cancer cell lines PC3 and Du 145 and in androgen-dependent LNCaP cells a significant decrease in K and a progressive increase in Mg and Na is observed. The changes in Na and K result in a progressively increasing Na/K ratio, significantly different from the control value from stage II onwards, when analysed in SEM. In the first stage of apoptosis a small increase in P is observed, with a small decrease in P in the second stage. The concentration of S decreases both in the first and the second stage, and this results in a significant increase of the P/S ratio in the first and second stage. In the third phase a marked increase in P is observed accompanied by a substantial increase in Ca. Sustained elevations in intracellular calcium are known to be related to the process. (Figure 14)

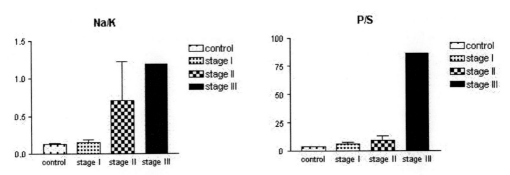

Figure 14. Na/K and P/S ratios show a progressive and significant (p< 0.001 compared to control) increase through the three stages of apoptosis.

Analysis in the STEM shows that in PC3 , Du145 and LNCaP cell lines, etoposide causes a decrease in K and an increase in Na. Etoposide treatment also causes an increase in the concentration of P and a decrease in S resulting in an increase in the ratio of P to S. The Na/K ratio increases markedly after exposure to etoposide. However, according to the results from analysis in the SEM the cellular P was decreased after etoposide, but the cellular S was even more decreased. Calculation of the P/S ratio in the SEM measurements confirms the etoposide-induced increase in the P/S ratio. The apparent discrepancy between the increase in P found in the STEM and the decrease in P found in the SEM analysis is explained by the fact that the size of the cells decreases after etoposide treatment . This means that in the SEM analysis, now also the substrate is excited due to overpenetration of the electron beam. The combination of STEM and SEM data can be interpreted as showing that the amount of P in the cell decreases as a consequence of a decrease in cell size, but that the local concentration of P itself increases relative to other components of the cell.

The cellular K concentration is closely correlated to the P concentration, both in control and in etoposide treated cells. Also the Mg concentration is positively correlated with the P concentration. For none of the other elements investigated, a consistent significant positive correlation between the concentration of that element and the concentration of P was found. The nature of the correlation between P and K did not change significantly after etoposide treatment, but the range of P and K concentrations was much larger than in the control cells, and the slope of the line describing the correlation decreased . [Salido 2001, Salido 2002, Salido 2004, Vilches 2004] (Figure 15)

Apoptosis is associated with a loss of K^+ ions from the cell [Bortner 1997, Dallaporta 1998, Dallaporta 1999]. The loss of K^+ accounts for the changes in cell volume that are universally associated with apoptosis and, interestingly, may actually activate enzymes in the apoptotic cascade and promote DNA degradation [Hughes 1999; Montague 1999]. Recent studies have suggested a primary role for intracellular ions, especially potassium (K^+) in the activation of caspases implicated in apoptosis [Hughes 1997, Kunzelman 2005], and K^+ loss seems to be crucial for activation of endonucleases in intact cells [Dallaporta 1998].

Apoptotic cells pass through morphologically identifiable changes in their pathway to death, with loss of cellular volume and cell shrinkage, while nuclei shrink until complete disintegration of the cell into apoptotic bodies.In addition, there are indications that K^+ loss is not just a consequence of apoptosis, but that it in itself may promote apoptosis, since caspases and nucleases are inhibited by high intracellular K^+, and inhibition of K^+ efflux results in inhibition of apoptosis. [Hughes, 1997].Moreover, K^+ efflux and apoptotic cell shrinkage may be triggered by various types of K^+ channels; some of them are even located in the mitochondrial membrane. [Kunzelman, 2005]. MitoK $_{ATP}$ has also been shown to play a role in protecting cardiomyocites against apoptosis in ischemic preconditioning. Pharmacological opening of MitoK $_{ATP}$ channels by diazoxide suppresses several markers of hydrogen peroxide-induced apoptosis, including TUNEL positivity, cytochrome c release, caspase-3 activation an PARPcleavage. In summary, the MitoK $_{ATP}$ channel has been demonstrated to play a key role in ischemic preconditioning and prevention of apoptosis. It remains unclear how the opening of MitoK $_{ATP}$ leads to protection against cell death. Three different mechanisms have been proposed [Ardehali and O'Rourke ,2005] These include changes in the levels of reactive oxygen species (ROS), mitochondrial matrix swelling, and changes in the mitochondrial Ca^{2+} [Ardehali 2005]. This mitoK$_{ATP}$ channel is modulated by a variety of K^+ channel openers and inhibitors. Furthermore, these modulators significantly influence mitochondrial function and cell survival, suggesting a link between mitoK$_{ATP}$ and protection against ischemic injury. This protective effect of mitoK$_{ATP}$ has been demonstrated in several tissues besides the heart, including brain, liver, kidney and gut. [Ardehali and O'Rourke ,2005]

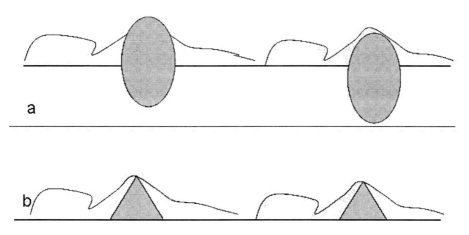

Figure 15. When the cell decreases in size, relatively more substrate is excited, and the spectrum from the cell is "diluted" by the contribution of the substrate. B) analysis in the STEM. Although the characteristic signal is reduced when the cell decreases in size, this is fully compensated for by the fact that also the background signal is reduced, in principle to the same extent.

The difference in concentrations of Na^+ and K^+ over the cell membrane is crucial for the maintenance of physiological intracellular gradients. The process requires adequate energy supply, an intact cell membrane and specific enzymatic activities. When the Na^+/K^+ pump in the cell membrane cannot maintain the ionic gradients, in the absence of a membrane potential, calcium ions flow through the voltage-dependent ion channels.

These changes may be due to activation of K^+ channels, or to inhibition of the Na^+/K^+-ATPase, presumably due to lack of ATP. Most likely the changes are due to both factors. Thus, the increase in Na/K ratio reflects the physiopathological conditions of the cell, and can be used as a very sensitive measure of various kinds of cell injury, and can be detected by X-ray microanalysis in cells and tissues undergoing programmed cell death. [Bowen 1988;

Sandstrom 1994, Warley 1997, Berger and Garnier 1999,Ingram 1999, Skepper 1999]. Voltage dependent K^+ channels have been described in most mammalian cells and their activation results in efflux of K^+ with a net decrease of intracellular K^+, as well as an imbalance in other ions such as Na^+ and Mg^{++} [Fernandez- Segura 1999, Fujii 1999; Skepper 1999;] Some studies have also pointed to the fact that K^+ channel function may be also required for tumor cell surface vesicle formation and shedding, changes in membrane permeability and nuclear fragmentation, i.e., the events leading to what we know by the term "programmed cell death" [Smith 1998]

Recent studies have coined the term apoptotic volume decrease , AVD, for the early-phase cell shrinkage associated with activation of K^+ and Cl^- channels in apoptosis [Barros 2001, Kunzelmann, 2005], and have shown that AVD induction precedes and is a prerequisite for cytochrome-c release, caspase-3 activation, DNA laddering and ultrastructural alterations [Maeno 2000]. Cells maintain and regulate their volume by controlling intracellular solute concentrations and water flux across the cell membrane. Alterations in cell volume are accompanied by the movement of water across the cell membrane and activation of volume-regulated ion transport channels, mainly K^+ and Na^+ channels. [Beauvais 1995, Mc Carthy and Cotter, 1997]

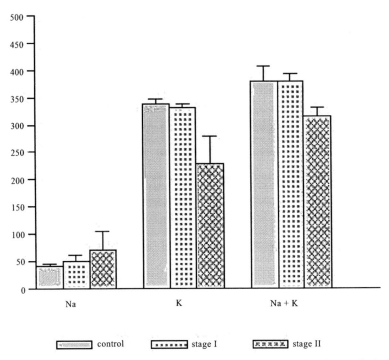

Figure 16. The shrinkage of the etoposide treated cells was correlated with microanalytical data when the sum of the Na and K content was used to assess the variation in cell volume.

The channel activation leads to a significant decrease in K content, and a significant increase in Na. The variation in cellular volume is evaluated using the sum of the Na and K content. In addition, a tendency to higher P can be noted. Hence, the increase in P/S ratio may be explained as a relative increase of P-containing macromolecules (mainly nucleotides) relative to S-containing macromolecules (mainly proteins).(Figure 16).

DNA digestion is mediated by either a Ca^{2+}/Mg^{2+}-dependent endonuclease or the acid-activated deoxyribonuclease II (DNase II), pH dependent, or both. X-ray microanalysis detects total Ca (in the millimolar range) whereas in the initial stages of apoptosis changes in free Ca^{2+} (in the nanomolar to micromolar range) are relevant. These cannot be measured by X-ray microanalysis. However, later in the apoptotic process also significant changes in total Ca occur, as described by us and others [Sandstrom 1994, Warley 1997, Gutierrez 1999, Salido 2001b].

With respect to the culture techniques employed, as described above, the method allows the rapid criofixation and adequate analysis by means of electron probe x-ray microanalysis. When cells are grown on a solid substrate as we describe for polypropylene filters, and analyzed in SEM, there is a theoretical possibility that the electron beam penetrates the cell entirely and excites the substrate even at low accelerating voltage. The likelihood for this increases when the cell shrinks, as is the case during apoptosis. This implies that a measured decrease in elemental concentration under such conditions can be (in part) due to a decrease in cell size. This possible artefact has to be taken into account in the interpretation of data obtained by analysis of cells on a solid substrate [Fernandez-Segura 1999; Roomans 2001]. To normalize the intensity counts of the different elements with respect to the mass of the cell analyzed, the phosphorus intensity signal is often taken as a measure of the analyzed mass and as a unit of reference for evaluating the peak intensity of the other elements [Abraham 1985]. However, this requires that the P content is constant during the experiment, and this is not always true during apoptosis, a process that affects cellular macromolecules, and progresses over a considerable period of time. Analysis of cells grown on thin plastic films on grids in the STEM avoids these problems, although the method as such is technically more difficult [Roomans 1996, Von Euler 1993]. A methodological comparison of the two methods [6] showed that systematic errors in the absolute concentrations could be easily introduced in both methods, but that elemental ratios would be relatively free of such artefacts. We have, therefore, also expressed our data as elemental ratios and the direction of the changes is the same in both types of measurement. Hence, the increase in P/S ratio may be explained as a relative increase of P-containing macromolecules (mainly nucleotides) relative to S-containing macromolecules (mainly proteins). The relative increase in P content matches with the etoposide-induced arrest in G_2/M which would result in an increased nucleotide content [Barbiero 1995,Salido 2001]

In androgen independent cell lines PC3 and Du 145, both bombesin and calcitonin inhibit fully the etoposide-induced increase in the Na/K ratio, Cl and Mg. Calcitonin, but not bombesin, also reverses the etoposide-induced changes in the P/S ratio. Because the change in the P/S ratio probably is multifactorial, [Grängsjö 2000, Grundin 1985; Smith and Cameron, 1999] an explanation for this difference cannot be provided currently but requires further study. (Figure 17)(Figure 18)(Figure19)(Figure 20)(Figure 21)

---

[6] Roomans GM. X-ray microanalysis of cultured cells in the scanning electron microscope and the scanning transmission electron microscope: a comparison. *Scanning Microsc* 1999; 13: 159-65.

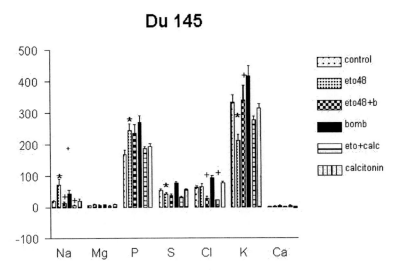

Figure 17. Elemental concentrations of cells from the Du 145 prostate carcinoma cell line were analyzed in the scanning transmission electron microscope. The mean values, (expressed in mmol/kg dry weight) for control cells, cells that were treated for 48 h with etoposide (eto48h), cells that were treated for 48h with etoposide in the presence of bombesin (eto48h+b), cells that were treated for 48 h with bombesin alone (bomb), cells treated for 48h with etoposide in the presence of calcitonin (eto+calc), and cells treated for 48 h with calcitonin alone (calcitonin) are shown. Statistically significant differences between control and etoposide-treated cells are denoted by an asterisk ($p<0.01$), and statistically significant differences between etoposide-treated cells and cells that were treated with etoposide and bombesin or calcitonin are indicated by a cross ($p<0.01$). Data based on 25-45 measurements per group from two separate experiments.

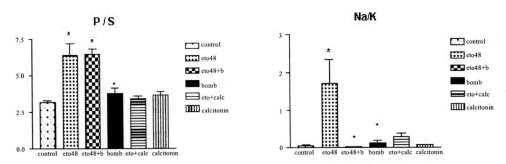

Figure 18. Elemental ratios in Du145 cells analyzed in the scanning transmission electron microscope. An increase in the P/S ratio (left) was observed after etoposide treatment that appeared to be blocked by calcitonin. The neuropeptides bombesin and calcitonin blocked the etoposide-induced increase in the Na/K ratio (right). Changes in the P/S ratio can be related to the increased nuclear-to-cytoplasmic ratio that occurs in apoptosis. The increase in intracellular concentrations of Na with respect to K reflects the loss of function of the Na/K pump and, thus, may be regarded as a reliable indicator of the loss of cell viability during apoptosis. Statistically significant differences between control and etoposide-treated cells are denoted by an asterisk ($p<0.01$), and statistically significant differences between etoposide-treated cells and cells that were treated with etoposide and bombesin or calcitonin are indicated by a cross ($p<0.01$). Data based on 25-45 measurements per group from two separate experiments. Control: control cells; etoposide: cells that were treated for 48 hours with etoposide; e+b: cells that were treated for 48 hours with etoposide in the presence of bombesin; bomb: cells that were treated for 48 hours with bombesin alone; eto+calc: cells that were treated for 48 hours with etoposide in the presence of cacitonin; calcitonin: cells that were treated for 48 hours with calcitonin alone.

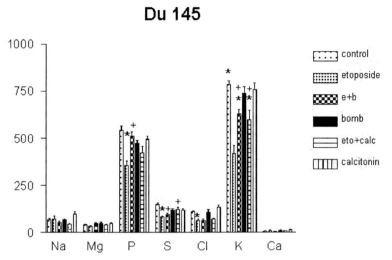

Figure 19. Elemental concentrations in Du145 cells analyzed in the SEM. Statistically significant differences between control cells and etoposide-treated cells are denoted by an asterisk (p<0.001), and statistically significant differences between etoposide-treated cells and cells that were treated with etoposide and bombesin or calcitonin are indicated by a cross (p<0.01).Data were based on 25-45 measurements per group from two separate experiments. Control: control cells; etoposide: cells that were treated for 48 hours with etoposide; e+b: cells that were treated for 48 hours with etoposide in the presence of bombesin; bomb: cells that were treated for 48 hours with bombesin alone; eto+calc: cells that were treated for 48 hours with etoposide in the presence of cacitonin; calcitonin: cells that were treated for 48 hours with calcitonin alone.

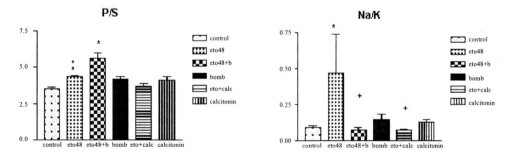

Figure 20. Elemental ratios in Du145 cells analyzed in the SEM ; left, the P/S ratio; right, the Na/K ratio. Statistically significant differences between control cells and etoposide-treated cells are denoted by an asterisk (p<0.001), and statistically significant differences between etoposide-treated cells and cells that were treated with etoposide and bombesin or calcitonin are indicated by a cross (p<0.01).Data were based on 25-45 measurements per group from two separate experiments. The discrepancy between scanning transmission electron microscope (STEM) data and SEM data for changes in the P concentration likely was due to overpenetration of the electron beam through the shrunken apoptotic cells; therefore, the P/S ratios are shown. Control: control cells; etoposide: cells that were treated for 48 hours with etoposide; e+b: cells that were treated for 48 hours with etoposide in the presence of bombesin; bomb: cells that were treated for 48 hours with bombesin alone; eto+calc: cells that were treated for 48 hours with etoposide in the presence of cacitonin; calcitonin: cells that were treated for 48 hours with calcitonin alone.

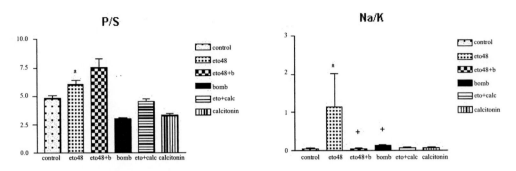

Figure 21. Elemental ratios in PC3 cells analyzed in the scanning electron microscope: left, the P/S ratio; right, the Na/K ratio. Similar results were obtained in scanning transmission electron microscope analysis. Statistically significant differences between control cells and etoposide-treated cells are denoted by an asterisk ($p<0.001$), and statistically significant differences between etoposide-treated cells and cells that were treated with etoposide and bombesin or calcitonin are indicated by a cross ($p<0.01$).Data based on 30-60 measurements per group, three separate experiments. Control: control cells; etoposide: cells that were treated for 48 hours with etoposide; e+b: cells that were treated for 48 hours with etoposide in the presence of bombesin; bomb: cells that were treated for 48 hours with bombesin alone; eto+calc: cells that were treated for 48 hours with etoposide in the presence of cacitonin; calcitonin: cells that were treated for 48 hours with calcitonin alone.

In the LNCaP androgen-sensitive cell line both bombesin and calcitonin block the etoposide-induced increase in Na/K ratio (Figure 22). The addition of the neuropeptides inhibits the etoposide-induced increase in Cl/K ratio and also increases the P/S ratio. (Figure 23)Bombesin, but not calcitonin blocks changes in chloride concentrations and both neuropeptides inhibit the changes observed in the rest of elements analyzed. Calcitonin was more effective in blocking changes in K, S, and P, whereas bombesin was more effective in blocking changes in Mg, Na and Cl. (Figure 24) Recently, calcitonin-induced resistance to etoposide induced apoptosis via the Akt/surviving pathway has been described. [Thomas and Shah, 2005]

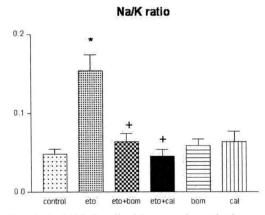

Figure 22. Elemental Na/K ratio in LNCaP cells. Means and standard error of the mean for control cells, cells treated for 48 h with etoposide (eto), cells treated for 48h with etoposide in the presence of bombesin (eto+bom), cells treated for 48 h with bombesin alone (bom), cells treated for 48h with etoposide in the presence of calcitonin (eto+cal), and cells treated for 48 h with calcitonin alone (cal) are given. Statistically significant differences between control and etoposide-treated cells are denoted

by an asterisk (*) (p<0.001), and statistically significant differences between etoposide-treated cells and cells treated with etoposide + bombesin or calcitonin are indicated by (+) (p<0.001). Data based on 30-60 measurements for group, two separate experiments.

Elemental ratios

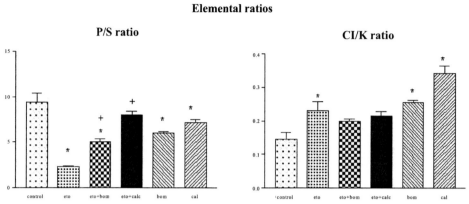

Figure. 23. Elemental concentrations in LNCaP cells analyzed in the STEM. Means and standard error of the mean for control cells, cells treated for 48 h with etoposide (eto), cells treated for 48h with etoposide in the presence of bombesin (eto+bom), cells treated for 48 h with bombesin alone (bom), cells treated for 48h with etoposide in the presence of calcitonin (eto+cal), and cells treated for 48 h with calcitonin alone (cal) are given. The data are expressed in mmol/kg dry weight. Statistically significant differences from control cells are denoted by an asterisk (*) (p<0.001), and statistically significant differences between etoposide-treated cells and cells treated with etoposide + bombesin or calcitonin are indicated by (+) (p<0.001). Data based on 25-45 measurements for group, two separate experiments.

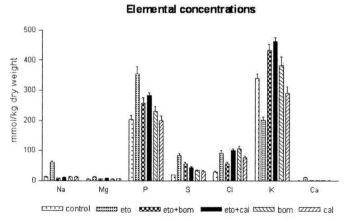

Figure. 24. Elemental ratios in LNCaP cells analyzed in the STEM (a) P/S ratio, (b) Cl/K ratio. Means and standard error of the mean for control cells, cells treated for 48 h with etoposide (eto), cells treated for 48h with etoposide in the presence of bombesin (eto+bom), cells treated for 48 h with bombesin alone (bom), cells treated for 48h with etoposide in the presence of calcitonin (eto+cal), and cells treated for 48 h with calcitonin alone (cal) are given. Statistically significant differences with control cells are denoted by an asterisk (*), and statistically significant differences between etoposide-treated cells and cells treated with etoposide + bombesin or calcitonin are indicated by (+) (p<0.01). Data based on 25-45 measurements per group, two separate experiments. The decrease in P/S ratio is blocked by calcitonin (p<0.001) and bombesin (p<0.01).

Activation of K^+ channels is one of the common features of apoptosis and prevention of etoposide induced apoptosis in thymocytes by blocking K channels has been reported earlier [Dallaporta 1998, Maeno 2000]. The addition of bombesin or calcitonin to etoposide treated cells reduces the percentages of apoptotic cells, with an increase in K and a decrease in Na, and subsequent decrease of the Na to K ratio, as an objective expression of the increase in the viability of the neoplastic cells in the presence of neuropeptides. (Figure 25)

Figure. 25. Scanning electron micrographs of LNCaP prostate carcinoma cells (a) etoposide treated cells and (b) etoposide plus bombesin treated cells. Apoptotic cells (*) are rounded, and detached from neighboring cells. In the etoposide-treated group most cells are in advanced stages of apoptosis as shown in the micrograph, with membrane alterations, such as blebs (↑). The presence of neuropeptides blocks apoptotic changes. Most cells do not pass to the latest stages of apoptosis, no blebs are observed, and cells remain larger and more attached to the substrate than in the etoposide-treated group.

Apoptosis is associated with an increase in both P and S, but more in S, which results in a decreased P/S ratio. The P signal mainly represents nucleotides and phosphorylated proteins, whereas the S signal mainly represents protein-bound sulfur. This correlates to the morphological changes associated with apoptosis, The fact that these changes are very marked, and can be inhibited to a large extent by neuropeptides, indicates that they are important in the process of apoptosis.

This protective effect on etoposide-induced apoptosis in PC3, Du 145 and LNCaP cells appears to be quite similar.[Salido 2000, Salido 2002, Salido 2004]. This confirms that neuropeptides confer antiapoptotic capabilities on non-neuroendocrine cells in close proximity to neuroendocrine cells and that this occurs both in androgen-dependent and androgen-independent cells. It can therefore be speculated that certain neuroendocrine peptides can increase the survival and further growth of neighboring cells [Ismail 2002] and thereby contribute to the aggressive clinical course of prostate tumors containing neuroendocrine elements [Hoosein, 1998, Quek 2006, Sciarra 2006, Slovin 2006].

According to our data, neuroendocrine cells, thus, have value as an indicator of poor prognosis in patients with prostate carcinoma, independently of hormonal status of the epithelial cell tumors.

There appears to be a direct relationship between the density of NE cells and enhancement of prostate cancer characteristics, such as increased Gleason grade, lose of androgen sensitivity, and autocrine/ paracrine activity. It has been suggested that NE cells provide paracrine stimuli for the propagation of local carcinoma cells and that NE differentiation is associated with the progression of prostate cancer toward an androgen-independent state [Cox 1999; Hansson and Abrahamsson 2001, Ito 2001, Segawa 2001, Goodin and Rutherford 2002, Quek 2006, Sciarra 2006, Slovin 2006].

The approach taken in the present study mimics the presence of neuroendocrine cells in prostatic carcinoma and makes it possible to correlate the morphological changes and elemental patterns that appear in neuropeptide-induced resistance to apoptosis.

The use of electron probe x-ray microanalysis makes it possible to evaluate alterations in total element composition in individual cells during apoptosis. The three morphologically identifiable stages of apoptosis are associated with alterations of intracellular ions, mainly sodium, potassium and chlorine, as well as with changes in the phosphorus/sulfur ratio. The use of x-ray microanalysis can, thus, be a helpful tool for further studies on cellular mechanisms involved in the control of programmed cell death of prostatic cancer cells. Induction of apoptosis is a process of high significance for the treatment of cancer. As we have shown, there is a strong correlation between apoptosis and elemental changes in the cell, and this could be used to control the apoptotic process in a better way. In particular, insight may be gained in the way in which neuropeptides such as calcitonin and bombesin, that can be secreted by neuroendocrine cells in close proximity to prostate carcinoma cells, inhibit etoposide-induced apoptosis in prostate cancer cells, and this knowledge may be used to overcome the stimulatory effect of these peptides on prostate tumor growth. Recently Thomas and coworkers have confirmed our data as describing Akt-mediated inhibition of apoptosis by calcitonin. [Thomas and Shah, 2005] The putative function of neuroendocrine cells in stimulating proliferation and/or inhibiting the apoptotic process, worsening the prostate cancer outcome, trough a paracrine hormonal mechanism, provides a rationale for the experimental use of drugs which are able to inhibit the secretion of neuroendocrine products, which the aim to counteract tumor progression.

APOPTOSIS, MITOCHONDRIAL MEMBRANE POTENTIAL (ΔΨ) AND NEUROENDOCRINE DIFFERENTIATION

Apoptosis is accompanied by major changes in ion compartmentalization and transmembrane potentials, with transient mitochondrial swelling and a subsequent loss of plasma membrane potential related to the loss of cytosolic K^+. [Salido et al., 2002].

Mitochondrial membrane potential (MMP), $\Delta\Psi m$, may control the permeability of the outer membrane and regulate cytochrome c release. When the mitochondria loss their $\Delta\Psi m$ undergo swelling, and release IMs proteins [Robertson 2000, Bouchier-Hayes 2005, Armstrong JS a 2006, Armstrong JS b 2006]. Furthermore, hyperpolarization appears to be linked to subsequent apoptosis changes, through disturbance of the electron transport system [Iijima 2003].

Alternatively, it may be that the release of cytochrome c involves a permeabilization of the outer membrane regulated by the pro-apoptotic Bcl-2 family proteins Bax or Bid.

In non-apoptotic cells Bax is predominantly found as a monomer in the cytosol. Bak is not found as a soluble form in the cells, it is already inserted in the outer mitochondrial membrane of normal cells. After apoptotic stimulation Bax specifically translocates to the mitochondria and is inserted into the outer mitochondrial membrane as large oligomers. Bcl-2 has been shown to inhibit Bax activation and oligomerization. The changes results in exposure of domains in the inactivated proteins, allowing the proteins to form multimeric complexes. Thus, these Bax channels would be large enough to release cytochrome c and other even larger proteins from the intermembrane space. Several studies have indicated that the mitochondrial structures are not damaged during the release of proteins from the intermembrane space during apoptosis. In Bax induced cytochrome c release no mitochondrial swelling is detected, demonstrating that opening of PTP is not involved. Cytochrome c triggers the activation of caspase 9 through formation of apoptosomes. Smac/DIABLO and HtrA2/ Omi bind to caspase inhibitors, thus preventing inhibition of activated caspases inside the cells.

The 'BH3 domain only' proteins act as an amplification loop of the receptor pathway. Caspase 8 activated through the receptor pathway cleaves Bid generating a C-terminal fragment (t cBid), which interacts with Bax and Bak triggering activation of these proteins and the mitochondrial pathway. Other 'BH3 domain only' proteins have been shown to activate the multidomain proteins in an indirect way, neutralizing their anti-apoptotic activity [Antonsson 2004, Armstrong JS a, 2006, Armstrong JS b].

Endonuclease G and AIF proteins released from the mitochondrial intermembrane space appear to activate additional well characterized caspase independent pathways. Endonuclease G is a nuclear encoded protein synthesized in the cytosol and subsequently imported to the mitochondria. During apoptosis endonuclease G is released from the mitochondria together with other apoptogenic factors. After the protein has been released into the cytosol it translocates to the nucleus and cleaves chromatin DNA into fragments .During apoptosis AIF, a mitochondrial intermembrane protein, is released into the cytosol. Then the protein translocates to the nucleus and induces DNA fragmentation and chromatin condensation. AIF does not exhibit DNAse activity and actually it remains unclear how the protein induce DNA fragmentation. [Li 2001, Armstrong JS a 2006, Armstrong JS b 2006]

Depolarization and subsequent permeabilization of the mitochondrial membrane are assumed to be determinant factors in cell death via necrosis or apoptosis. Mitochondrial membrane potential (MMP) , $\Delta\psi m$, may control the permeability of the outer membrane and regulate cytochrome c release. [Petronili 2001, James 2002, Lim 2002, Iijima 2003, Armstrong JS b 2006]

Detection of the mitochondrial permeability transition event (PT) provides an early indication of the initiation of cellular apoptosis. This process is tipically defined as a collapse in the electrochemical gradient across the mitochondrial membrane, as measured by the change in the mitochondrial membrane potential ($\Delta\psi$). Changes in the mitochondrial $\Delta\psi$ lead to the insertion of proapoptotic proteins into the membrane and possible oligomerization of BID, BAK, BAX or BAD. This could create pores, which dissipate the transmembrane potential, thereby releasing cytochrome c into the cytoplasm. Loss of mitochondrial $\Delta\psi$, indicative of apoptosis, can be detected by a unique fluorescent cationic dye, 5,5',6,6'-tetrachloro-1-1',3,3'-tetraethyl-benzamidazolocarbocyanin iodide, commonly known as JC1.

JC-1 is a cationic dye that exhibits potential-dependent accumulation in mitochondria, indicated by a fluorescence emission shift from green (~525 nm) to red (~590 nm). Consequently, mitochondrial depolarization is indicated by a decrease in the red/ green fluorescence intensity ratio. The potential-sensitive color shift is due to concentration-dependent formation of red fluorescent J-aggregates. [Salvioli 1997]

JC-1 is more specific for mitochondrial versus plasma membrane potential, and more consistent in its response to depolarization, than other cationic dyes such as DiOC6 and rhodamine 123. [Salvioli 1997, Dedov 1999, Mathur 2000, Dedov 2001]

The ratio of green to red fluorescence is dependent only on the membrane potential and not on other factors such as mitochondrial size, shape, and density that may influence single-component fluorescence signals. Use of fluorescence ratio detection therefore allows researchers to make comparative measurements of membrane potential and determine the percentage of mitochondria within a population that respond to an applied stimulus. Subtle heterogeneity in cellular responses can be discerned in this way. [Cosarizza, 1996]

We have demonstrated that bombesin and calcitonin prevent etoposide induced apoptosis of prostate cancer cells in vitro. [Salido 2000, Salido 2002, Salido 2004] A recent study from our laboratory demonstrated the capability of etoposide in the induction of alterations in $\Delta\Psi$m, as the first step in mitochondrial disfunction. (Figure 26)(Figure 27) (Figure 28) (Figure 29). However, the mechanisms mediating resistance to etoposide-induced apoptosis have not yet been elucidated. Etoposide inhibits topoisomerase II at the nuclear level poisoning the enzyme and induces it to generate DNA breaks that are lethal to the cell. Topo II-targeted drugs show a limited sequence preference, triggering double-stranded breaks throughout the genome, inducing cell cycle arrest and p53 activation and it has been described that also cleaves caspase 8 and 9, and is involved in Phase I trials for solid tumors, and phase II trials for ovarian carcinoma and gliomas. [Bouchier-Hayes 2005, Wang, 2005, Duca, 2006].

Figure 26. Control cells as observed in the confocal microscope. J-aggregates are predominant (upper right) as indicative of well preserved mitochondria with high $\Delta\Psi$. Cells observed in a Leica TCS SL confocal microscope, with excitation laser of 488 nm and emission at 525 nm (green) and 590 nm (red).(x63)

Δψm ETOPOSIDE-TREATED CELLS

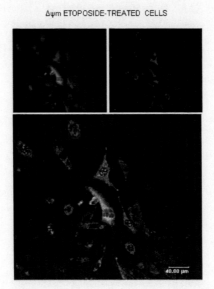

Figure 27. Alterations of ΔΨ in etoposide-induced apoptosis in PC-3 cancer cells in different stages of the apoptotic process after 24 h treatment with etoposide as observed in the confocal microscope. (JC-1 fluorescence) Upper Left: Negatively charged mitochondria appear as red particles in the cytoplasm. Right : mitochondrial ΔΨ collapse produces green fluorescence distributed throughout the cell. Down The overlay image shows how cells undergoing apoptosis contain less red dye aggregate spots than non-apoptotic cells and the entire cell appears more and more green as the mitochondrial ΔΨ disintegrates. (x 63)

Δψm ETOPOSIDE PLUS BOMBESIN -TREATED CELLS

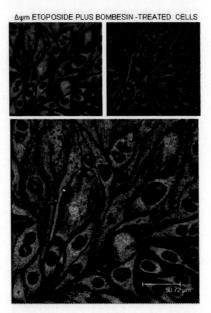

Figure 28. PC-3 cells after 24 h.treatment with etoposide in the presence of bombesin as observed in the confocal microscope after JC-1 staining. (x 63) J-aggregates are present in most cells as an indicative of well preserved mitochondrial function. Some of the cells also show green fluorescence in the cytosol, but in a significatively lower proportion than in etoposide treated cells.

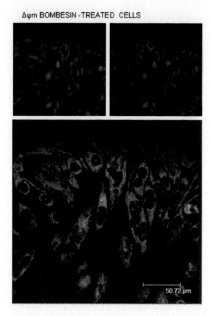

Δψm BOMBESIN-TREATED CELLS

Figure 29. PC3 cells after 24 h treatment with bombesin as observed in the confocal microscope after JC-1 staining. (x 63) Cells appear red, even in higher proportion than those observed for control group, as indicative of high Δψ due to the presence of the neuropeptide. Only residual signal for green (monomeric) form of the dye is observed in the green channel.

The SMase can be directly stimulated by caspase 3. Ceramide can be formed, aside from the novo synthesis, by the action of sphingomielinase ,SMase. Ceramide, and its soluble analogues, itself can induce apoptosis, and etoposide stimulates SMase. The proposed role for ceramide on mitochondrial function results on cytochrome c release, dependent of Δψm, caspase activation and DNA fragmentation. [Lawen, 2003, van Blitterswijk, 2003]

Cellular loss of K^+ appears to be an important trigger of apoptosis in a wide variety of cells [Hughes 1997, Hughes and Cidlowski 1999, Montague 1999, Maeno 2000,Lang, 2005, Yurinskaya 2005a, Yurinskaya 2005b]. Activation of K^+ channels leads to hyperpolarization of the cell membrane, thus increasing the electrical driving force for Cl^- exit into the extracellular space. Thus, if K^+ channel activity is paralleled by Cl^- channel activity, it leads to cellular loss of KCl, with osmoticaly obliged water and hence to cell shrinkage, a hallmark of apoptosis. [Lang, 2005].

As we have described previously, analysis in the STEM showed that in PC3, Du 145 and LNCaP cell lines, etoposide causes a decrease in intracellular K^+ and Cl^-. [Salido 2002, Salido 2004, Vilches 2004]

Bombesin acts as a mitogen in prostate cancer and may do so through activation of transcription factor Elk-1 and c-fos. In in vitro assays, the neuropeptide stimulates androgen-independent growth and the invasiveness of prostate cancer cell lines. Calcitonin affects growth and migration of certain prostate cancer cell lines and may play a role in the regulation of prostate cell growth and metastasis, especially to the bone. [di Sant'Agnese 2004]

The growth factors have been shown to activate K^+ channels. This activation appears to be particularly important for the early G1 phase of the cell cycle. [Lang 2005] Reports on the role of K^+ channels in apoptosis are conflicting. Inhibition of K channels appears to favour and activation of K^+ channels to inhibit apoptosis.

It is possible to speculate that, as a consequence of our previous descriptions of the loss of K^+ and Cl^- in etoposide-induced apoptosis, the neuropeptides bombesin and calcitonin could induce an activation of K channels for inhibiting apoptosis in vitro in androgen-dependent and androgen- independent prostatic cells. This is a relevant fact that would explain why carcinomas with neuroendocrine differentiation have worse prognostic together with a higher therapeutical failure.

CONCLUSION

Our evidence shows that the presence of neuroendocrine cells and their secretory products confers anti-apoptotic capabilities on non-neuroendocrine cells in androgen-sensitive and androgen-insensitive prostatic cancer cells. On the other hand, the interaction between neuroendocrine cells and carcinoma prostatic cells is a novel model for the study of basic mechanisms of apoptosis. New therapeutic protocols and trials need to be developed to test drugs acting through the neutralization of antiapoptotic intracellular pathways mediated by neuroendocrine hormones. Hopefully, this will lead to the development of entirely new therapeutic approaches in hormone refractory prostate cancer.

REFERENCES

Allard, P., Beaulieu, P., Aprikian, A. and Chevalier, S. Bombesin modulates the association of Src with a nuclear 110-kd protein expressed in dividing prostate cells. *J Androl* 2000, 21: 367-75.

Antonsson B. Mitochondria and the Bcl-2 family proteins in apoptosis signaling pathways. Mol.Cell. *Biochem.* 2004, 256/257, 141-155

Aprikian AG, Han K, Guy L, Landry F, Begin LR, Chevalier S. Neuroendocrine differentiation and the bombesin/gastrin-releasing peptide family of neuropeptides in the progression of human prostate cancer. *Prostate Suppl* 1998, 8: 52-61.

Ardehali H, O'Rourke B. Mitochondrial K(ATP) channels in cell survival and death. *J Mol Cell Cardiol.* 2005;39(1):7-16.

Ardehali H. Cytoprotective channels in mitochondria. *J Bioenerg Biomembr.* 2005, 37(3):171-7.

Armstrong JS [a]. Mitochondria: a target for cancer therapy.*British J. Pharmacol.*2006, 147: 239-248.

Armstrong JS [b]. Mitochondrial membrane permeabilization: the sine qua non for cell death. *BioEssays* 2006, 28: 253-260.

Arrebola F, Canizares J, Cubero MA, Crespo PV, Warley A, Fernandez-Segura E. Biphasic behavior of changes in elemental composition during staurosporine-induced apoptosis. *Apoptosis* 2005, 10(6):1317-31.

Arrebola F, Fernandez-Segura E, Campos A, Crespo PV, Skepper JN, Warley A. Changes in intracellular electrolyte concentrations during apoptosis induced by UV irradiation of human myeloblastic cells.*Am J Physiol Cell Physiol.* 2006, 290(2):C638-49.

Arrebola F, Zabiti S, Canizares FJ, Cubero MA, Crespo PV, Fernandez-Segura E. Changes in intracellular sodium, chlorine, and potassium concentrations in staurosporine-induced apoptosis.*J Cell Physiol.* 2005b, 204(2):500-7.

Barbiero G., Duranti F., Bonelli G., Amenta JS, Baccino F.M. Intracellular ionic variations in the apoptotic death of L cells by inhibitors of cell cycle progression. *Exp.Cell. Res.* 1995, 217, 410-418.

Baretton G.B., Klenk U., Diebold J., Schmeller N., Lohrs U. Proliferation and apoptosis associated factors in advanced prostatic carcinomas before and after androgen deprivation therapy: prognostic sinificance of p21/waf1/CIP1 expression. *Br. J. Cancer* 1999, 80, 546-555.

Barros LF., Hermosilla T.,Castro J. Necrotic volume increase and the early physiology of necrosis. Comp. Biochem. Physiol. A Mol. Integr. *Physiol.*; 2001,130, 401-409.

Berchem, G.J., Bosseler, M., Sugars, L.Y., Voeller, H.J., Zeitkin, S., Gelman, E.P. Androgens induce resistance to bcl-2 mediated apoptosis in LNCaP prostate cancer cells. *Cancer Res.* 1995, 55: 735-738.

Bonkhoff, H. Endocrine-paracrine cell types in the prostate and prostatic adenocarcinoma are postmitotic cells. *Hum. Pathol.* 2005, 26: 167-170.

Bonkoff H., Stein U., Remberger K. Androgen receptor status in endocrine-paracrine cell types of the normal, hyperplastic and neoplastic human prostate. Virchows Arch. A. *Pathol. Anat.* 1993, 423, 291-294.

Borner, M.M., Myers, C.E., Sartor, O., Sei, Y., Toko, T., Trepel, J.B., et al. Drug induced apoptosis is not necessarily dependent on macromolecular synthesis or proliferation in the p53 negative human prostate cancer cell line PC-3. *Cancer Res.* 1995, 55: 2122-2128.

Bortner C.D., Hughes FM. Jr, Cidlowski JA. A primary role for K+ and Na+ efflux in the activation of apoptosis. *J. Biol. Chem.* 1997 ,272: 32436-32442.

Bouchier-Hayes L, Lartigue L, Newemyer D. Mitochondria: pharmacological manipulation of cell death. *J. Clin.Invest.* 2005, 115 (10): 2640-2647.

Chien J, Ren Y, Qing Wang Y, Bordelon W, Thompson E, Davis R, Rayford W, Shah G. Calcitonin is a prostate-epithelium-derived growth stimulator peptide. *Mol Cell Endocrinol.* 2001, 181: 69-79.

Clegg N, Ferguson C., True L, Arnold H, Moorman A, Quinn J, Vessella RL, Nelson PS. Molecular characterization of prostatic small cell carcinoma. *Prostate* 2003, 55: 55-64.

Cosarizza A, Cecarelli D, Masini A. Functional heterogeneity of isolated mitochondrial population revealed by cytofluorimetric análisis at the single organelle level. *Exp. Cell. Res.* 1996, 222: 84-94.

Cox ME, Deeble PD., Lakhnai S., Parsons SJ. Acquisition of neuroendocrine characteristics by prostate tumor cells is reversible: implications for prostate cancer progression. *Cancer Res.* 1999, 59: 3821-3830.

Dallaporta B., Hirsch T., Susin S.A., Zamzami N., Larochette N., Brenner C., Marzo I. et al. Potassium leakage during the apoptosis degradation phase. *J. Immunol.* 1998, 160: 5605-15.

Dallaporta B., Marchetti P., de Pablo M.A., Maisse C., Duc H.T., Metivier D., Zamzami N., et al. Plasma membrane potential in thymocyte apoptosis. *J. Immunol.* 1999 ,162: 6534-42.

Dedov VN, Cox GC, Roufogalis BD. Visualisation of mitochondria in living neurons with single and two-photon fluorescence laser microscopy. Micron 2001, 32: 653-660.

Dedov VN, Roufogalis BD. Organisation of mitochondria in living sensory neurons. *FEBS Lett.* 1999, 456: 171-174.

Di Sant'Agnese PA, Hung J, Yao J.. The Neuroendocrine Prostate.[online] [2004 11 02] Available from:URL:www.urmc.rochester.edu/path/neuroendo/

Duca M, Guianvarc'h D, Oussedik K, Halby L, Garbesi A, Dauzonne D, Monneret C, et al. Molecular basis of the targeting of topoisomerase II-mediated DNA cleavage by VP16 derivatives conjugated to triplex-forming oligonucleotides. *Nucleic Acids Res.* 2006, 34(6):1900-11

Fairbairn, L.L., Cowling, G.J., Reipert, B.M., Dexter, T.M. Suppression of apoptosis allows differentiation and development of a multipotent hemopoietic cell line in the absence of added growth factors. *Cell* 1993, 74: 823-832,

Fernandez-Segura E., Cañizares FJ., Cubero MA., Warley A., Campos A. Changes In Elemental Content During Apoptotic Death Studied By Electron Probe X-Ray Microanalysis. *Exp. Cell Res.* 1999, 253, 454-462.

Festuccia C, Guerra F, D'Ascenzo S, Giunciuglio D, Albini A, Bologna M. In vitro regulation of pericellular proteolysis in prostatic tumor cells treated with bombesin. *Int J Cancer* 1998, 75: 418-31.

Fixemer T, Remberger, K, Bonkhoff H. Apoptosis resistance of neuroendocrine phenotypes in prostatic adenocarcinoma. *Prostate* 2002, 53: 118-123

Goodin J.L., Rutherford C.L. Identification of differentially expressed genes during cyclic adenosine monophosphate-induced neuroendocrine differentiation in the human prostatic adenocarcinoma cell line LNCaP. Mol. *Carcinogenesis* 2002, 33: 88-98.

Grundin TG, Roomans GM, Forslind B, Lindberg M, Werner Y. X ray microanalysis of psoriatic skin. *J. Invest. Dermatol.*1985, 85, 378-380.

Gutierrez AA., Arias JM., Garcia L., Mas-Oliva J., Guerrero-Hernandez A. Activation of Ca2+ permeable cation channel by two different inducers of apoptosis in a human prostatic cancer cell line, *J. Physiol.* (London) 1999,15:95-107.

Han K., Viallet J., Chevalier S., Zheng W., Bazinet M., Aprikian A.G. Characterization of intracellular calcium mobilization by bombesin-related neuropeptides in PC-3 human prostate cancer cells. *Prostate* 1997, 31, 53-60.

Hansson J, Abrahamsson PA. Neuroendocrine pathogenesis in adenocarcinoma of the prostate. *Ann Oncol.* 2001, 12 supp 2: 145-152.

Hoosein N.M. Neuroendocrine and immune mediators in prostate cancer progression, Front. *Biosci.* 1998, 3: 1274-1279.

Hughes FM. Jr., Bortner CD., Purdy GD., Cidlowski JA. Intracellular K+ suppresses the activation of apoptosis in lymphocytes. *J Biol Chem.* 1997, 272: 30567-76.

IijimaT., MishimaT., Akagawa K., Iwao Y.. Mitochondrial hyperpolarization after transient oxygen-glucose deprivation and subsequent apoptosis in cultured rat hippocampal neurons. *Brain Res.* 2003, 993: 140-145.

Ingram P., Shelburne JD., Le Furgey A. Principles And Instrumentation. In: Ingram, P., Shelburne, J., Roggli, V., Le Furgey, A., eds. *Biomedical Applications Of Microprobe Analysis*. London:Academic Press;1999;1-58.

Ismail H, Landry F, Aprikian AG, Chevalier S.. Androgen ablation promotes neuroendocrine cell differentiation in dog and human prostate. *Prostate* 2002,51: 117-125.

Ito T, Yamamoto S, Ohno Y, Namiki K, Aizawa T., Akiyama A, Tachibana M. Up-regulation of neuroendocrine differentiation in prostate cancer after androgen deprivation therapy, degree and androgen independence. *Oncol Rep* 2001, 8: 1221-1224

James Am, Murphy MP. How mitochondrial damage affects cell function. *J Biomed Sci.* 2002, 9: 475-487.

Jongsma J, Oomen MH, Noordzij MA, Romijn JC, van der Kwast TH, Schröder FH, van Steenbrugge GJ. Androgen-independent growth is induced by neuropeptides in human prostate cancer cell lines. *Prostate* 2000, 42: 34-44.

Kunzelman K. Ion channels and cancer. *J. Membrane Biol.* 2005, 105: 159-173.

Lang F, Föller M, Lang KS, Ritter M, Gulbins E, Vereninov A, Huber SM. Ion channels in cell proliferation and apoptotic cell death. *J. Membrane Biol* 2005, 205:147-157.

Lawen A.. Apoptosis an introduction. *BioEssays* 2003,25:888-896.

Li L.Y., Luo X., Wang X. Endonuclease G is an apoptotic DNase when released from mitochondria. *Nature* 2001, 412: 95-99.

Lim MLR, Lum MG, Hansen TM, Roucou X. On the release of cytochrome c from mitochondria during cell death signalling. *J. Biomed Sci.* 2002, 9: 488-506.

Maeno E, Ishizaki Y, Kanaseki T, Hazama A, Okada Y. Normotonic cell shrinkage because of disordered volume regulation is an early prerequisite to apoptosis. *Proc. Natl. Acad. Sci. USA* 2000, 97: 9487-92.

Mason RP. Calcium channel blockers, apoptosis and cancer: is there a biologic relationship? J Am Coll Cardiol 1999, 34: 1857-66.

Mathur A, Hong Y, Kemp BK, Alvarez A, Erusalimsky JD. Evaluation of fluorescent dyes for the detection of mitochondrial membrana potencial changes in cultured cardiomyocytes. *Cardiovasc. Res.* 2000, 46: 126-138.

Montague JW, Bortner CD, Hughes FM Jr, Cidlowski JA. A necessary role for reduced intracellular potassium during the DNA degradation phase of apoptosis. Steroids 1999, 64: 563-69.

Nilius B. Chloride channels go cell cycling *J. Physiol.* 2001:32, 581.

Oberhammer, F., Wilson, J., Dive, C., Morris, I.D., Hickman, J.A., Wakeling, A.G., et al. Apoptotic death in epithelial cells: cleavage of DNA prior to or in the absence of internucleosomal fragmentation. *EMBO J.* 1993, 12(9): 3679-3684.

Petronili V, Penzo D, Scorrano L, Bernardi P, Di Lisa F. The mitochondrial permeabilty transition, release of cytochrome c and cell death. *J Biol Chem.* 2001, 276: 12030-12034.

Robertson J.D. , Orrenius S.. Molecular mechanisms of apoptosis induced by cytotoxic chemicals. *Crit. Rev. Toxicol.* 2000, 30: 609-627.

Roomans GM, Hongpaisan J, Jin Z, Mörk AC, Zhang A. In vitro systems and cultured cells as specimens for X-ray microanalysis. *Scanning Microsc* 1996, Suppl 10: 359-74.

Roomans GM., Vilches J., López A., Salido M. Neuropeptides Bombesin and calcitonin inhibit apoptosis-related elemental changes in a prostate cancer cell line. In: *Proceedings*

SCANDEM 2001. Stockholm: Scandinavian Society for Electron Microscopy, 2001: 132-133.

Salido M, Vilches J, Lopez A, Roomans GM. Neuropeptides bombesin and calcitonin inhibit apoptosis related elemental changes in prostate cancer cell lines. *Cancer* 2002, 94: 368-377.

Salido M, Vilches J, Lopez A. Neuropeptides bombesin and calcitonin induce resistance to etoposide induced apoptosis in prostate cancer cell lines. *Histol Histopathol.* 2000,15: 729-38.

Salido M, Vilches J, Roomans GM. Changes in elemental concentrations in LNCaP cells are associated with a protective effect of neuropeptides on etoposide-induced apoptosis. *Cell Biol.Int.* 2004, 28 (5): 397-402

Salido M., Vilches J., López A., Roomans G.M. X-ray microanalysis of etoposide induced apoptosis in the PC-3 prostatic cancer cell line. *Cell Biol. Int.* 2001, 25, 499-508.

Salido M., Vilches J., Moreno M.A., López A., Roomans G.M. Morphological observation and x-ray microanalysis on apoptotic cell death in an androgen sensitive prostate cancer cell line.*Biol.Cell.* 2001b,93: 403-404.

Salido,M.; Larran,J.; Lopez,A.; Vilches,J.; Aparicio,J. Etoposide sensitivity of human prostatic cancer cell lines PC-3, DU 145 and LNCaP. *Histol.Histopathol.*1999, 14: 125-134.

Salvioli S, Ardizzoni A, Franceschi C, Cossarizza A. JC-1, but not $DiOC_6$ (3) or rhodamine 123, is a reliable fluorescent probe to asses $\Delta\psi$ changes in intact cells: implications for studies on mitochondrial functionality during apoptosis. *FEBS Lett.* 1997, 411:77-82.

Sciarra A, Cardi A, Dattilo C, Mariotti G, Di Monaco F, Di Silverio F. New perspective in the management of neuroendocrine differentiation in prostate adenocarcinoma.*J.Clin.Pract.* 2006, 60:462-470.

Segawa N, Mori I, Utsunomiya H, Nakamura M, Nakamura Y, Shan L, Kakudo K, Katsuoka Y. Prognostic significance of neuroendocrine differentiation, proliferation activity and androgen receptor expression in prostate cancer. *Pathol. Int.* 2001, 51:452- 459.

Skepper JN., Karydis I., Garnett MR., Hegyi L., Hardwick S.J., Warley A., Mitchinson MJ.,et al. Changes in elemental concentrations are associated with early stages of apoptosis in human monocyte-macrophages exposed to oxidized low-density lipoprotein: an X-ray microanalytical study. *J. Pathol.* 1999, 188: 100-106.

Smith NKR, Cameron IL. Ionic regulation of proliferation in normal and cancer cells. In: Ingram P, Shelburne JD, Roggli VL, LeFurgey A, eds. *Biomedical applications of microprobe analysis.* San Diego: Academic Press.1999;445-459.

Tang D.G., Li L, Chopra D.P., Porter A.T. Extended survivability of prostate cancer cells in the absence of trophic factors: increased proliferation, evasion of apoptosis, and the role of apoptosis proteins. *Cancer Res.* 1998, 58: 3466-3479.

Thomas S, Shah G. Calcitonin induces apoptosis resistance in prostate cancer cell lines against cytotoxic drugs via the Akt/survivin pathway.*Cancer Biol Ther.* 2005, 4(11):1226-33.

Van Blitterswijk WJ, van der Luit AH, Veldman RJ, Verheij M, Borst J. Ceramide, second messenger or modulator of membrane structure and dinamics ? *Biochem J* 2003, 369 : 199-211.

Vilches J, Salido M, Fernandez-Segura E, Roomans GM Neuropeptides, Apoptosis and Ion changes in Prostate Cancer. Methods of study and recent developments.. *Histol. Histopathol.*2004: 19: 951-961.

Vilches J, Salido M. Mitochondria and cell death. *Histol. Histopathol.* 2005, supp 1, S58-59.

Von Euler A, Pålsgård E, Vult von Steyern C, Roomans GM. X-ray microanalysis of epithelial and secretory cells in culture. *Scanning Microsc* 1993, 7: 191-202.

Wang P, Song JH, Song DK, Zhang J, Hao C. Role of death receptor and mitochondrial pathways in conventional chemotherapy drug induction of apoptosis. *Cell.* Signal 2006 18(9):1528-35

Warley A. X-Ray Microanalysis for biologists. Cambridge:Portland Press. 1997.

Wasilenko W.J., Cooper J., Palad A.J., Somers K.D., Blackmore P.F., Rhim J.S., Wright GL Jr., et al. Calcium signaling in prostate cancer cells: evidence for multiple receptors and enhanced sensitivity to bombesin/GRP. *Prostate* 1997,30, 167-173.

Wilson EM, Oh Y, Hwa V, Rosenfeld RG. Interaction of IGF-binding protein-related protein 1 with a novel protein, neuroendocrine differentiation factor, results in neuroendocrine differentiation of prostate cancer cells. *J Clin. Endocrinol. Metab.* 2001, 86: 4504-4511.

Wright, G.L., Grob, B.M., Haley, C., Grossman, K., Newhall, K., Petrylak, D., et al. Upregulation of prostate specific membrane antigen after androgen-deprivation therapy. *Urology* 1996, 48(2): 326-334,.

Yurinskaya VE, Moshkov AV, Rozanov YM, Yu M, Shirokova AV, Vassilieva IO, Shumilina EV et al. Thymocyte K^+, Na^+,and water balance during dexametasone and etoposide-induced apoptosis. *Cell. Physiol. Biochem* 2005a, 16:15-22

Yurinskaya VE, Goryachaya TS, Guzhova TV, Moshkov AV, Rozanov YM, Sakuta GA, Shirokova AV et al. Potassium and sodium balance in U937 cells during apoptosis with and without cell shrinkage. *Cell. Physiol. Biochem* 2005b, 16: 155-162.

FIGURE PERMISSION LIST

Figure 1 and 4b are reprinted with permission from Histology and Histopathology (2000). M Salido, J Vilches, A Lopez. Neuropeptides bombesin and calcitonin induce resistance to etoposide induced apoptosis in prostate cancer cell lines. Vol 15:729-738.

Figures 2, 4, 5, 6 are reprinted with permission from Histology and Histopathology (2004). M Salido, J Vilches, E Fernandez Segura and GM Roomans. Neuropeptides apoptosis and ion changes. Methods of study and recent developments. Vol 19:951-961.

Figures 12, 13, 14, 16 reprinted with permission from Cell Biology International vol 25. M Salido, J. Vilches, A. Lopez, GM Roomans. X-ray microanalysis of etoposide-induced apoptosis in the PC3 prostatic cancer cell line. Pages 499-508. Copyright 2001, with permission from Elsevier.

Figs 17-21 reprinted from SalidoM, Vilches J, Lopez A, Roomans GM. Neuropeptides bombesin and calcitonin inhibit apoptosis-related elemental changes in prostate carcinoma cell lines. Cancer 2002; 94: 368-77. With permission from John Wiley and Sons Inc.

Figures 22,23,24,25 reprinted with permission from Cell Biology International vol 28. M Salido, J. Vilches, Roomans. Changes in elemental concentrations in LNCaP cells are associated with a protective effect of neuropeptides on etoposide-induced apoptosis.Pages 397-402. Copyright 2004, with permission from Elsevier

In: Cell Apoptosis Research Trends
Editor: Charles V. Zhang, pp. 151-168

ISBN: 1-60021-424-X
© 2007 Nova Science Publishers, Inc

Chapter V

The Role of IAP as a Novel Diagnostic and Therapeutic Target for Prostate Cancer

Takeo Nomura and Hiromitsu Mimata

Department of Oncological Science (Urology), Oita University Faculty of Medicine,
1-1 Idaigaoka, Hasama-machi, Yufu, Oita 879-5593, Japan

ABSTRACT

Prostate cancer is the most frequently diagnosed malignancy and the second leading cause of cancer death among men in the United States. The progression of prostate cancer from the androgen-dependent to the androgen-independent state is the main obstacle to improving the survival and quality of life in patients with advanced prostate cancer. Prostate cancer progression and the development of the androgen-independent state have been related to genetic abnormalities that influence not only the androgen receptor but also crucial molecules involved in apoptosis. Apoptosis is a programmed cell death executed by a family of intracellular cysteine proteases known as caspases. Activation of the caspase cascade is associated with proteolytic cleavage of diverse structural and regulatory proteins that contribute to apoptosis. Several proteins that inhibit apoptosis, including p53 and the Bcl-2 family, are involved in the promotion of tumorigenesis and drug-resistance in several cancers. In addition, the inhibitor of apoptosis proteins (IAPs) have been identified in baculovirus. The IAPs are a family of caspases inhibitors that block the apoptotic pathway via binding pro- and/or active forms of caspase-3, -7 and -9 and inhibiting their processing. To date, eight human IAPs have been recognized: X-chromosome-linked IAP (XIAP), human IAP-1 (hIAP-1), hIAP-2, survivin, neuronal apoptosis inhibitory protein (NAIP), apollon, livin, and IAP-like protein-2 (ILP-2). IAPs are overexpressed in prostate cancer and their expression correlates with tumor progression and drug-sensitivity, therefore therapies that target IAPs have the potential to improve outcomes for patients with prostate cancer. This review focuses on the experimental evidence that associates IAPs expression with prostate carcinogenesis, and analyzes the roles of IAPs in chemotherapy in order to develop novel therapeutic strategies.

INTRODUCTION

Prostate cancer is the most frequently diagnosed malignancy and the second leading cause of cancer-related death among men in the United States [1]. Since the early 1990s around the time prostate-specific antigen (PSA) first became available, prostate cancer incidence increased dramatically [2]. Prostate cancers are relatively slow growing malignancies with doubling times for local tumors estimated at 2 to 4 years. The current therapeutic regimens for treatment of prostate cancer include surgery, radiation therapy, hormonal therapy, and chemotherapy, as well as a combination of these regimens, depending on specific situations of the patient. Although localized prostate cancer may be successfully treated by surgery or radiation therapy, 15% of patients relapse after apparently successful treatment, eventually progress and develop metastasis [3]. Prostate cancer is characterized by an initial period during which the tumor growth is androgen dependent. Androgen-deprivation procedures by medical or surgical castration result in a temporary regression of the disease, but the effect is usually palliative and temporary. The reason for this is that decreased levels of androgens favor the growth of tumor cells that are androgen independent. The development of a hormone-independent state is an irreversible phenomenon observed in the majority of patients and occurs within several years after initiation of androgen deprivation. The progression of prostate cancer from the androgen-dependent to the androgen-independent state is the main obstacle to improving the survival and quality of life in patients with advanced prostate cancer. Therefore, much attention has been focused on the evolution from androgen-dependent to androgen-independent prostate cancer. However, there has been no effective therapy for the treatment of androgen-independent prostate cancer, and the treatment has been far from satisfactory. Although progress in cell and molecular biology has enhanced our understanding of the mechanisms involved in prostate cancer growth, differentiation, and metastasis, many of the biologic processes leading to an overwhelmingly hormone-independent state remain unclear because of its unusual biologic features. An increasing number of biologic events have been identified during the process of prostate cancer progression. These are genetic alterations such as mutations of tumor suppressor genes [4], expression of various oncogenes affecting the cell growth, proliferation, and cell death [5], enhanced angiogenesis [6], and a variety of gene mutations that result in a functionally altered expression of the androgen receptor (AR) [7]. It has been reported that insufficient programmed cell death, termed apoptosis, represents the explanation for the accumulation of prostate cancer cells [8, 9]. Cancer cells have lost a balance between mitosis and cell death, resulting in a gradual accumulation of cells with the potential deviated from the normal cell homeostasis. That is to say, progression of androgen-dependent cancer to hormone-refractory disease is related to genetic abnormalities that influence not only the AR but also crucial molecules involved in apoptosis.

Cell death can be accidental or programmed in a multicellular organism [10, 11]. The former is called necrosis and the latter is apoptosis. Necrosis is always accompanied by extensive inflammations resulting in uncontrolled cell death, with related cell swelling and rupturing. On the other hand, apoptosis is a series of morphologically and biochemically related processes, and its features are controlled cell death in the absence of inflammation, resulting in the bundling of cellular organelles into membrane-enclosed apoptotic vesicles. The main morphological characteristics are nuclear fragmentation and cellular collapse in apoptotic vesicles, with resulting rapid condensation and budding of the cell, and apoptotic cells can be effectively digested by surrounding resident macrophages or neighboring epithelial cells [12]. It seems that there is an inherent suicide program which can be activated when the cell death is desirable for the advantage of the rest of the community. The commonly used term "apoptosis" stems from the old Greek word " *falling off like leaves from a tree* ", and its essential function is a cell deletion during early development and growth of normal tissues [13-16]. Interestingly, apoptosis also occurs spontaneously in malignant cells, often markedly retarding tumor growth, and it is increased in tumors responding to several therapies including irradiation, chemotherapy, thermal stress, and hormone ablation [9]. Molecularly, apoptosis is executed by the activation of caspases, a family of intracellular cysteine proteases that cleave substrates at aspartic acid residues [17, 18]. In addition, the apoptotic pathways are highly regulated in the presence of proapoptotic and antiapoptotic proteins by both extrinsic and intrinsic stimuli [19]. Unfortunately, cancer fails to respond to treatment in varying degrees. In part, the failure of cell death is caused by failure of the apoptosis and caspase activation pathways. The inhibitor of apoptosis proteins (IAPs) are a family of antiapoptotic proteins that block cell death by inhibiting the downstream of the caspase activation pathway [20-23]. It has been reported that there is a positive correlation between IAP expression and tumor progression in prostate cancer [24, 25].

Understanding the machinery of IAPs function can potentially allow for the development of novel therapeutic strategies targeting caspases and IAPs for prostate cancer. In this review, we will focus on the experimental evidence of the roles of caspases and their negative regulators, IAPs, and discuss how this evidence is being translated into the clinical field as the development of new diagnostic and prognostic markers.

PROSTATE CANCER AND ANDROGEN

Although the great advance in cell and molecular biology has enhanced our understanding of the multiple mechanisms involved in prostate cancer development, the exact etiology of prostate cancer remains unknown. However, it seems valid that prostate cancer initiation and progression are influenced by androgens, even if we have been impressed by the high incidence of development of prostate cancer in aged men with low levels of androgen. Androgen is a necessary growth factor for prostate cancer. The action of androgens is mediated by a specific receptor protein, androgen receptor (AR), which is located on the human X-chromosome. The AR, a transcription factor belonging to the nuclear receptor superfamily, is activated by phosphorylation, and this activation promotes nuclear localization and binding of AR to androgen response elements in the androgen target genes such as PSA. AR mediates testosterone and dihydrotestosterone (DHT) activity by initiating transcription of androgen-responsive genes. In addition to androgens, AR also plays a crucial role in several stages of male development and the progression of prostate cancer [26, 27]. It has been also reported that many of the growth factors such as insulin-like growth factor (IGF), epidermal growth factor (EGF), and keratinocyte growth factor (KGF) activate AR-related gene transcription [28].

CHARACTERISTICS OF APOPTOSIS IN PROSTATE CANCER

Even the patients with advanced prostate cancer respond to androgen ablation by medical or surgical castration and by blocking the effects of residual androgen with competitive androgen receptor antagonists, and serum PSA levels decrease in almost all patients. After androgen ablation, androgen-responsive prostate cancer cells cease to proliferate and fall into cell death. This cell death occurs as apoptosis [29]. The irreversible genomic DNA fragmentation takes place and results from activation of Ca^{2+}/Mg^{2+} dependent endonuclease activity in the cancer cell nucleus. This activation is because of sustained increase of intracellular Ca^{2+} induced after androgen ablation. However, as the tumor progresses, the threshold of apoptosis progressively drops to a point at which cell proliferation exceeds apoptosis [30]. This increase of proliferating cells is due to the accumulation of androgen-independent cells that eventually relapse and metastasize, resulting in a short survival of the patients with advanced prostate cancer. Progression to androgen independence is a multifactorial process by which cells acquire the ability to both survive in the absence of androgens and grow using androgen non-related stimuli for mitogenesis. Several mechanisms have been proposed to lead androgen independence, including the hypersensitivity of AR, inappropriate activation of AR by other ligands, and alterations in the regulators of the cell death pathway [31, 32]. This stage of cancer is also characterized by the emergence of apoptosis-resistant cells resulting from various genetic mutations and overexpression of antiapoptotic genes [33]. In general, there is a strong tendency in gene expression after androgen ablation. Clusterin, AR, Bcl-2, Bcl-xL, Hsp27, IGFBP-2 and IGFBP-5 are upregulated by androgen ablation and remain overexpressed in androgen-independent cancer [34]. We previously reported that the expression of cellular IAP-1 (cIAP-1) and cIAP-2 was upregulated in patients treated with androgen ablation [35]. In addition, it has been reported that the androgen-independent prostate cancer cell lines PC3 and DU145 cells are highly resistant to apoptosis and overexpressed cIAP-1, cIAP-2, XIAP, and NAIP [36]. The aspect of prostate cancer cell biology of these antiapoptotic molecules is developing rapidly, and IAPs may prove to be the importance of antiapoptotic action in prostate cancer progression.

PROAPOPTOTIC AND ANTIAPOPTOTIC SIGNALING PATHWAYS

Apoptosis is characterized by morphological (cytoplasmic condensation, nuclear condensation, cell shrinkage, and budding) and biochemical (DNA fragmentation and degradation of apoptotic protein) features. The molecular signaling pathways involved in apoptosis are also well recognized. Two main apoptotic pathways that activate caspases, triggered by physiological and pharmacological agents have been identified (Fig. 1). Apoptosis in prostate cancer is initiated by either extrinsic (death receptor-mediated) pathway or intrinsic (mitochondria-mediated) pathway or a combination of both pathways. The extrinsic pathway is activated by ligand-bound death receptors of the tumor necrosis factor (TNF) family such as Fas, TNFR1, or the tumor necrosis factor-related apoptosis-inducing ligand (TRAIL) receptors DR4 and DR5. After ligand binding to the death receptor, the death-inducing signaling complex (DISC) is formed and mediates the apoptotic process. Although death receptors can promote cell growth under some situation [37], the main ability of these receptors is to induce apoptosis. Cell death from these activated receptors occurs due to recruitment of an adaptor protein called Fas associated death domain (FADD), which recruits pro-caspase-8 directly or via another adaptor protein called TNFR-associated death domain (TRADD), which binds to TNFR1. Recruitment of pro-caspase-8 through FADD leads to its autocatalytic activation [38]. Active caspase-8 subsequently activates effector caspases such as caspase-3 and proteolysis of substrate proteins resulting in induction of apoptosis. On the other hand, the intrinsic pathway, triggered by radiation or by agents that elevate intracellular Ca^{2+} is mediated through mitochondria. Both proapoptotic and antiapoptotic members of Bcl-2 family proteins that are localized on the outer mitochondrial membrane influence permeability and regulate mitochondrial initiation associated with cytochrome c release into the cytosol [39]. Released cytochrome c interacts with apoptotic protease-activating factor-1 (Apaf-1) and pro-caspase-9 to form the apoptosome [40]. This complex activates caspase-9, which in turn activates effector caspases to induce apoptosis. The death receptor-mediated pathway and mitochondria-mediated pathway converge at the level of downstream effector caspase-3. In addition, a fragment of Bid (t-Bid) which is cleaved by caspase-8 translocates to the mitochondria and induces the release of cytochrome c to activate the intrinsic pathway [41]. Finally, these two caspase activation pathways can be connected and amplify the death receptor apoptotic signal. In addition to Bcl-2 family, IAPs that are direct inhibitor of caspases are known as antiapoptotic factor in caspase cascade. Furthermore, second mitochondrial derived activation of caspase/direct IAP binding protein with low pI (Smac/DIABLO), high temperature requirement A (HtrA2/Omi), and XIAP-associated factor 1 (XAF1) exist as blockers of IAPs [42, 43]. In this review, we focused on caspases, effector protein of apoptosis, and IAPs, recently identified negative regulator of caspases.

CASPASES

Caspases are a unique family of conserved cysteine proteases that usually cleave after an aspartate residue in their substrates, the downstream target caspases, to activate them [44]. To date, 14 caspases involved in apoptosis have been identified and classified into two groups as the initiator caspases and the effector caspases. All caspases are synthesized in cells as catalytically inactive zymogens, and undergo an activation process. The initiator caspases, such as caspase-2, -8, -9, and -10 are activated by various apoptotic signals, and they subsequently activate downstream effector caspases, such as caspase-3, -6, and -7, through an internal cleavage to separate the large and small subunits.

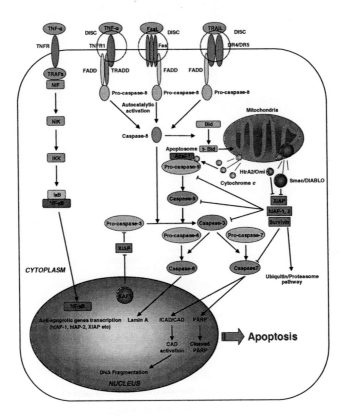

Figure 1. Apoptotic pathways and mechanisms of the inhibition of apoptosis by IAP family proteins. Binding of ligand (TNF- α, FasL, TRAIL) to their respective death receptors (TNFR1, Fas, DR4/DR5) triggers the formation of a DISC involving the recruitment of the adaptor protein FADD and caspase-8. Activation of caspase-8 from this complex initiates a caspase cascade resulting in the activation caspase-3 and subsequent cleavage of proteins leading to the execution of apoptosis. Activated caspase-8 also triggers the mitochondrial pathway through caspase-8-mediated Bid cleavage, translocation of t-Bid into the mitochondria and release of cytochrome c. Cytochrome c binds directly Apaf-1, which subsequently oligomerizes to form the apoptosome. This complex recruits and activates caspase-9, and apoptosome-activated caspase-9 acts on downstream effector caspases, such as caspase-3, -7, and –6, triggering DNA fragmentation and oligonucleosome formation through CAD. IAPs (XIAP, hIAP-1, hIAP-2, survivin) inhibit apoptosis by directly inhibiting effector caspase-3 and –7, and by interfering with caspase-9 activity/processing. Smac/DIABLO and HtrA2/Omi released from mitochondria inhibit all IAPs directly, and XAF1 exclusively inhibits XIAP.

On the other hand, the initiator caspases are autoactivated under apoptotic situations forming multicomponent complexes, such as apoptosome, which is responsible for the activation of caspase-9. In addition to cleaving other members of the caspase family, effector caspases can cleave various structural and functional proteins such as oncoproteins, cytoplasmic or nuclear substrates, which mark morphologic features of apoptotic cell death [44]. For example, caspase-3 activates endonucleases, responsible for nucleosomal DNA fragmentation. This event is initiated by cleavage of inhibitor of caspase-activated DNase

(ICAD) that cleaves DNA into the oligomeric fragments. It is also noteworthy that the activation of caspases depends on the type of cells and the type of apoptosis-inducing agent.

CASPASE EXPRESSION AND PROSTATE CANCER

The possibility exists that understanding the regulation of caspase activation and its activity may contribute to the development of therapy for prostate cancer. Therefore, it is necessary to determine caspase-3 expression, which functions ultimately downstream in the apoptotic pathway, and its correlation with the apoptosis in prostate cancer. Sohn et al reported that positive staining for caspase-3 by immunohistochemistry (IHC) was lower in prostate cancer than in benign nodular hyperplasia (BPH) and there was no relationship between caspase-3 expression and Gleason score, however, apoptosis index was significantly higher in prostate cancer than in BPH [45]. Some reports showed that caspase-3 expression was inversely related to Gleason score [46, 47]. Inhibition of caspases by cytokine response modifier A (CrmA) blocked androgen ablation-mediated prostate cancer cell death in xenografted LNCaP-Crm A tumors on nude mice [48]. Taken together, altered caspase-3 expression may contribute to the progression of prostate cancer and the acquisition of resistance to androgen ablation.

INHIBITOR OF APOPTOSIS PROTEINS (IAPs)

The IAPs represent the only known family of direct endogenous caspase inhibitors [20-23, 49]. To date, eight IAP-encoding genes have been recognized in the human genome, including X-chromosome-linked IAP (XIAP), human IAP-1 (hIAP-1, cIAP-2), hIAP-2 (cIAP-1), survivin, neuronal apoptosis inhibitory protein (NAIP), apollon (BRUCE), livin (ML-IAP, KIAP), and IAP-like protein-2 (ILP-2), which are evolutionarily conserved with apparent homologies identified in flies, worms, yeast and several mammalian species including mice, rats, chickens, pigs, and humans [21, 50-54]. All the IAPs show varying degrees of antiapoptotic ability, depending on the different mechanisms of action for each IAP homology. The IAPs are characterized by a highly conserved domain of ~70 amino acids termed the baculoviral IAP repeat (BIR) domain and really interesting new gene (RING)-zinc binding domain [23, 55]. Some of BIR domains can bind to and potently inhibit activated caspase-3, -7, and -9 [23]. A RING domain has ubiquitin protease activity; it can bind to ubiquitin-conjugating enzymes that promote autoubiquitination and degradation of IAP-caspase complexes after apoptosis stimulus [55]. The requirement of RING domain for inhibition of apoptotic pathway seems to be dependent on the type of cells. IAPs can protect cells from various triggers of intrinsic and extrinsic pathways. All the IAPs except NAIP can bind to and inhibit caspase-3 and –7 [21, 50]. XIAP, hIAP-1, hIAP-2, and survivin were also shown to bind to and inhibit caspase-9, but not caspase-1, -6, -8, or –10 [21]. Although IAPs cannot bind to or inhibit caspase-8, they bind to and inhibit its substrate caspase-3, thus providing protection from Fas/TNF family death receptor mediated apoptosis [21, 50]. In contrast, XIAP, hIAP-1, and hIAP-2 bind to directly pro-caspase-9, and prevent its processing and activation induced by cytochrome c released from mitochondria, thus they can prevent the proteolytic processing of pro-caspase-3, -6, and –9 [50]. It has been reported that XIAP mainly binds to active caspase-3, but also partially to the unprocessed pro-caspase-3 [56]. We transiently cotransfected with XIAP and pro-caspase-3 cDNAs into LNCaP cells, and observed the strong interaction between XIAP and pro-caspase-3 by immunoprecipitation and immunoblot analysis [57]. XIAP may block a common downstream by directly inhibiting pro- and active caspase-9, and by interfering with caspase-3 activity and/or processing of pro-caspase-3. Moreover, it has been demonstrated that the antiapoptotic characteristics of IAPs are related to the nuclear factor kappa B (NF-κB) pathway and mitogen-activated protein (MAP) kinase signaling pathway. The hIAP-1 and hIAP-2 also bind to TNF receptor-associated factor (TRAF) heterocomplexes through their N-terminal BIR domain, interfering with the upstream activation of caspase-8 [58, 59]. TRAF-1, TRAF-2, XIAP, hIAP-1, and hIAP-2 are identified as gene targets of NF-κB transcription activity [59, 60]. The activation of NF-κB and the induction of IAPs are an essential part in the process that protects cells from apoptotic signals caused by TNF-α. In addition, XIAP, NAIP, and ML-IAP bind to TGF β-activated kinase (TAK-1) and activate TAK-1/c-Jun N-terminal kinase (JNK) signaling cascade, with resulting inhibition of apoptosis [61]. TAK-1 dependent JNK activation also plays an important role in antiapoptotic efficacy of IAPs. In addition to regulation of apoptosis, IAP members such as survivin have been found to be a potent regulator of cell cycle progression and mitosis [62]. Survivin may regulate cell death by not only an antiapoptotic mechanism but also a caspase cascade-independent mechanism. Interestingly, it has been reported that survivin may contribute to tumor angiogenesis via angiopoietin-1 stabilization [63]. Taken together, the possibility exist that IAPs have unknown mechanisms of cancer growth and protection from cell death.

ANTAGONISTS OF THE IAP FAMILY

Currently, the following three negative regulators are known to bind to IAPs and inhibit their activity; Smac/DIABLO, HtrA2/Omi, and XAF1 [64]. HtrA2/Omi, which belongs to the shock response serine protease-chaperone HtrA family, is released together with Smac/DIABLO from mitochondria [65, 66]. Smac/DIABLO promotes apoptosis by binding to XIAP and disrupting its association with active caspase-9 [67]. Some reports suggested that Smac/DIABLO selectively interacts with hIAP-1 and hIAP-2 and reduces the protein level of them through ubiquitin/proteasome pathway in Hela cells [68, 69]. XAF1 is known as the other negative regulator of XIAP. In contrast to Smac/DIABLO and HtrA2/Omi, XAF1 is localized in the nucleus and can affect a redistribution of XIAP from the cytosol to the nucleus, resulting in inactivation of XIAP [70]. Taken together, these antagonists of the IAP family may play a role as a potential tumor suppressor.

IAPs EXPRESSION AND PROSTATE CANCER

The upregulation of IAPs expression has been considered one of mechanisms for escape from elimination by apoptosis. To investigate the IAPs expression in prostate cancer is essential for the development of novel therapeutic strategies targeting IAPs for prostate cancer. Some of IAP family protein are overexpressed in various cancers and related to progression and poor survival, including breast cancer [71], esophageal cancer [72], gastric cancer [73], colorectal cancer [74], neuroblastma [75], non-small cell lung cancer [76], urinary bladder cancer [77], and epithelial ovarian cancer [78]. These results suggest that IAPs may contribute to tumor progression and drug resistance. However, there have been few studies on the expression of IAPs in prostate cancer. Kishi et al reported that survivin mRNA expression was positively correlated with the progression (T-stage, lymph node metastasis, vessel invasion, surgical margin, and Gleason score) and aggressiveness (proliferative activity) of prostate cancer [25]. Shariat et al showed that survivin expression was associated with higher Gleason score and positive lymph node metastasis [79]. In contrast, Krajewska et al reported that expression levels of survivin, hIAP-1, hIAP-2, and XIAP by IHC on the microarrays elevated in prostate cancer, but IAPs expression did not correlate with Gleason score and PSA levels [24].

Considerable controversy exists as to the association between IAPs and androgen response. Some reports showed that the IAPs are overexpressed in commonly used prostate cancer cell lines, including androgen-dependent LNCaP cells, androgen-independent DU145 and PC3 cells [36, 80]. McEleny et al confirmed the expression of NAIP, hIAP-1, hIAP-2, XIAP, and survivin in LNCaP, DU145, and PC3 cells at the level of the mRNA and the protein. They also showed an increased expression of hIAP-1, hIAP-2, and XIAP in DU145 and PC3 cells compared with LNCaP cells, and this expression is correlated with resistance to apoptosis [36]. Another study identified the expression of hIAP-2 and XIAP in DU145 and PC3 cells, but identified hIAP-1 expression only in DU145 cells, and did not identify the expression of NAIP in these cell lines [80]. Interestingly, there are poor relationships between mRNA and protein expression for survivin [36], hIAP-1, hIAP-2, and XIAP [80], suggesting that these proteins are post-transcriptionally regulated. We previously showed that the expression of cIAP-1 and cIAP-2 was upregulated in patients treated with androgen ablation by IHC, suggesting that the advent of residual cancer cells after androgen ablation was due to the induction of these IAPs [35]. Zhang et al indicated that androgen stimulation with DHT increased survivin expression and antiandrogen therapy with flutamide decreased its expression in LNCaP cells, suggesting that survivin played a potentially important role in androgen sensitivity and resistance to antiandrogen therapy [81]. In contrast, another study indicated that IAP expression in LNCaP cells was unaffected by charcoal-stripped medium [36]. They also confirmed that androgen-supplemented medium did not influence IAP expression in LNCaP cells [36]. These results suggest that IAP expression is unrelated to an androgen environment, then androgen ablation does not affect IAP expression and the acquisition of androgen independence is not due to the expression of IAP in prostate cancer. Taken together, IAPs, particularly survivin, are thought to have an important role in tumor progression, and may be of great value in clinical and prognostic markers in patients with prostate cancer, but a relationship between IAP and androgen response remains to be elucidated.

ARE IAPs TARGETS FOR THERAPY IN THE TREATMENT OF PROSTATE CANCER?

There appear to be more studies in the recent literature focusing on survivin and XIAP as potential therapeutic targets. The reason for this is that among IAPs, XIAP is the most potent inhibitor of caspases and apoptosis [21], and that survivin plays an important role in mitosis by interacting with other chromosomal passenger protein as well as an inhibitor of caspases [82]. In addition, the most important feature of these two molecules is that there are upregulated in various cancers and high levels of survivin and/or XIAP are associated with poor prognosis. There have been confirmed correlations of survivin and XIAP expression levels with responses to irradiation and chemotherapeutic drugs.

Radiation triggers the intrinsic pathway, resulting in apoptotic cell death in cancer [83]. Although radiation therapy is an effective treatment for localized prostate cancer, the development of radioresistance may occur in some cases. It seems that MAP kinase is activated by radiation, and cancer cells escape radiation-induced apoptosis in prostate cancer [84]. It is reported that a low dose of γ-irradiation in non-small cell lung cancer resulted in upregulation of XIAP, and cancer cells acquired the resistance to γ-irradiation [85]. Another study showed that an inverse relationship between survivin expression and radiosensitivity in pancreatic cancer [86]. There are no reports on the expression of IAPs in prostate cancer patients receiving radiation therapy. Studies explaining the elusive mechanisms following radiation are expected to constitute a rational approach for molecular targeting treatment.

There have been no satisfactory chemotherapeutic strategies for the treatment of both hormone-sensitive and hormone-resistant prostate cancer. The first standard chemotherapy consisting mainly of mitoxantrone/prednisone has no survival advantage for patients with hormone-resistant prostate cancer. Recently, docetaxel (taxotere) or paclitaxel (taxol) based combination chemotherapy demonstrated significant antitumor activity and improvement in overall survival in advanced prostate cancer [87, 88]. Although the relationship between IAPs expression and chemosensitivity is still unknown in prostate cancer, several experimental studies showed inhibition of apoptosis by IAPs in response to chemotherapeutic agents in various cancers [57, 80, 89-91]. Overexpression of IAPs such as XIAP and survivin confers resistance to chemotherapy and stimuli that trigger the intrinsic and extrinsic pathways of caspase cascade. We previously showed that overexpression of XIAP by stable transfection in LNCaP cells inhibited taxol- and cisplatin-induced apoptosis [57, 90]. Another study reported that XIAP suppressed apoptosis following treatment with some genotoxic agents or after irradiation in myeloid leukemia cells [92]. In addition to experimental studies, the relationship between increased IAPs expression and chemosensitivity was clinically reported in several cancers [72, 93]. Survivin expression is a significant predictor of chemosensitivity and survival in esophageal cancer [72] and lymphoma [93]. These results also suggest that IAPs have the potential as a novel determinant of chemosensitivity and therapeutic target.

Cisplatin, a most effective and widely used chemotherapeutic agent, is reported as a negative regulator of XIAP in several cancers including ovarian cancer [89, 94], oral cancer [95], and glioblastoma [96]. We reported that cisplatin induced apoptosis by the inhibition of XIAP expression in LNCaP cells [90]. We also reported that cisplatin-resistant LNCaP cells overexpressed hIAP-2, XIAP, and survivin, resulting in cross-resistance to several chemotherapeutic agents [97]. Although the current regimens of chemotherapy for prostate cancer have no satisfactory advantage, therapeutic strategies interfering with XIAP expression by cisplatin may confer a novel insight to develop a XIAP-targeted therapy.

DEVELOPMENT OF IAPS INHIBITORS

Two approaches have been taken to develop IAP inhibitors; antisense oligonucleotides and small molecule inhibitors. The antisense molecule contains DNA and RNA oligonucleotides. The antisense inhibits its target by forming a duplex with native mRNA, and in turn, this duplex recruits RNAase H that cleaves the native mRNA. In short, antisense oligonucleotides target the entire protein via the degradation of native mRNA, therefore this is the most efficient approach to the inhibition of IAPs. On the other hand, small molecules bind a single domain and blocks active sites on the protein. Until recently, antisense oligonucleotides against XIAP and survivin are in clinical trials, while practical application of small molecule XIAP inhibitors is in progress. In contrast to small molecule XIAP inhibitor, attempts to develop this agent for survivin present difficulties due to a lack of knowledge about structure of survivin. It has been reported that abrogation of XIAP expression by XIAP antisense induced apoptosis and enhanced sensitivity to cisplatin and TRAIL in DU145 cells [98]. Another study showed that an antisense to cIAP-1 sensitized PC3 cells to Fas antibody and TNF-mediated apoptosis [99]. These results suggest that antisense strategy to downregulate IAPs provides an effective therapeutic approach to hormone refractory prostate cancer.

Currently, it is generally accepted that gene targeting using small-interfering RNA (siRNA) has become a common strategy to explore gene function because of its marked efficacy and specificity. RNA interference (RNAi) is a mechanism whereby double-stranded RNA post-transcriptionally silences a specific gene. We confirmed that LNCaP cells transfected with synthetic, double-stranded siRNA against XIAP are enhanced to suppress cell growth and induce apoptosis when cells are treated with taxol (unpublished observation). It may be proven that the application of siRNA technique to gene therapy is effective.

Several novel approaches to interference of IAP expression have not only the potential for overcoming the antiapoptotic mechanism of IAP in prostate cancer but also an insight into the function of IAP in tumor progression and drug-resistance.

CONCLUSION

Prostate cancer initiation and progression has been associated with various genetic alterations. Therefore, the development of novel therapeutic strategies to overcome these alterations should be preceded by the identification of genetic and molecular profiles of tumors for each patient. We discussed the regulators of the apoptotic pathway including IAP family proteins. These molecules may play an important role not only as a new diagnostic and prognostic marker but also as a therapeutic target in prostate cancer. Downregulation of IAP expression by antisense oligonucleotides, small molecules, and siRNA, may have important practical implications in the management of prostate cancer. In addition, antagonists of the IAP family may be of a great value as a new target gene. Further studies of mechanisms that regulate IAP expression in prostate cancer are expected to lead to the development of novel therapeutic strategies.

ACKNOWLEDGMENTS

We thank Ms. N. Hamamatsu, Ms. Y. Ayaki and numerous members of our laboratory for their stimulating comments.

REFERENCES

[1]   Greenlee, R. T., Hill-Harmon, M. B., Murray, T., and Thun, M. Cancer statistics, 2001. *CA Cancer J Clin, 51:* 15-36, 2001.

[2]   Hankey, B. F., Feuer, E. J., Clegg, L. X., Hayes, R. B., Legler, J. M., Prorok, P. C., Ries, L. A., Merrill, R. M., and Kaplan, R. S. Cancer surveillance series: interpreting trends in prostate cancer--part I: Evidence of the effects of screening in recent prostate cancer incidence, mortality, and survival rates. *J Natl Cancer Inst, 91:* 1017-1024, 1999.

[3]   Hull, G. W., Rabbani, F., Abbas, F., Wheeler, T. M., Kattan, M. W., and Scardino, P. T. Cancer control with radical prostatectomy alone in 1,000 consecutive patients. *J Urol, 167:* 528-534, 2002.

[4]   Thompson, T. C., Kadmon, D., Timme, T. L., Merz, V. W., Egawa, S., Krebs, T., Scardino, P. T., and Park, S. H. Experimental oncogene induced prostate cancer. *Cancer Surv, 11:* 55-71, 1991.

[5]   Berges, R. R., Furuya, Y., Remington, L., English, H. F., Jacks, T., and Isaacs, J. T. Cell proliferation, DNA repair, and p53 function are not required for programmed death of prostatic glandular cells induced by androgen ablation. *Proc Natl Acad Sci U S A, 90:* 8910-8914, 1993.

[6]   Weidner, N., Semple, J. P., Welch, W. R., and Folkman, J. Tumor angiogenesis and metastasis--correlation in invasive breast carcinoma. *N Engl J Med, 324:* 1-8, 1991.

[7]   Taplin, M. E., Bubley, G. J., Shuster, T. D., Frantz, M. E., Spooner, A. E., Ogata, G. K., Keer, H. N., and Balk, S. P. Mutation of the androgen-receptor gene in metastatic androgen-independent prostate cancer. *N Engl J Med, 332:* 1393-1398, 1995.

[8]   Carson, D. A. and Ribeiro, J. M. Apoptosis and disease. *Lancet, 341:* 1251-1254, 1993.

[9]   Kerr, J. F., Winterford, C. M., and Harmon, B. V. Apoptosis. Its significance in cancer and cancer therapy. *Cancer, 73:* 2013-2026, 1994.

[10] Cohen, J. J. Apoptosis. *Immunol Today, 14:* 126-130, 1993.

[11] Gerschenson, L. E. and Rotello, R. J. Apoptosis: a different type of cell death. *FASEB J, 6:* 2450-2455, 1992.

[12] Savill, J. S., Wyllie, A. H., Henson, J. E., Walport, M. J., Henson, P. M., and Haslett, C. Macrophage phagocytosis of aging neutrophils in inflammation. Programmed cell death in the neutrophil leads to its recognition by macrophages. *J Clin Invest, 83:* 865-875, 1989.

[13] Cross, M. and Dexter, T. M. Growth factors in development, transformation, and tumorigenesis. *Cell, 64:* 271-280, 1991.

[14] Strasser, A., Harris, A. W., and Cory, S. bcl-2 transgene inhibits T cell death and perturbs thymic self-censorship. *Cell, 67:* 889-899, 1991.

[15] Nagata, S. Apoptosis by death factor. *Cell, 88:* 355-365, 1997.

[16] Vaux, D. L., Haecker, G., and Strasser, A. An evolutionary perspective on apoptosis. *Cell, 76:* 777-779, 1994.

[17] Cryns, V. and Yuan, J. Proteases to die for. Genes Dev, *12:* 1551-1570, 1998.

[18] Stennicke, H. R. and Salvesen, G. S. Properties of the caspases. *Biochim Biophys Acta, 1387:* 17-31, 1998.

[19] Cho, S. G. and Choi, E. J. Apoptotic signaling pathways: caspases and stress-activated protein kinases. *J Biochem Mol Biol, 35:* 24-27, 2002.

[20] Wright, C. W. and Duckett, C. S. Reawakening the cellular death program in neoplasia through the therapeutic blockade of IAP function. *J Clin Invest, 115:* 2673-2678, 2005.

[21] Roy, N., Deveraux, Q. L., Takahashi, R., Salvesen, G. S., and Reed, J. C. The c-IAP-1 and c-IAP-2 proteins are direct inhibitors of specific caspases. *EMBO J, 16:* 6914-6925, 1997.

[22] Deveraux, Q. L., Stennicke, H. R., Salvesen, G. S., and Reed, J. C. Endogenous inhibitors of caspases. *J Clin Immunol, 19:* 388-398, 1999.

[23] Deveraux, Q. L. and Reed, J. C. IAP family proteins--suppressors of apoptosis. *Genes Dev, 13:* 239-252, 1999.

[24] Krajewska, M., Krajewski, S., Banares, S., Huang, X., Turner, B., Bubendorf, L., Kallioniemi, O. P., Shabaik, A., Vitiello, A., Peehl, D., Gao, G. J., and Reed, J. C. Elevated expression of inhibitor of apoptosis proteins in prostate cancer. *Clin Cancer Res, 9:* 4914-4925, 2003.

[25] Kishi, H., Igawa, M., Kikuno, N., Yoshino, T., Urakami, S., and Shiina, H. Expression of the survivin gene in prostate cancer: correlation with clinicopathological characteristics, proliferative activity and apoptosis. *J Urol, 171:* 1855-1860, 2004.

[26] Avila, D. M., Zoppi, S., and McPhaul, M. J. The androgen receptor (AR) in syndromes of androgen insensitivity and in prostate cancer. *J Steroid Biochem Mol Biol, 76:* 135-142, 2001.

[27] Montgomery, J. S., Price, D. K., and Figg, W. D. The androgen receptor gene and its influence on the development and progression of prostate cancer. *J Pathol, 195:* 138-146, 2001.

[28] Culig, Z., Hobisch, A., Cronauer, M. V., Radmayr, C., Trapman, J., Hittmair, A., Bartsch, G., and Klocker, H. Androgen receptor activation in prostatic tumor cell lines

by insulin-like growth factor-I, keratinocyte growth factor, and epidermal growth factor. *Cancer Res, 54:* 5474-5478, 1994.

[29]   Isaacs, J. T., Lundmo, P. I., Berges, R., Martikainen, P., Kyprianou, N., and English, H. F. Androgen regulation of programmed death of normal and malignant prostatic cells. *J Androl, 13:* 457-464, 1992.

[30]   Berges, R. R., Vukanovic, J., Epstein, J. I., CarMichel, M., Cisek, L., Johnson, D. E., Veltri, R. W., Walsh, P. C., and Isaacs, J. T. Implication of cell kinetic changes during the progression of human prostatic cancer. *Clin Cancer Res, 1:* 473-480, 1995.

[31]   Feldman, B. J. and Feldman, D. The development of androgen-independent prostate cancer. *Nat Rev Cancer, 1:* 34-45, 2001.

[32]   Visakorpi, T. The molecular genetics of prostate cancer. *Urology, 62:* 3-10, 2003.

[33]   Amanatullah, D. F., Reutens, A. T., Zafonte, B. T., Fu, M., Mani, S., and Pestell, R. G. Cell-cycle dysregulation and the molecular mechanisms of prostate cancer. *Front Biosci, 5:* D372-390, 2000.

[34]   Gleave, M., Miyake, H., and Chi, K. Beyond simple castration: targeting the molecular basis of treatment resistance in advanced prostate cancer. *Cancer Chemother Pharmacol, 56 Suppl 1:* 47-57, 2005.

[35]   Mimata, H., Satoh, F., Hanada, T., Kasagi, Y., Sakamoto, S., Hamada, Y., and Nomura, Y. Expression of the IAP gene family in prostate cancer. *Eur Urol, 37 Suppl 2:* 104, 2000.

[36]   McEleny, K. R., Watson, R. W., Coffey, R. N., O'Neill, A. J., and Fitzpatrick, J. M. Inhibitors of apoptosis proteins in prostate cancer cell lines. *Prostate, 51:* 133-140, 2002.

[37]   Algeciras-Schimnich, A., Barnhart, B. C., and Peter, M. E. Apoptosis-independent functions of killer caspases. *Curr Opin Cell Biol, 14:* 721-726, 2002.

[38]   Salvesen, G. S. and Dixit, V. M. Caspase activation: the induced-proximity model. *Proc Natl Acad Sci U S A, 96:* 10964-10967, 1999.

[39]   Tsujimoto, Y. Bcl-2 family of proteins: life-or-death switch in mitochondria. *Biosci Rep, 22:* 47-58, 2002.

[40]   Li, P., Nijhawan, D., Budihardjo, I., Srinivasula, S. M., Ahmad, M., Alnemri, E. S., and Wang, X. Cytochrome c and dATP-dependent formation of Apaf-1/caspase-9 complex initiates an apoptotic protease cascade. *Cell, 91:* 479-489, 1997.

[41]   Luo, X., Budihardjo, I., Zou, H., Slaughter, C., and Wang, X. Bid, a Bcl2 interacting protein, mediates cytochrome c release from mitochondria in response to activation of cell surface death receptors. *Cell, 94:* 481-490, 1998.

[42]   van Gurp, M., Festjens, N., van Loo, G., Saelens, X., and Vandenabeele, P. Mitochondrial intermembrane proteins in cell death. *Biochem Biophys Res Commun, 304:* 487-497, 2003.

[43]   Leaman, D. W., Chawla-Sarkar, M., Vyas, K., Reheman, M., Tamai, K., Toji, S., and Borden, E. C. Identification of X-linked inhibitor of apoptosis-associated factor-1 as an interferon-stimulated gene that augments TRAIL Apo2L-induced apoptosis. *J Biol Chem, 277:* 28504-28511, 2002.

[44]   Thornberry, N. A. and Lazebnik, Y. Caspases: enemies within. *Science, 281:* 1312-1316, 1998.

[45] Sohn, J. H., Kim, D. H., Choi, N. G., Park, Y. E., and Ro, J. Y. Caspase-3/CPP32 immunoreactivity and its correlation with frequency of apoptotic bodies in human prostatic carcinomas and benign nodular hyperplasias. *Histopathology, 37:* 555-560, 2000.

[46] Winter, R. N., Kramer, A., Borkowski, A., and Kyprianou, N. Loss of caspase-1 and caspase-3 protein expression in human prostate cancer. *Cancer Res, 61:* 1227-1232, 2001.

[47] O'Neill, A. J., Boran, S. A., O'Keane, C., Coffey, R. N., Hegarty, N. J., Hegarty, P., Gaffney, E. F., Fitzpatrick, J. M., and Watson, R. W. Caspase 3 expression in benign prostatic hyperplasia and prostate carcinoma. *Prostate, 47:* 183-188, 2001.

[48] Srikanth, S. and Kraft, A. S. Inhibition of caspases by cytokine response modifier A blocks androgen ablation-mediated prostate cancer cell death in vivo. *Cancer Res, 58:* 834-839, 1998.

[49] Reed, J. C. The Survivin saga goes in vivo. *J Clin Invest, 108:* 965-969, 2001.

[50] Deveraux, Q. L., Roy, N., Stennicke, H. R., Van Arsdale, T., Zhou, Q., Srinivasula, S. M., Alnemri, E. S., Salvesen, G. S., and Reed, J. C. IAPs block apoptotic events induced by caspase-8 and cytochrome c by direct inhibition of distinct caspases. *EMBO J, 17:* 2215-2223, 1998.

[51] Tamm, I., Wang, Y., Sausville, E., Scudiero, D. A., Vigna, N., Oltersdorf, T., and Reed, J. C. IAP-family protein survivin inhibits caspase activity and apoptosis induced by Fas (CD95), Bax, caspases, and anticancer drugs. *Cancer Res, 58:* 5315-5320, 1998.

[52] Chen, Z., Naito, M., Hori, S., Mashima, T., Yamori, T., and Tsuruo, T. A human IAP-family gene, apollon, expressed in human brain cancer cells. *Biochem Biophys Res Commun, 264:* 847-854, 1999.

[53] Kasof, G. M. and Gomes, B. C. Livin, a novel inhibitor of apoptosis protein family member. *J Biol Chem, 276:* 3238-3246, 2001.

[54] Richter, B. W., Mir, S. S., Eiben, L. J., Lewis, J., Reffey, S. B., Frattini, A., Tian, L., Frank, S., Youle, R. J., Nelson, D. L., Notarangelo, L. D., Vezzoni, P., Fearnhead, H. O., and Duckett, C. S. Molecular cloning of ILP-2, a novel member of the inhibitor of apoptosis protein family. *Mol Cell Biol, 21:* 4292-4301, 2001.

[55] Yang, Y., Fang, S., Jensen, J. P., Weissman, A. M., and Ashwell, J. D. Ubiquitin protein ligase activity of IAPs and their degradation in proteasomes in response to apoptotic stimuli. *Science, 288:* 874-877, 2000.

[56] Deveraux, Q. L., Takahashi, R., Salvesen, G. S., and Reed, J. C. X-linked IAP is a direct inhibitor of cell-death proteases. *Nature, 388:* 300-304, 1997.

[57] Nomura, T., Mimata, H., Takeuchi, Y., Yamamoto, H., Miyamoto, E., and Nomura, Y. The X-linked inhibitor of apoptosis protein inhibits taxol-induced apoptosis in LNCaP cells. *Urol Res, 31:* 37-44, 2003.

[58] Rothe, M., Pan, M. G., Henzel, W. J., Ayres, T. M., and Goeddel, D. V. The TNFR2-TRAF signaling complex contains two novel proteins related to baculoviral inhibitor of apoptosis proteins. *Cell, 83:* 1243-1252, 1995.

[59] Wang, C. Y., Mayo, M. W., Korneluk, R. G., Goeddel, D. V., and Baldwin, A. S., Jr. NF-kappaB antiapoptosis: induction of TRAF1 and TRAF2 and c-IAP1 and c-IAP2 to suppress caspase-8 activation. *Science, 281:* 1680-1683, 1998.

[60] Stehlik, C., de Martin, R., Kumabashiri, I., Schmid, J. A., Binder, B. R., and Lipp, J. Nuclear factor (NF)-kappaB-regulated X-chromosome-linked iap gene expression protects endothelial cells from tumor necrosis factor alpha-induced apoptosis. *J Exp Med, 188:* 211-216, 1998.

[61] Sanna, M. G., da Silva Correia, J., Ducrey, O., Lee, J., Nomoto, K., Schrantz, N., Deveraux, Q. L., and Ulevitch, R. J. IAP suppression of apoptosis involves distinct mechanisms: the TAK1/JNK1 signaling cascade and caspase inhibition. *Mol Cell Biol, 22:* 1754-1766, 2002.

[62] Reed, J. C. and Bischoff, J. R. BIRinging chromosomes through cell division--and survivin' the experience. *Cell, 102:* 545-548, 2000.

[63] Papapetropoulos, A., Fulton, D., Mahboubi, K., Kalb, R. G., O'Connor, D. S., Li, F., Altieri, D. C., and Sessa, W. C. Angiopoietin-1 inhibits endothelial cell apoptosis via the Akt/survivin pathway. *J Biol Chem, 275:* 9102-9105, 2000.

[64] Wrzesien-Kus, A., Smolewski, P., Sobczak-Pluta, A., Wierzbowska, A., and Robak, T. The inhibitor of apoptosis protein family and its antagonists in acute leukemias. *Apoptosis, 9:* 705-715, 2004.

[65] Gray, C. W., Ward, R. V., Karran, E., Turconi, S., Rowles, A., Viglienghi, D., Southan, C., Barton, A., Fantom, K. G., West, A., Savopoulos, J., Hassan, N. J., Clinkenbeard, H., Hanning, C., Amegadzie, B., Davis, J. B., Dingwall, C., Livi, G. P., and Creasy, C. L. Characterization of human HtrA2, a novel serine protease involved in the mammalian cellular stress response. *Eur J Biochem, 267:* 5699-5710, 2000.

[66] Hegde, R., Srinivasula, S. M., Zhang, Z., Wassell, R., Mukattash, R., Cilenti, L., DuBois, G., Lazebnik, Y., Zervos, A. S., Fernandes-Alnemri, T., and Alnemri, E. S. Identification of Omi/HtrA2 as a mitochondrial apoptotic serine protease that disrupts inhibitor of apoptosis protein-caspase interaction. *J Biol Chem, 277:* 432-438, 2002.

[67] Ekert, P. G., Silke, J., Hawkins, C. J., Verhagen, A. M., and Vaux, D. L. DIABLO promotes apoptosis by removing MIHA/XIAP from processed caspase 9. *J Cell Biol, 152:* 483-490, 2001.

[68] Hu, S. and Yang, X. Cellular inhibitor of apoptosis 1 and 2 are ubiquitin ligases for the apoptosis inducer Smac/DIABLO. *J Biol Chem, 278:* 10055-10060, 2003.

[69] Yang, Q. H. and Du, C. Smac/DIABLO selectively reduces the levels of c-IAP1 and c-IAP2 but not that of XIAP and livin in HeLa cells. *J Biol Chem, 279:* 16963-16970, 2004.

[70] Liston, P., Fong, W. G., Kelly, N. L., Toji, S., Miyazaki, T., Conte, D., Tamai, K., Craig, C. G., McBurney, M. W., and Korneluk, R. G. Identification of XAF1 as an antagonist of XIAP anti-Caspase activity. *Nat Cell Biol, 3:* 128-133, 2001.

[71] Tanaka, K., Iwamoto, S., Gon, G., Nohara, T., Iwamoto, M., and Tanigawa, N. Expression of survivin and its relationship to loss of apoptosis in breast carcinomas. *Clin Cancer Res, 6:* 127-134, 2000.

[72] Kato, J., Kuwabara, Y., Mitani, M., Shinoda, N., Sato, A., Toyama, T., Mitsui, A., Nishiwaki, T., Moriyama, S., Kudo, J., and Fujii, Y. Expression of survivin in esophageal cancer: correlation with the prognosis and response to chemotherapy. *Int J Cancer, 95:* 92-95, 2001.

[73] Lu, C. D., Altieri, D. C., and Tanigawa, N. Expression of a novel antiapoptosis gene, survivin, correlated with tumor cell apoptosis and p53 accumulation in gastric carcinomas. *Cancer Res, 58:* 1808-1812, 1998.

[74] Kawasaki, H., Altieri, D. C., Lu, C. D., Toyoda, M., Tenjo, T., and Tanigawa, N. Inhibition of apoptosis by survivin predicts shorter survival rates in colorectal cancer. *Cancer Res, 58:* 5071-5074, 1998.

[75] Adida, C., Berrebi, D., Peuchmaur, M., Reyes-Mugica, M., and Altieri, D. C. Anti-apoptosis gene, survivin, and prognosis of neuroblastoma. *Lancet, 351:* 882-883, 1998.

[76] Monzo, M., Rosell, R., Felip, E., Astudillo, J., Sanchez, J. J., Maestre, J., Martin, C., Font, A., Barnadas, A., and Abad, A. A novel anti-apoptosis gene: Re-expression of survivin messenger RNA as a prognosis marker in non-small-cell lung cancers. *J Clin Oncol, 17:* 2100-2104, 1999.

[77] Swana, H. S., Grossman, D., Anthony, J. N., Weiss, R. M., and Altieri, D. C. Tumor content of the antiapoptosis molecule survivin and recurrence of bladder cancer. *N Engl J Med, 341:* 452-453, 1999.

[78] Sui, L., Dong, Y., Ohno, M., Watanabe, Y., Sugimoto, K., and Tokuda, M. Survivin expression and its correlation with cell proliferation and prognosis in epithelial ovarian tumors. *Int J Oncol, 21:* 315-320, 2002.

[79] Shariat, S. F., Lotan, Y., Saboorian, H., Khoddami, S. M., Roehrborn, C. G., Slawin, K. M., and Ashfaq, R. Survivin expression is associated with features of biologically aggressive prostate carcinoma. *Cancer, 100:* 751-757, 2004.

[80] Tamm, I., Kornblau, S. M., Segall, H., Krajewski, S., Welsh, K., Kitada, S., Scudiero, D. A., Tudor, G., Qui, Y. H., Monks, A., Andreeff, M., and Reed, J. C. Expression and prognostic significance of IAP-family genes in human cancers and myeloid leukemias. *Clin Cancer Res, 6:* 1796-1803, 2000.

[81] Zhang, M., Latham, D. E., Delaney, M. A., and Chakravarti, A. Survivin mediates resistance to antiandrogen therapy in prostate cancer. *Oncogene, 24:* 2474-2482, 2005.

[82] Adams, R. R., Carmena, M., and Earnshaw, W. C. Chromosomal passengers and the (aurora) ABCs of mitosis. *Trends Cell Biol, 11:* 49-54, 2001.

[83] Zhivotovsky, B., Joseph, B., and Orrenius, S. Tumor radiosensitivity and apoptosis. *Exp Cell Res, 248:* 10-17, 1999.

[84] Hagan, M., Wang, L., Hanley, J. R., Park, J. S., and Dent, P. Ionizing radiation-induced mitogen-activated protein (MAP) kinase activation in DU145 prostate carcinoma cells: MAP kinase inhibition enhances radiation-induced cell killing and G2/M-phase arrest. *Radiat Res, 153:* 371-383, 2000.

[85] Holcik, M., Yeh, C., Korneluk, R. G., and Chow, T. Translational upregulation of X-linked inhibitor of apoptosis (XIAP) increases resistance to radiation induced cell death. *Oncogene, 19:* 4174-4177, 2000.

[86] Asanuma, K., Moriai, R., Yajima, T., Yagihashi, A., Yamada, M., Kobayashi, D., and Watanabe, N. Survivin as a radioresistance factor in pancreatic cancer. *Jpn J Cancer Res, 91:* 1204-1209, 2000.

[87] Tannock, I. F., de Wit, R., Berry, W. R., Horti, J., Pluzanska, A., Chi, K. N., Oudard, S., Theodore, C., James, N. D., Turesson, I., Rosenthal, M. A., and Eisenberger, M. A.

Docetaxel plus prednisone or mitoxantrone plus prednisone for advanced prostate cancer. *N Engl J Med, 351:* 1502-1512, 2004.

[88] Trivedi, C., Redman, B., Flaherty, L. E., Kucuk, O., Du, W., Heilbrun, L. K., and Hussain, M. Weekly 1-hour infusion of paclitaxel. Clinical feasibility and efficacy in patients with hormone-refractory prostate carcinoma. *Cancer, 89:* 431-436, 2000.

[89] Li, J., Feng, Q., Kim, J. M., Schneiderman, D., Liston, P., Li, M., Vanderhyden, B., Faught, W., Fung, M. F., Senterman, M., Korneluk, R. G., and Tsang, B. K. Human ovarian cancer and cisplatin resistance: possible role of inhibitor of apoptosis proteins. *Endocrinology, 142:* 370-380, 2001.

[90] Nomura, T., Mimata, H., Yamasaki, M., and Nomura, Y. Cisplatin inhibits the expression of X-linked inhibitor of apoptosis protein in human LNCaP cells. *Urol Oncol, 22:* 453-460, 2004.

[91] Chandele, A., Prasad, V., Jagtap, J. C., Shukla, R., and Shastry, P. R. Upregulation of survivin in G2/M cells and inhibition of caspase 9 activity enhances resistance in staurosporine-induced apoptosis. *Neoplasia, 6:* 29-40, 2004.

[92] Datta, R., Oki, E., Endo, K., Biedermann, V., Ren, J., and Kufe, D. XIAP regulates DNA damage-induced apoptosis downstream of caspase-9 cleavage. *J Biol Chem, 275:* 31733-31738, 2000.

[93] Schlette, E. J., Medeiros, L. J., Goy, A., Lai, R., and Rassidakis, G. Z. Survivin expression predicts poorer prognosis in anaplastic large-cell lymphoma. *J Clin Oncol, 22:* 1682-1688, 2004.

[94] Sasaki, H., Sheng, Y., Kotsuji, F., and Tsang, B. K. Down-regulation of X-linked inhibitor of apoptosis protein induces apoptosis in chemoresistant human ovarian cancer cells. *Cancer Res, 60:* 5659-5666, 2000.

[95] Matsumiya, T., Imaizumi, T., Yoshida, H., Kimura, H., and Satoh, K. Cisplatin inhibits the expression of X-chromosome-linked inhibitor of apoptosis protein in an oral carcinoma cell line. *Oral Oncol, 37:* 296-300, 2001.

[96] Roa, W. H., Chen, H., Fulton, D., Gulavita, S., Shaw, A., Th'ng, J., Farr-Jones, M., Moore, R., and Petruk, K. X-linked inhibitor regulating TRAIL-induced apoptosis in chemoresistant human primary glioblastoma cells. *Clin Invest Med, 26:* 231-242, 2003.

[97] Nomura, T., Yamasaki, M., Nomura, Y., and Mimata, H. Expression of the inhibitors of apoptosis proteins in cisplatin-resistant prostate cancer cells. *Oncol Rep, 14:* 993-997, 2005.

[98] Amantana, A., London, C. A., Iversen, P. L., and Devi, G. R. X-linked inhibitor of apoptosis protein inhibition induces apoptosis and enhances chemotherapy sensitivity in human prostate cancer cells. *Mol Cancer Ther, 3:* 699-707, 2004.

[99] McEleny, K., Coffey, R., Morrissey, C., Williamson, K., Zangemeister-Wittke, U., Fitzpatrick, J. M., and Watson, R. W. An antisense oligonucleotide to cIAP-1 sensitizes prostate cancer cells to fas and TNFalpha mediated apoptosis. *Prostate, 59:* 419-425, 2004.

In: Cell Apoptosis Research Trends   ISBN: 1-60021-424-X
Editor: Charles V. Zhang, pp. 169-206   © 2007 Nova Science Publishers, Inc

Chapter VI

Different Response of Normal and Cancer Human Colon Epithelial Cells to Dietary Fatty Acids and Endogenous Apoptotic Regulators of the TNF Family

Jiřina Hofmanová, Alena Vaculová,
Martina Hýžďalová, and Alois Kozubík

Laboratory of Cytokinetics, Institute of Biophysics, Academy of Sciences of the Czech
Republic, Královopolská 135, 612 65 Brno, Czech Republic

ABSTRACT

In the colon, signals from nutritional compounds (like dietary lipids) and endogenous factors regulating cell growth, differentiation and apoptosis are integrated within the cell and have a substantial impact determining the final phenotype, metabolism and kinetics of colon epithelial cell populations. Moreover, altered response of transformed cell to these signals is supposed during colon carcinogenesis.

Using human colon epithelial cells *in vitro*, derived either from normal human fetal colon tissue (FHC) or from adenocarcinoma (HT-29) the response to (i) TNF-α, anti-Fas antibody and TRAIL and (ii) polyunsaturated fatty acids of n-6 (arachidonic acid, AA) or n-3 (docosahexaenoic acid, DHA) series, short chain fatty acid - sodium butyrate (NaBt) and their mutual interactions have been investigated and compared. Generally, even if both cell lines showed some similar characteristics (morphology, extensive proliferation, not full differentiation), they differ in their ability to respond to endogenous inductors of apoptosis like TRAIL as well as to exogenous dietary factors like PUFAs or NaBt produced from fiber.

The response of HT-29 and FHC cells to TNF-α, anti-Fas antibody and TRAIL was compared. Both cell lines did not respond to TNF-α, they expressed a limited sensitivity to anti-Fas antibody and different response to TRAIL. Some molecular determinants of limited sensitivity of HT-29 and very low sensitivity of FHC cells to TRAIL were determined. This could be attributed to a lower activity of FHC cell mitochondrial pathway components followed by decreased caspase activities and PARP cleavage.

PUFAs and NaBt, working individually or together, initiated higher apoptotic response in FHC than in HT-29 cells. Moreover, NaBt also induced higher differentiation response in FHC cells, although the proliferation of both cell types was similarly affected. Combined treatment of cells with either AA or DHA and NaBt resulted in potentiation of apoptotic effects especially in FHC cells, DHA being more effective. The relationship between these effects and different lipid and oxidative metabolism of normal and cancer cells (changes of lipids in plasma as well as mitochondrial membrane, accumulation of lipid droplets in cytoplasm associated with changes of MMP and ROS production) is supposed.

INTRODUCTION

Colorectal cancer (CRC) is one of the main causes of cancer-related death in western countries. Most of CRC are derived from pre-malignant adenomas (polyps), which are derived from normal colonic epithelium. The transition from normal state to malignancy includes accumulation of mutation in many oncogenes and tumour suppressor genes accompanied by epigenetic changes and promoted by factors operating by non-genotoxic way. These events together result in dysregulation of cellular proliferation, differentiation and apoptosis, impaired tissue homeostasis, and development of cancer.

It is now clear that in addition to endogenous regulators maintaining cell and tissue homeostasis, exogenous dietary factors interacting with colonic epithelium play their role in the etiology of CRC. Individual components of the lumenal contents are being investigated to clarify the role and possible mechanisms by which dietary factors may influence colonic health. An important topic represents possible different effects of endogenous as well as dietary factors on gene expression, proliferation, differentiation and apoptosis of normal colonic cells and cells in various stages of malignant transformation.

ENDOGENOUS REGULATORS OF COLONIC CELL APOPTOSIS

The epithelium of the mammalian intestine is a continuously renewing tissue serving a number of critical physiological functions. Dynamic balance between cell production at the base and cell death at the surface of the colonic crypts is precisely regulated by a number of naturally occurring endogenous factors, such as hormones and cytokines. Apoptosis and its special form - anoikis (detachment-induced apoptosis) may represent the final differentiation step for colonic enterocytes, may be initiated by depletion of survival factors or may represent a mean of the eliminating damaged cells. Suppression of apoptosis can cause cellular transformation and favour progression at every stage of the adenoma-carcinoma sequence [1].

Cytokines from the tumour necrosis factor (TNF) family have been identified as participants in the regulation of apoptosis [2] despite the fact that their role in regulating epithelial cell turnover is not fully understood and their effect on colon cancer cells and the associated molecular and cellular mechanisms have yet to be elucidated [3]. TNF-α, Fas ligand and TNF-related apoptosis-inducing ligand (TRAIL) induce apoptosis by binding to their respective death receptors (DR) possessing intracellular death domains, which recruit certain adaptor molecules to form the "death-inducing signaling complex" (DISC) activating the apoptotic caspase cascade. Apoptosis is induced by trimerisation of these receptors by their natural ligands constituting a complementary TNF family [4]. By caspase-8 activation, subsequent downstream signals are started, either through direct activation of effector caspases (type I cells - extrinsic pathway) or by transferring a signal to mitochondria (type II cells - intrinsic pathway) mediated by cleavage of Bid protein [5]. Changes in mitochondria (reactive oxygen species - ROS - production, decrease of mitochondrial membrane potential - MMP) are associated with the activity of pro- and anti-apoptotic proteins of the Bcl-2 family and initiate subsequent events (release of cytochrome c, Smac/DIABLO) leading to activation of caspase-9, effector caspases, death-substrate cleavage, and finally cell death [6].

TNF-α plays an important role in immune responses in inflammatory conditions of the intestinal mucosa. In spite of the fact that it generates potent antitumour activity both *in vivo* and *in vitro* [7], many cancer cells are resistant to TNF-mediated killing. The cell surface receptor CD95 (APO-1/Fas) initiates an apoptotic signal to apoptosis-sensitive cells when oligomerised by natural ligand, CD95L, or anti-Fas antibody [8]. Whereas CD95 is expressed in every colonocyte of normal colon mucosa, it is down-regulated or lost in the majority of colon carcinomas. Normal cells are relatively sensitive to Fas-mediated apoptosis, but the biological significance of CD95 expression in the gut is still under discussion [9]. In contrast to the normal colonic epithelium, many colon carcinoma cell lines are relatively resistant to CD95 cross-link [10]. The therapeutic use of the FasL/Fas or the TNF-α /TNFR system in cancer treatment has been hampered by severe side effects. The systemic administration of TNF-α causes a septic shock-like response possibly mediated by NF-kB activation, and the injection of agonist Ab to Fas can be lethal.

In contrast to TNF-α and FasL, TRAIL induces apoptosis in a wide variety of transformed cell lines, but seems to have little or no cytotoxic effect on most normal cells *in vitro* and *in vivo.* Moreover, TRAIL induces apoptosis irrespective of p53 status when the wild type of DR4 and 5 is expressed [11]. However, despite the fact that an increase in the sensitivity to TRAIL-induced apoptosis during adenoma to colon carcinoma transition was detected [12], many types of cancer cells become resistant to TRAIL [13]. Induction of apoptosis by TRAIL is mediated by interaction with the two death receptors TRAIL-R1 (DR4) and TRAIL–R2 (DR5) and the mechanism of induction of apoptosis is believed to be similar to that of TNF-α and FasL [4]. TRAIL-R1 showed a tissue distribution very similar to that of CD95. It was expressed in epithelial cells all along the crypt axis and in some interstitial mononuclear cells. Interestingly, DR4 and DR5 were co-expressed mainly in epithelial cells of the crypt mouth and the surface epithelium where senescent cells undergo apoptosis. However, isolated crypt cells are completely resistant to TRAIL *in vitro* [14]. Some cells seem to be protected from TRAIL-induced apoptosis by their expression of decoy receptors (DcR1, DcR2) which do not transduce apoptotic signals as well as at the level of some molecules involved in intracellular signaling pathway [11]. Since TRAIL may be useful as a therapeutic agent in cancer, particular attention to different sensitivity of normal and cancer cells and to its molecular determinants needs to be considered to optimize such therapy.

Dietary Lipid Components

Polyunsaturated fatty acids (PUFAs)

A lot of experimental and epidemiological studies support the idea that lipid nutritional components play an important role in colon cancer etiology [15]. It was shown that in addition to increased supply of calories, quantitative and qualitative content of PUFAs and the ratio of n-3 a n-6 type is highly important [16]. In spite of much contradiction in the literature, it is generally thought that high calorie and fat intake represents a risk factor in cancer development and that n-6 PUFAs (from plant oils rich in linoleic acid) can be positively correlated with this process. On the other hand, PUFAs of the n-3 series, alpha-linolenic acid and especially eicosapentaenoic (EPA) and docosahexaenoic (DHA) acids (rich in fish oil), were shown to be protective against cancer development [17]. However, the epidemiological data searching association between fish and fish oil consumption and cancer risk and incidence seem to be inconsistent. Recently great attention has been paid to the reasons for this discrepancy [18]. From these reports necessity arises for further studies to elucidate and verify cell and molecular mechanisms of n-6 and n-3 PUFA action for better understanding of their role in human cancer risk [19].

Cell culture systems *in vitro* allow to investigate relationship between diet and cancer and detail mechanisms of actions of individual dietary components. When the cells are incubated with PUFAs, PUFAs are incorporated into cell membrane phospholipids and affect cellular functions including cell proliferation, differentiation and apoptosis. However, the susceptibility to individual fatty acids varies with the fatty acid concentration, exposure, and the cellular system investigated.

As a part of membrane phospholipids and due to their structural and physical properties, PUFAs can modulate structure and function of cell membranes. Changes of fatty acid spectrum in membranes may influence membrane fluidity, especially ligand-receptor interactions and other functions mediated by membrane [20]. PUFAs play an important role in transduction of signals from extracellular space and function as inter- and intracellular mediators and modulators of cellular signaling network [21]. In addition to changes on membrane level, the research is focused to changes of oxidative metabolism, i. e. reactive oxygen (ROS) and nitrogen species (RNS) production and lipid peroxidation, modulation of production of biologically active mediators – eicosanoids and control of the events in the nucleus, i. e. interaction with intracellular receptors and effects on gene transcription [22]. The effects of PUFAs on various levels of cell organization and their interaction with other endogenous or exogenous factors can finally significantly influence cell proliferation, differentiation and apoptosis of colon epithelial cells.

Once inside the cell, PUFAs and their metabolites are involved in many intracellular signaling pathways activating or inhibiting various components of this machinery and modulate gene expression [23]. Moreover, they may alter the levels and signaling pathways of other mediators, such as eicosanoids and cytokines [24].

In response to elevated fatty acids, lipid droplets (LDs) are rapidly formed in cytoplasm of cultured cells and *in vivo*. Conversion to neutral lipids and subsequent sequestration in LDs serve for stockpiling fatty acids and sterols that are used for metabolism, membrane synthesis (phospholipids and cholesterol) and steroid synthesis [25]. LDs appear to be complex, metabolically active organelles directly involved in membrane traffic and phospholipid recycling. They consist of a core of neutral lipids, predominantly triacylglycerols (TAG) or cholesteryl esters that are surrounded by a monolayer of PL and associated proteins. They are fundamental component of intracellular lipid homeostasis and serve as an important reservoir of signal molecules and inflammatory mediators. It has been observed that in addition to adipocytes, all other cell types have the ability to generate LDs and that abnormal accumulation of LDs occurs in a variety of pathological conditions. Lipid accumulation is thought to be important modulator of immune cell functions. Recently, increased LD formation was correlated with reduced cell number and its association with both differentiation and apoptosis was reported. Accumulation of cytosolic LDs during differentiation as well as apoptosis of human myeloid cells, inflammatory cells and lymphocytes was also reported [26] [27] [28]. However, the storage of fatty acids in LDs may protect cells from exposure to high concentrations of non-esterified fatty acids. Therefore more studies are needed to clarify the role of LDs and their association with cytokinetics of various cell types.

Susceptibility of double bonds to oxidation is the reason for production of various types of biologically active metabolites like eicosanoids, ROS and RNS, and lipid peroxides (LPs). Antiproliferative and apoptotic effects of PUFAs were shown to be associated with induction of oxidative stress. Supplementation of cultured cells with PUFAs enhanced production of hydrogen peroxide and superoxide and increased level of lipid peroxidation products [29]. LPs were shown to be involved in regulation of the cell cycle and proliferation [30].

Enzymatic conversion of PUFAs (especially arachidonic acid, AA, 20:4, ω-6) to local mediators eicosanoids plays an important role in a variety of physiological and pathological processes. Transformation of colon epithelial cells is characterized by elevated AA liberation and metabolism especially by overexpression of cyclooxygenase-2 (COX-2) [31]. Results of our investigation also suggest an important role of modulation of AA metabolism in cytokinetics in various cell systems using specific inhibitors of COXs, lipoxygenases or cytochrome P450 monooxygenase and their interaction with both cytokines and differentiation agents [32-35].

Mitochondria are affected early in the apoptotic process and are considered as strategic centers of cell death regulation. In mitochondria, fatty acids increase proton conductance and can promote opening of the permeability transition pores, which may cause release of mitochondrial apoptogenic proteins into cytosol [36]. These events are often associated with dissipation of MMP, increased ROS production and are regulated by proteins from Bcl-2 family [37]. PUFAs physically interact with mitochondrial membranes and alter their permeability by opening the permeability transition pores and decreasing MMP. Next, in colonocytes DHA is incorporated into cardiolipin, a mitochondrial inner membrane PL, which coincided with increasing unsaturation, induction of oxidative stress, release of cytochrome c and apoptosis [38, 39]. PUFAs were shown to significantly modulate the level of several Bcl-2 family proteins (e.g. Bid, Bcl-2) that interact with mitochondrial membrane lipids regulating apoptosis and cardiolipin has been proposed as a „lipid receptor" to Bid [40].

Mammalian organism is not able to synthesize PUFAs. Their availability depends on external supply in the diet and thus the changes of cell membrane composition become independent on the genome. PUFAs may assert in the initiation stage of carcinogenesis especially through enzymatic systems activating or metabolizing procarcinogens like prostglandin hours synthase and cytochrome P-450 and they play the main role in promotion stage of carcinogenesis as well as in metastatic process. However, specific role of n-6 and n-3, FAs in carcinogenesis and mechanisms of their effects on various cell types remain to be elucidated.

SHORT-CHAIN FATTY ACIDS (SCFA)

Epidemiological studies confirm a protective role of dietary fiber against CRC development. However, better understanding of changes in the luminal environment and the effects on colonic epithelium induced by fiber and/or its constituents is crucial for therapeutic intervention [41]. SCFAs are produced in gastrointestinal tract by anaerobic fermentation of dietary fiber. The most important is butyrate, which can significantly influence colonic cell behaviour [42]. It is generally thought that it serves as the main oxidative fuel for normal colonic epithelium, increases normal cell proliferation and its withdrawal results in apoptosis. On the other hand, sodium butyrate (NaBt) can decrease the proliferation of neoplastic colonocytes *in vitro* and *in vivo*, and induces their differentiation as well as apoptosis. Thus, butyrate could be effective in prevention and therapy of CRC when the appropriate cells would be exposed to its adequate dose. However, the data on butyrate effects on CRC are rather confusing. Butyrate can promote or inhibit proliferation as well as apoptosis in dependence upon the cell type, stage of differentiation, concentration, physiological conditions during the study, and the presence of other factors. However, the mechanisms of butyrate effects are still far from being clarified. A recent gene array and proteome analysis of human colon cancer cell lines revealed that the genes (mostly transcription factors) and proteins linked to the cell growth, apoptosis and oxidative metabolism are most significantly affected [43].

Together with others we have shown that mitochondria play an important role in NaBt induced apoptosis of colon cancer cell lines [44]. A necessity of mitochondrial function for initiation and maintenance of SCFA-induced cell cycle arrest and the role of MMP in the induction of apoptosis was demonstrated. Mitochondrial function was required for butyrate-induced cell cycle arrest and apoptosis of colonic carcinoma cells. Initiation of at least some apoptotic pathway has been linked to dissipation of MMP and plays a pivotal role in coordinating these processes. Heerdt et al. [45] [46] showed that butyrate induces mitochondrial gene expression, mitochondrial enzymatic activity, a transient, simultaneous arrest of cells in G0/G1 and G2/M of the cell cycle affecting cyclin D1 and cdk inhibitors like p21 and p27. Moreover, it has been shown that the effects of butyrate are tightly linked to its mitochondrial beta-oxidation and dependent upon mitochondrial function. Butyrate-initiated dissipation of the MMP is required for progression through apoptotic cascade and intact MMP is essential for the initiation and maintenance of the early p21 induction and the subsequent arrest of cells in G0/G1. Thus there is evidence that mitochondria plays a critical role in integrating multiple signals that coordinate inherent p53-independent proliferation and apoptosis pathways in human colonic epithelial cells [47].

Butyrate was shown to exert modulatory effects on nuclear proteins such as selective inhibition of histone phosphorylation, hypermethylation of cytosine residues in DNA, and histone hyperacetylation. The most commonly reported mechanism by which it modulates gene expression involves an alteration of chromatin structure subsequent to increased histone acetylation [48]. Butyrate differentiation effects are associated with PPARγ activation and are enhanced by inhibition of PI3K. During stress, inflammatory and immune responses, cells rapidly activate a transcription factor NF-κB. NaBt prevention of NF-κB activation is mediated by suppressing proteasome activity [49]. However, some effects are independent on histone acetylation, e. g. effects on nitric oxide synthase or COX-2 transcriptional activity [50]. Further studies are necessary to determine the precise mechanisms of butyrate effects and its applicability to the clinical situation in order to have an impact in the incidence of CRC.

It can be summarized that dietary lipid compounds can significantly affect maintenance of cell population homeostasis and thus they are involved in the processes leading to pathological states including cancer. They may function mainly by non-genotoxic mechanisms during cancer promotion and progression [51]. However, it is difficult to establish their direct clinical effects. Therefore, the better understanding of the pathways and mechanisms by which lipids affect cell functions, the better possibilities are to use these agents in prevention and therapy.

Our previous studies using butyrate, PUFAs and inhibitors of their metabolism in cell lines derived from human leukemia, colon adenocarcinoma or mouse fibrosarcoma demonstrated significant effects of these compounds on proliferation, differentiation and apoptosis and modulation of significant signaling pathways. Particularly effective were the effects after combined treatments. Our results showed potentiation of apoptosis in leukemia cell lines after combined treatment with inhibitors of AA metabolism and TNF [33] and potentiation of differentiation induced by retinoic acid or DMSO [52]. As demonstrated by us and others, the effects of TNF family molecules on cancer cells could be significantly potentiated by NaBt [32] [53] or polyunsaturated fatty acids [34] [35].

Summarizing our experience and research results we recognized that cell response to molecules regulating cytokinetics as well as lipid compounds is cell type specific and may depend on the level of cell transformation. Therefore, the investigation of differences in the response of normal and cancer colon epithelial cells could be helpful in finding molecular mechanisms responsible for cell resistance to TNF family apoptotic inducers and in clarifying the effects of various dietary lipid components on colonic epithelium. This type of research helps to design right approaches in cancer prevention and therapeutic application of these compounds.

Here, we report the results obtained by comparison of the response of cell lines derived from human colon adenocarcinoma (HT-29) and normal human fetal colon (FHC) to i) TNF-α, Fas antibody (anti-Fas), and TRAIL, ii) AA, DHA, and iii) NaBt. Moreover, the possibility of mutual interaction of PUFAs and NaBt is demonstrated.

MATERIAL AND METHODS

Cell cultures. Human colon adenocarcinoma HT-29 cells (ATCC; Rockville, MD, USA) were cultured in 25 mm^2 flasks (TPP; Trasadigen, Switzerland) in McCoy's 5A medium (Sigma Aldrich; Prague, Czech Rep.) supplemented with gentamycin (50 mg/l; Serva Electrophoresis GmbH; Heidelberg, Germany) and 10% fetal calf serum (FCS) (PAN Biotech GmbH; Aidenbach, Germany). The normal human colon FHC cells (CRL-1831; ATCC) were cultured in a 1:1 mixture of Ham's F12 and Dulbecco's modified Eagle's mediums (Sigma Aldrich) containing HEPES (25 mM), cholera toxin (10 ng/ml; Calbiochem-Novabiochem Corp.; La Jolla, CA, USA), insulin (5 μg/ml), transferrin (5 μg/ml) and hydrocortisone (100 ng/ml; all Sigma Aldrich), and supplemented with 10% fetal calf serum. The cultures were maintained at 37 °C in 5 % CO$_2$ and 95 % humidity. They were passaged twice a week after exposure to trypsin/EDTA (0.05/0.02 %; PAN Biotech, GmbH).

Application of the Agents

Cytokines: 24 hours after seeding the cells were treated with TNF-α (15 ng/ml; Sigma Aldrich), anti-Fas antibody (CH-11; 200 ng/ml; Immunotech; Marseille, France) or TRAIL (100 ng/ml; provided by Dr. L. Anděra, IMG Prague, Czech Rep.). For new protein synthesis inhibition, cycloheximide (CHX; 5 μg/ml; Sigma Aldrich) was used. MEK 1/2 kinases were inhibited by U0126 (10 μM; Cell Signaling Technology, Inc.; Beverly, MA, USA).

PUFAs and butyrate: 72 hours after seeding, the medium was changed and the cells were treated with AA or DHA (50 μM) or sodium butyrate (3 mM) for 48 or 72 h. The resulting concentration of FCS during treatments was 5%. AA and DHA (Sigma-Aldrich Corp., St. Louis, MO, USA) were dissolved in 96% ethanol and stored as stock solution (100 mM) under nitrogen at -80 °C. For the experiments, fatty acids were diluted with the growth medium. The control cells were treated with an appropriate concentration of ethanol. This type of treatment did not influence any of the parameters tested. NaBt (Sigma) was dissolved in PBS and then diluted to growth medium.

Cell counts, floating cell quantification and viability assays. Floating and adherent cells were counted separately using a Coulter Counter (model ZM; Beckman Coulter, Inc.; Fullerton, CA, USA), and the amount of floating cells was expressed as a percentage of the total cell number. Cell viability was determined microscopically by eosin (0.15 %) dye exclusion assay.

Cell cycle analysis. Fixed cells (70 % ethanol) were washed with PBS, low-molecular-weight fragments of DNA were extracted in citrate buffer (Na_2HPO_4, $C_6H_3O_7$, pH 7.8), RNA was removed by ribonuclease A (5 Kunitz U/ml), and DNA was stained with propidium iodide (PI; 20 µg/ml PBS) for 30 min in the dark. Fluorescence was measured using a flow cytometer (FACSCalibur; Becton Dickinson; San Jose, CA, USA) equipped with an argon ion laser at 488 nm wavelength for excitation. A total of $2x10^4$ cells were analysed in each sample. The ModFit 2.0 (Verity Software House; Topsham, CA, USA) and CellQuest (Becton Dickinson) software was used to generate DNA content frequency histograms and quantify the amount of the cells in individual cell cycle phases including subG0/G1 population.

Fluorescence microscopy. The cells were stained with a 4,6-diamidino-2-phenyl-indol (DAPI; Fluka; Buchs, Switzerland) solution (1 µg DAPI/ml ethanol) at room temperature in the dark for 30 min. They were then mounted in Mowiol, and the percentage of apoptotic cells (with chromatin condensation and fragmentation) was determined using a fluorescence microscope (Olympus IX-70; Olympus; Prague, Czech Rep.) from a total number of 200 cells.

Expression of mitochondrial membrane Apo2.7 protein. The cells ($1x10^6$) were washed in PBS and permeabilised by digitonin (100 µg/ml; Sigma Aldrich) in PBSF (2.5% fetal calf serum and 0.1% NaN_3 in PBS) on ice for 20 min and then washed in PBS. The Apo2.7 protein was marked using anti-Apo2.7 antibody conjugated with phycoerythrin (Immunotech). The cells were incubated with 20 µl of antibody and 80 µl of PBSF at room temperature for 15 min in the dark. Fluorescence was measured using a flow cytometry (a total of $2x10^4$ cells in each sample) and analysed using the CellQuest software.

Production of reactive oxygen species (ROS). The intracellular production of ROS was detected by FCM analysis using dihydrorhodamine-123 (DHR-123, Fluka, Switzerland), which reacts with intracellular hydrogen peroxide. The cells treated with the appropriate agent were harvested, washed twice in PBS, and resuspended in Hanks' buffered saline solution (HBSS). DHR-123 was added in a final concentration of 0.2 µM. The samples ($2x10^4$ cells per sample) were then incubated for 15 min at 37 °C in 5% CO_2. Fluorescence was detected with a 530/30 (FL-1) optical filter. Forward and side scatters were used to gate the viable population of cells.

Detection of mitochondrial membrane potential (MMP). The changes of MMP were analysed by FCM using tetramethylrhodamine ethyl ester perchlorate (TMRE; Molecular Probes, Eugene, OR, USA). The cells were washed with HBSS, resuspended in 100 nM of TMRE in HBSS, and incubated for 20 min at room temperature in the dark. The cells were then washed twice with HBSS, resuspended in 500 µl of the total volume, and analysed ($2x10^4$ cells per sample). Fluorescence was detected with a 585/42 (FL-2) optical filter. Forward and side scatters were used to gate the viable population of cells. The data were evaluated (Cell Quest Software) as a percentage of the cells with decreased MMP.

Detection of lipid droplets. Accumulation of lipid droplets was detected by FCM after staning of cells with Nile red. The cells treated with the appropriate agent were harvested, washed twice and resuspended in PBS. Nile red (stock solution 1mg/ml in aceton) was added in a final concentration of 0.1 µg/ml. The samples ($5x10^5$ cells per sample) were then incubated for 5 min at RT in the dark. Fluorescence was detected in 2×10^4 cells per sample with a 530/30 (FL-1) optical filter.

ALP activity determination. The cells grown on 40×10 mm dishes were treated with NaBt (3 mM), TNF-α (0.5, 5 or 15 ng/ml) or their combinations for 24, 48 or 72 hours. The cells were trypsinised and counted using a haemocytometer. 5×10^5 cells were resuspended in 500 μl of substrate buffer (10% diethanolamine; 5 mM $MgCl_2$; pH 9.7) and lysed by sonication for 5×10 seconds on Branson Sonifier B-12 at a power of 30 watts (output 1). The cell lysate and ALP (Sigma) in several concentrations (15.6-1000×10^{-6} U/well) for a calibration curve were incubated in 37°C with ALP substrate (4-p-nitrophenylphosphate; Fluka) in a 96-well plate (4 parallel wells in each group) for 30 min. The reaction was stopped by adding 3 M NaOH (50 μl/well) and the optical densities were measured at 405 nm (DigiScan Reader). The reading values (units$\times 10^{-6}$/50 000 cells) were converted to the percentage of control.

Expression of carcinoembryonic antigen (CEA). The cells treated with the appropriate agent were harvested, washed twice in azide buffer (PBS with 0.1% sodium azide), resuspended in 100μl of this buffer and 5 μl anti-CD66e-FITC monoclonal antibody (SEROTEC) was added. As isotype control mouse IgG1-FITC (Pharmingen) was used. Samples were incubated 30 min on ice in the dark, washed twice with azide buffer, resuspended in 0.5 ml of it. Fluorescence was detected in 2×10^4 cells per sample with a 530/30 (FL-1) optical filter.

Caspase activities. The cells were lysed in lysis buffer (250 mM HEPES, 25 mM CHAPS, 25 mM DTT, 40 μM protease inhibitor cocktail; Sigma Aldrich) on ice for 20 min and then centrifuged at 15,000 g for 15 min in 4 °C. The acquired proteins (equal concentrations) were incubated with caspase-3 (Ac-DMQD-AMC; 50 μM; Alexis; Carlsbad, CA, USA), caspase-8 (Ac-IETD-AMC; 50 μM; Sigma) or caspase-9 (Ac-LEHD-AMC; 50 μM; Alexis) substrates overnight in assay buffer (40 mM HEPES, 20 % glycerol, 4 mM DTT). Fluorescence was measured (355/460 nm) using a Fluostar Galaxy fluorometer (BMG Labtechnologies GmbH; Offenburg, Germany). Caspase-9 activity was inhibited by specific inhibitor (Z-LEDH-FMK; Bio Vission, Inc.; New Minas, Canada).

Immunoblotting. The cells were lysed in Laemmli sample buffer (100 mM Tris, pH 7.4; 1 % sodium dodecyl sulphate; 10 % glycerol), diluted to an equal concentration, mixed with bromphenol blue (0.01 %) and mercaptoethanol (1 %), and subjected (20 μg) to SDS-PAGE. The polyacrylamide gels were transferred to polyvinylidene difluoride membranes (Immobilon-P; Millipore Corp.; Bedford, MA, USA) electrophoretically in a buffer containing 192 mM glycine, 25 mM Tris, and 15 % methanol. The membranes were blocked overnight in 5 % non-fat milk in wash buffer (0.05 % Tween-20 in 20 mM Tris; pH 7.6; 100 mM NaCl), and then probed with anti-PARP (1:500, SC-7150, Santa Cruz Biotechnology; Santa Cruz, CA, USA), anti-Bid (1:500, SC-6538, Santa Cruz Biotechnology), anti-FLIP (1:250, AAP-440, StressGen Biotechnologies Corp.; Victoria, BC, Canada), anti-DR5 (1:1000, D3938, Sigma Aldrich) or anti-DcR2 (1:1000, 68861N, Pharmingen, BD Biosciences; San Jose, CA, USA) antibodies. The recognised proteins were detected using horseradish peroxidase-labelled rabbit (1:6000, #NA934, Amersham Biosciences; Buckinghamshire, UK) or mouse (1:3000, #NA931, Amersham Biosciences) anti-IgG secondary antibody and an enhanced chemiluminescence kit (ECL; Amersham Biosciences). An equal loading was verified using β-actin quantification.

Statistical analysis. The results of at least three independent experiments were expressed as the means + S.E.M. Statistical significance ($P < 0.05$) was determined by one-way ANOVA followed by Tukey test or non-parametric Mann-Whitney test.

RESULTS

The Effects of TNF-α, Anti-Fas, And TRAIL

Parameters reflecting cell proliferation and cell death were detected in HT-29 and FHC cells after 24 hours treatment with TNF-α, anti-Fas or TRAIL. Moreover, modulation of these parameters after combined treatment of the cells with apoptotic inducers and protein synthesis inhibitor cycloheximide (CHX) or inhibitor of MEK1/2 kinase activity (U0126) was evaluated. The concentrations of the drugs used were selected on the basis of our previous experiments [34, 54, 55] and doses routinely used for this type of cells.

Effects of TNF-a, Anti-Fas, and TRAIL on Cell Proliferation

In selected concentrations, none of the TNF-family apoptotic inducers caused significant changes in the total cell number and cell cycle parameters of both HT-29 and FHC cells after 24 hours of treatment (data not shown).

Effects of TNF-α, Anti-Fas, and TRAIL on Cell Death

As apparent from Table 1, neither HT-29 nor FHC cells responded to TNF-α in the used concentration (15 ng/ml) and incubation time (24 h). An increase of parameters reflecting the cell death (cell detachment, subG0/G1 population and cells with apoptotic nuclear morphology) was observed after anti-Fas treatment with a slightly stronger response of HT-29 cells. The two cell lines studied differed significantly in their susceptibility to TRAIL. The values of all three parameters detected were about twofold higher in HT-29 than in FHC cells. Both cell types were markedly sensitised to all apoptotic inducers by addition of CHX (Table 1).

The differences in the apoptotic effects of TNF-α, anti-Fas, and TRAIL were further confirmed by detection of PARP cleavage (Fig 1). PARP was not cleaved after TNF-α treatment in both HT-29 and FHC cells. After treatment of cells with anti-Fas and particularly TRAIL, a more intensive PARP cleavage in HT-29 than in FHC cells was detected. Furthermore, an enhancement of a number of cells with decreased MMP was detected after treatment of both types of cells with anti-Fas or TRAIL (Fig. 2).

Differences In HT-29 and FHC Cell Response to TRAIL

Besides the parameters evaluated above, our attention was focused to other plausible determinants of different sensitivity of HT-29 and FHC cells to TRAIL effects.

Table 1. Quatification of cell death after 24 hour treatment of HT-29 and FHC cells with TNF-α (TNF; 15 ng/ml), anti-Fas (CH11; 200 ng/ml), TRAIL(100 ng/ml), or their combinations with CHX (5 µg/ml) or U0126 (10 µM)

Treatment groups	% of floating cells		% of cells in subG0-population		% of apoptotic cells (DAPI staining)	
	HT-29	FHC	HT-29	FHC	HT-29	FHC
Control	10.0±5.4	7.7±3.3	2.5±1.2	2.8±1.0	1.8±0.9	1.4±0.6
TNF	12.4±6.5	10.6±4.7	5.1±3.8	4.5±3.0	2.3±0.9	2.1±0.9
TNF+CHX	27.5±4.6 *§	17.5±5.2 *	15.1±3.9 *	17.0±7.9 **§	7.6±1.4 **¶	7.5±3.5 **§
TNF+U0126	13.3±2.9 #	7.0±1.2	6.5±3.4 *	4.8±1.1 *	1.2±0.6	1.3±1.1
CH11	21.8±8.3 *	12.3±4.9	15.8±7.6 **	12.5±7.8 *	6.9±2.6 **	3.8±1.4 *
CH11+CHX	38.9±2.0 **§	49.2±17.5 **	18.5±3.1 *	21.8±2.4 **	11.8±3.4 **	11.1±4.7 **§
CH11+U0126	27.8±5.9 * #	14.1±0.8 *	13.9±3.4 *	14.6±1.1 *	4.8±0.2 * #	3.0±0.7 *
TRAIL	41.6±5.4 ** #	24.0±10.6 **	15.4±4.7 **	9.8±4.6 **	12.4±5.8 **	6.5±1.9 **
TRAIL+CHX	54.6±2.6 **§	63.5±14.9 **§	21.2±4.1 *	27.7±4.0 **¶	17.5±4.3 **	16.6±6.2 **§
TRAIL+U0126	51.6±2.3 ** #	27.0±2.0 *	18.9±6.4 *	18.1±10.3 *	9.2±1.6 *	5.7±1.9 *

The values are means ± S.E.M. of at least three independent experiments; * - P < 0.05 (** - P < 0.01) versus untreated control; § - P<0.05 (¶ - P<0.01)versus TNF-α, anti-Fas or TRAIL alone treatment; # - P<0.05 between HT-29 and FHC cells.

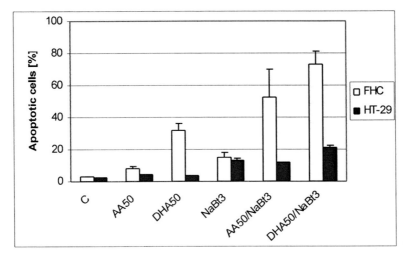

Figure 1. .Cleavage of PARP and β-actin quantification in HT-29 and FHC cells treated with TNF-α (TNF; 15 ng/ml), anti-Fas (200 ng/ml) or TRAIL (100 ng/ml) for 24 hours, detected by Western blotting. The results are representative of three independent experiments.

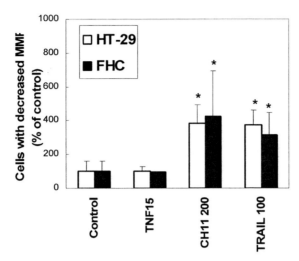

Figure 2. Cells (HT-29, FHC) with decreased mitochodrial membrane potential (MMP) after treatment with TNF-α (TNF; 15 ng/ml), anti-Fas (CH-11; 200 ng/ml) or TRAIL (100 ng/ml); P<0.05 (*) versus untreated control.

Expression of APO2.7 Mitochondrial Protein

The APO2.7 antigen is specifically expressed on the mitochondrial membrane of cells undergoing apoptosis. After 24 hours of TRAIL treatment, the number of cells with APO2.7 positivity confirmed a higher effect of TRAIL on HT-29 than on FHC cells. CHX strongly potentiated this effects in both cell lines, shifting positivity to about 95 % of the cells. Moreover, in HT-29 cells, but not in FHC cells, percentage of APO2.7 cells was significantly increased by co-treatment with U0126 (Fig. 3).

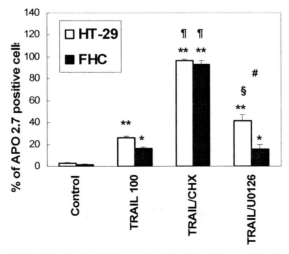

Figure 3. Expression of APO2.7 mitochodrial protein (% of positive cells) after 24 hour treatment of HT-29 and FHC cells with TRAIL (100 ng/ml) and its combination with CHX (5 μg/ml) or U0126 (10 μM); P<0.05 (*) versus untreated control, (§) versus TRAIL alone, (#) between HT-29 and FHC cells; P<0.01 (**) versus untreated control, (¶) versus TRAIL alone.

Caspase Activities

TRAIL induced an increase in caspase activities in both cell lines. Higher activities of caspases -8, -9, -3 after 4 hours of TRAIL treatment were detected in HT-29 compared to FHC cells (Fig. 4). These differences between the two cell lines were further enhanced by co-treatment of the cells with TRAIL and U0126. Combined treatment of cells with TRAIL and CHX did not significantly increase caspase-8 and -9 activities in comparison with TRAIL alone.

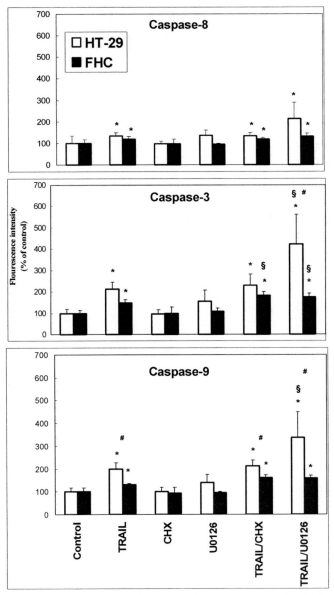

Figure 4. Caspase-8, -3 and -9 activities (% of control) after 4h treatment of HT-29 and FHC cells with TRAIL (100 ng/ml), CHX (5 μg/ml), U0126 (10 μM), or their combinations; P<0.05 (*) versus untreated control; (§) versus TRAIL alone; (#) between HT-29 and FHC cells.

PARP Cleavage

The results of PARP cleavage (Fig. 5) showed a more intensive effect of TRAIL in HT-29 then in FHC cells. Further potentiation of PARP cleavage by CHX as well as by U0126 was detected in both cell lines. However, combined treatment with TRAIL and U0126 resulted in a different cleavage pattern in FHC cells compared to HT-29 cells (lack of 89 kD cleavage product).

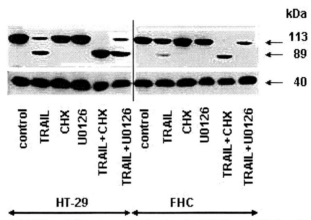

Figure 5. Cleavage of PARP and β-actin quantification in HT-29 and FHC cells treated with TRAIL (100 ng/ml), CHX (5 µg/ml), U0126 (10 µM) or their combinations for 24 hours, detected by Western blotting. The results are representative of three independent experiments.

Inhibition of Caspase-9

Co-treatment of cells with TRAIL (or TRAIL+CHX) and caspase-9 inhibitor Mch6 presented no significant changes of APO2.7 expression (not shown), but decreased the percentage of cells with apoptotic nuclear morphology in HT-29 cells after both TRAIL or TRAIL with CHX, and in FHC cells after treatment with TRAIL and CHX (Table 2).

Table 2. Quatification of apoptosis after 24 hour treatment of HT-29 and FHC cells with TRAIL (100 ng/ml), CHX (5 µg/ml) or their combination with/without presence of caspase-9 inhibitor Mch6 (2 µM)

	% of apoptotic cells (DAPI staining)	
Treatment	HT-29	FHC
Control	0.8±0.4	0.7±0.2
TRAIL	6.0±2.3 *	4.3±1.2 *
CHX	2.2±0.4	0.7±0.2
TRAIL+CHX	14.7±5.9 *	10.3±0.9 **
Mch6	1.3±1.4	1.3±0.8
TRAIL+Mch6	2.7±0.2 *	3.2±1.6
CHX+Mch6	1.2±0.6	1.2±0.4
TRAIL+CHX+Mch6	6.8±0.8 * ¶	4.3±0.8 * §

The values are means ± S.E.M. of three and more independent experiments; * - P < 0.05(** - P < 0.01) versus untreated control (Mch6 alone); § - P<0.05 (¶ - P<0.01) versus TRAIL alone (TRAIL+Mch6)

Expression of DR5, DCR2, and FLIP

No significant differences in the expression of DR5 protein detected by Western blotting between HT-29 and FHC cells were observed. As shown in Fig. 6A, co-treatment of cells with TRAIL and CHX resulted in its loss in both cell types.

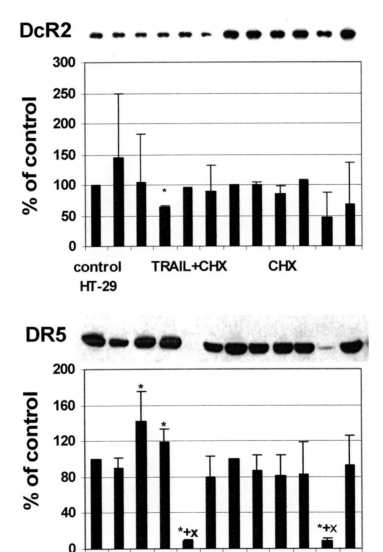

Figure 6. The level of DR5 and DcR2 proteins in HT-29 and FHC cells treated with TRAIL (100 ng/ml), CHX (5 µg/ml), U0126 (10 µM) or their combinations for 24 hours. Representative Western blots and the means ± S.E.M. of optical densities (related to β-actin content) are presented; P<0.05 (*) versus untreated control; (+) versus TRAIL; (x) versus CHX.

In comparison with HT-29 cells, FHC cells presented a higher expression of DcR2 protein, which was decreased by combined treatment with TRAIL and CHX (Fig. 6B).

The expression of FLIP was decreased after TRAIL treatment in both cell lines. This effect was significantly potentiated by CHX. Compared to TRAIL alone, a slight additional decrease of FLIP expression was observed after co-treatment with TRAIL and U0126 in both cell lines (Fig. 7 A).

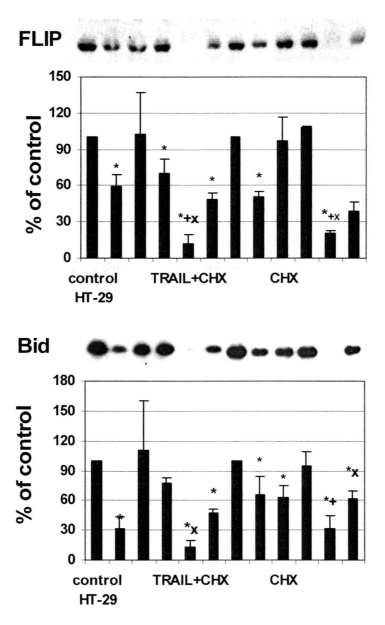

Figure 7. The levels of $FLIP_L$ (FLIP) and Bid protein in HT-29 and FHC cells treated with TRAIL (100 ng/ml), CHX (5 μg/ml), U0126 (10 μM) or their combinations for 24 hours. Representative Western blots and the means ± S.E.M. of optical densities (related to β-actin content) are presented; P<0.05 (*) versus untreated control; (+) versus TRAIL; (x) versus CHX.

Expression of BCL-2 Family Proteins

Expression of Bcl-2 family proteins was examined 24 hours after treatment with TRAIL and its combination with CHX and U0126. HT-29 cells expressed no Bcl-2 protein. While no changes of Bax and Bak expression were observed after any type of treatment (data not shown), a significant cleavage of Bid was detected. In HT-29 cells, Bid protein was more intensively cleaved after 24 hours TRAIL treatment in comparison with FHC cells. A strong potentiation of this effect (almost complete cleavage) by co-treatment with CHX was detected in both cell lines (Fig. 7 B).

Table 3.The effects of arachidonic acid (AA, 50 μM), docosahexaenoic acid (DHA, 50 μM), and sodium butyrate (NaBt, 3 mM) on FHC or HT-29 cell growth, cell death and cell cycle after 48 hour treatment

Treatment		Total cell number (x10⁵)	Floating cells (%)	Viable cells (%)	SubG0/G1 (%)	Cell cycle phase (%)		
						G0/G1	S	G2/M
Control	FHC	32.9±6.6	4.5±1.5	94.8±1.5	4.6±1.3	63.4±4.9	22.3±1.3	11.3±1.4
	HT-29	34.5±6.1	3.9±1.0	93.5±0.5	3.3±1.3	63.5±4.9	24.8±3.6	11.6±3.1
Ethanol	FHC	30.2±6.8	4.1±0.9	88.2±3.0 *	5.5±1.5	62.8±3.9	23.2±2.0	11.6±1.3
	HT-29	36.8±4.4	3.1±1.2	89.8±2.9	3.4±1.2	63.2±3.2	23.9±3.4	12.8±4.3
AA 50	FHC	27.2±6.1	33.7±12.1 **	75.4±8.3 *	15.1±4.6 **	58.1±7.5	25.0±3.6	9.9±3.9
	HT-29	34.1±3.1	4.1±1.6 ##	86.5±5.3	4.0±1.9 ##	61.4±2.2	25.1±4.4	13.5±5.2
DHA 50	FHC	24.2±5.1	27.9±13.7 **	64.8±11.8 **	26.7±9.0 **	71.6±2.6	19.6±2.0	7.3±0.8 **
	HT-29	33.0±7.4	4.2±1.7 #	86.8±2.4 #	3.9±1.5 ##	64.7±3.5	23.0±3.4 #	12.3±4.2 #
NaBt 3	FHC	10.2±2.0 **	19.4±2.2 **	74.6±13.9 *	26.3±10.1 **	74.4±5.4	13.1±5.5 **	10.5±3.1
	HT-29	14.3±2.2 **	17.8±6.5 **	83.3±3.3 *	13.8±4.8 *	73.6±4.1	19.8±3.6 ##	3.1±1.3 *

Furthermore, accumulation of lipid droplets (LDs) in cell cytoplasm (Nile red), changes of MMP (TMRE techniques) and ROS production (DHR-123) were detected using flow cytometry. Significant increase of Nile red fluorescence (LD accumulation) was detected in FHC cells after 48 hours of treatment with both AA and DHA compared to control. After DHA treatment, LD accumulation was also enhanced in HT-29 cells, but less than in FHC cells (Fig. 8). Both AA and DHA enhanced number of cells with decreased MMP to a similar extent in FHC cells. Slight but not significant tendency was also apparent in HT-29 cells (Fig. 9). After 48 hours of treatment with both AA and DHA, the ROS production was markedly increased in FHC cells and only slightly elevated in HT-29 cells (Fig. 10).

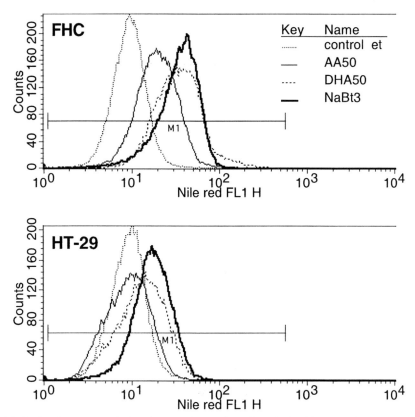

Figure 8. Lipid droplets accumulation in cytoplasm of FHC and HT-29 cells after 48 hour treatment with arachidonic acid (AA; 50 μM), docosahexaenoic acid (DHA; 50 μM) or sodium butyrate (NaBt; 3 mM). Cells were stained with Nile red and analyzed using flow cytometry. Mean of fluorescence (FL-1) was evaluated.

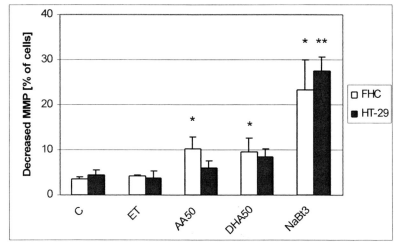

Figure 9. Number of FHC and HT-29 cells with decreased mitochondrial membrane potential (MMP) after 48 hour treatment with arachidonic acid (AA; 50 μM), docosahexaenoic acid (DHA; 50 μM) and sodium butyrate (NaBt; 3 mM). Both intact cells (C) and cells with ethanol vehicle (ET) were considered as appropriate controls. P<0.05 (*), P<0.01(**): compared to appropriate control.

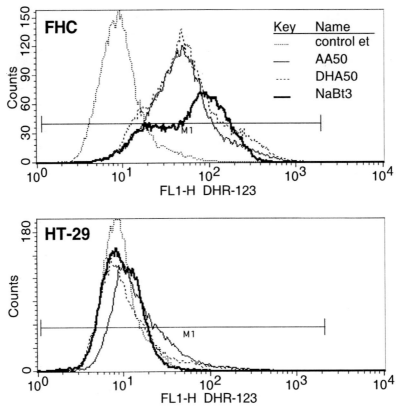

Figure 10. Production of reactive oxygen species in FHC and HT-29 cells treated 48 hours with arachidonic acid (AA; 50 µM), docosahexaenoic acid (DHA; 50 µM) and sodium butyrate (NaBt; 3 mM). Cells were stained with dihydrorhodamine-123 and analyzed using flow cytometry. Mean of fluorescence (FL-1) was evaluated.

For detection of final stages of apoptosis, the nuclear morphology of cells treated with PUFAs for 72 hours was analyzed by fluorescence microscopy after DAPI staining (Fig. 11). An increase of number of apoptotic FHC cells after AA (8%) and especially DHA (32%) treatment was observed compared to control (3%). In HT-29 cells, no differences in apoptotic cell numbers compared to control were detected.

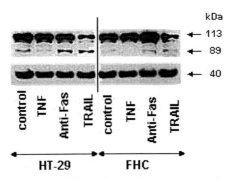

Figure 11. The amount of FHC and HT-29 cells with apoptotic nuclear morphology after 72 hour treatment with with arachidonic acid (AA; 50 µM), docosahexaenoic acid (DHA; 50 µM), sodium butyrate (NaBt; 3 mM) and their combinations. Cells were stained with DAPI and evaluated by fluorescence microscopy (200 cells per sample were counted). C: control.

No changes in differentiation markers, i. e. ALP activity (Fig. 12) and CEA expression (not shown) were observed either in HT-29 or FHC cells after 48 hours of treatment with PUFAs.

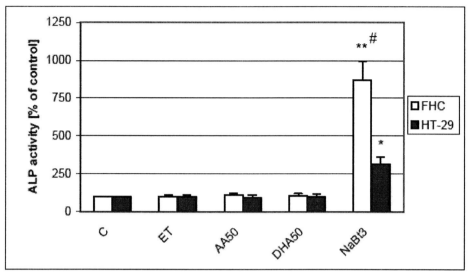

Figure 12. Activity of alkaline phosphatase (ALP) of FHC and HT-29 cells treated 48 hours with arachidonic acid (AA; 50 μM), docosahexaenoic acid (DHA; 50 μM) and sodium butyrate (NaBt; 3 mM).). Both intact cells (C) and cells with ethanol vehicle (ET) were considered as appropriate controls: $P<0.05$ (*), $P<0.01$(**); $P<0.01$ (#) compared FHC vs. HT-29 cells.

Sodium Butyrate (Nabt)

The effects of NaBt (3 mM) on cell growth and death are presented in Table 3. After 48 hours, NaBt significantly decreased total number of both FHC (30% of control) and HT-29 (40% of control) cells and enhanced the number of floating cells similarly in both cell lines. Significantly decreased FHC and HT-29 cell viability and increased percentage of cells in subG0/G1 populations accompanied these changes. After 48 hours of NaBt treatment, increased number of cells in G0/G1 population was detected in both cell types (74% compared to 64% in control), accompanied by decrease of cells in S-phase, significant for FHC cells.

NaBt also caused similar significant decrease of MMP in both cell lines tested (Fig. 9). Nile red fluorescence (LDs) increased about three-fold in FHC and two-fold in HT-29 cells (Fig. 8). While NaBt induced high production of ROS in FHC cells, only slight increase was observed in HT-29 cells (Fig. 10). The amount of apoptotic cells detected 72 hours after NaBt treatment was the same (about 15%) in both FHC and HT-29 cells (Fig. 11).

NaBt in 3 mM concentration induced differentiation of both cell lines. However, differentiation markers measured in 48 hours interval showed higher ability of FHC cells to differentiate in response to NaBt. Compared to control, the activity of ALP increased nine-fold in FHC cells and three-fold in HT-29 cells (Fig. 12). While untreated control HT-29 cells expressed CEA only slightly, much higher expression was detected in FHC cells. The basal expression was increased by NaBt about 2.5-fold in FHC and 1.5-fold in HT-29 cells (Fig. 13).

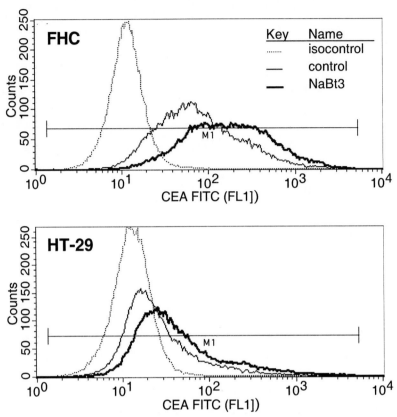

Figure 13. Expression of carcinoembryonic antigen (CEA) in FHC and HT-29 cells. Cells were untreated (control) or treated with sodium butyrate (NaBt; 3mM) for 48 hours. Binding of anti CD66e monoclonal antibody with FITC was evaluated using flow cytometry (FL 1). As isocontrol a mouse IgG1-FITC was used.

Combined Effects of PUFAs and Nabt

To investigate the interaction of PUFAs with NaBt, the cells were treated with 50 µM of AA or DHA together with NaBt (3 mM) for 48 hours and 72 hours. The potentiation of the effects (compared to PUFAs or NaBt alone) on most of the parameters tested was clearly apparent in FHC cells (Table 4), while in HT-29 cells only less profound effects of DHA combined with NaBt were observed (data not shown). In comparison with the DHA and NaBt used alone, combined treatment of FHC cells resulted in significant increase of floating cells, cells in subG0/G1 population and decreased cell viability (Table 4). Surprisingly, DHA attenuated the suppressive effect of NaBt on FHC cell growth, enhancing the amount of cells in S-phase by about 10% and simultaneously decreasing percentage of cells in G0/G1 phase.

Table 4. The effects of arachidonic acid (AA; 50 μM), docosahexaenoic acid (DHA; 50 μM), sodium butyrate (NaBt; 3 mM) and their combination on FHC cell growth, cell death and cell cycle after 48 hour treatment.

Treatment	Total cell number (x10^5)	Floating cells (%)	Viable cells (%)	SubG0/G1(%)	Cell cycle phase (%)		
					G0/G1	S	G2/M
Control	30.2±6.8	4.1±0.9	88.2±3.0	5.5±1.5	62.8±3.9	23.2±2.0	11.6±1.3
AA 50	27.2±6.1	33.7±12.1 **	75.4±8.3 *	15.1±4.6 **	58.1±7.5	25.0±3.6	9.9±3.9
DHA 50	24.2±5.1	32.9±12.2 **	64.8±11.8 **	26.7±9.0 **	71.6±2.6	19.6±2.0	7.3±0.8 **
NaBt 3	9.6±2.2 **	23.0±4.8 **	82.6±1.7 *	20.7±3.7 **	75.6±4.0	12.8±5.1 **	9.4±3.2
AA 50 + NaBt 3	13.2±3.4 ** xx	34.9±5.5 **	61.4±9.9 ** ++	34.2±7.2 ** xx	77.8±6.8 * xx	13.4±7.0 x	7.1±2.2 **
DHA 50 + NaBt 3	14.2±4.0 ** xx	47.1±6.3 ** ++	48.4±9.6 ** ++	52.6±8.4 ** xx +	67.3±7.6	22.6±10.2	9.8±3.9

Values are means ± S.E.M.; P<0.05 (single symbol) or 0.01 (double symbol): (*) compared to control; (x) compared to AA or DHA alone; (+) compared to NaBt alone; (#) compared FHC vs. HT-29 cells.

In Table 5, the combined effects of PUFAs and NaBt on LD accumulation, and changes of MMP are presented. A strong potentiation of Nile red fluorescence (LD accumulation) after combined treatment with PUFAs (particularly DHA) and NaBt was detected in both FHC and HT-29 compared to the agents used alone. Significantly higher effects were achieved in FHC compared to HT-29 cells. After combined treatment, about two-fold enhancement of cell number with decreased MMP (48 hours) and massive accrual of apoptotic cells (up to 52% with AA and 73% with DHA after 72 hours) was detected in FHC cells (Fig. 11). In HT-29 cell line, combined treatment of NaBt with DHA enhanced number of apoptotic cells only from 13 to 22 %.

Table 5. Per cent of cells with decreased mitochodrial membrane potential (MMP) and lipid droplet accumulation (mean Nile red fluorescence) of FHC cells treated with arachidonic acid (AA; 50 μM), docosahexaenoic acid (DHA; 50 μM), sodium butyrate (NaBt; 3 mM) or their combination for 48 hours.

		Control	AA 50	DHA 50	NaBt 3	AA 50 + NaBt 3	DHA 50 + NaBt 3
MMP	FHC	4.2±0.4 *	10.1±2.8 *	9.6±2.9 *	25.9±6.1 *	50.6±6.4 * x +	45.6±1.3 * x +
	HT-29	3.8±1.6	6.0±1.5	8.5±1.7	27.8±4.8 **	28.1±0.8 ** xx	24.3±1.3 ** xx
NILE RED	FHC	10.1±0.3	23.7±0.3 *	38.2±10.6 *	36.7±4.4 *	44.8±4.5 * x	84.2±10.8 * x +
	HT-29	9.5±0.2	11.3±1.1 * #	14.9±2.5 * #	18.1±2.0 * #	24.7±0.8 * + x #	30.3±8.8 * + x #

Values are means ± S.E.M.; P<0.05 (single symbol) or 0.01 (double symbol): (*) compared to control; (x) compared to AA or DHA alone; (+) compared to NaBt alone; (#) compared FHC vs. HT-29 cells.

DISCUSSION

Metabolism and turnover of colon epithelial cells are regulated by various endogenous factors regulating cell growth, differentiation and apoptosis. Dietary components such as PUFAs and butyrate can influence signaling pathways modulating cytokinetics [56]. Moreover, signals from both nutritional compounds and endogenous factors come into mutual interaction and this may have a substantial impact determining the final phenotype, metabolism and kinetics of colon epithelial cell population. Since significant changes in cell signaling occur during cell transformation, different response of normal and cancer cells may be assumed.

Cell culture systems of colonic cells *in vitro* can simulate well the *in vivo* situation by the ability of these cells to be induced to differentiation as well as to detachment-induced apoptosis - anoikis (shedding into medium) by specific factors. They allow studying in more detail relationship between specific cell types (normal vs. transformed cells, cells with various genetic background etc.) and the effects of endogenous as well as dietary components during colon carcinogenesis. Particularly important is modulation of oxidative metabolism, membrane properties, inter- and intracellular signaling pathways, activation of transcription factors and changes in gene expression, the processes which all can regulate cell proliferation, differentiation, and apoptosis. In this chapter, the effects of (1) endogenous regulators of TNF family, (2) AA, DHA, and NaBt and their combination were investigated using human colon epithelial cell lines. HT-29 are non-invasive, well-differentiated cells derived from colon adenocarcinoma, and FHC cells are derived from normal fetal colon tissue.

DIFFERENT RESPONSE OF NORMAL AND CANCER COLON CELLS TO TNF FAMILY MOLECULES

Our results showed that triggering of the TNF family receptors can induce different degree of response in cancer HT-29 and non-cancer FHC human colon cells. While both cell types did not respond to TNF-α in the concentration used, they showed recognizable response to anti-Fas antibody and markedly different response to TRAIL-mediated apoptosis. Generally, our results showed different sensitivity of normal and cancer colon cells to TNF family molecules and imply that synthesis of specific proteins involved in intracellular signaling pathways determines the cell response.

Relatively slow nature of death receptor-induced death in HT-29 cells has been reported [57]. In our previous experiments, TNF-α initially induced dose-dependent growth arrest and cytotoxicity was enhanced only by cultivation for longer time intervals [54], even though activation of caspase-3 and PARP cleavage were detected earlier [55]. In the experiments presented here, no effects on proliferation as well as cell death were detected either in HT-29 or FHC cells after 24 hours treatment with TNF-α. In general, TNF-α elicited acute apoptotic effects only when administered along with metabolic inhibitors such as CHX.

HT-29 cells seem to be slightly more sensitive to the effects of anti-Fas antibody than FHC cells. Similarly to TNF-α, cell sensitivity can be increased by CHX. Colon carcinoma cell lines usually show relative resistance to CD95-inducible cell death and malignant transformation of colon epithelium is often accompanied by abnormally low levels of CD95 protein [10]. We detected only a low expression of the CD95 receptor in both normal and cancer colon cell lines (data not shown). It is supposed that the apoptotic responsiveness to CD95 crosslinking might be more complex and remains to be further elucidated [14].

As supposed, HT-29 cells presented higher sensitivity to TRAIL than FHC cells, and sensitivity of both cell types was increased by CHX. Molecular determinants of response to TRAIL on various levels of signaling pathway are suggested and they vary among different cell types [58]. At the receptor level, high expression of DR4 and DR5 was reported in various types of colorectal neoplastic disorders [59]. Cells may become resistant through regulation of the DR cell surface transport and/or their relocalisation in the plasma membrane [60, 61]. Our results showed that there are no significant differences in level of DR5 between HT-29 and FHC cells. Interestingly, DR5 was lost after co-treatment with TRAIL and CHX, which implies the role of new protein synthesis in the regulation of the level of this receptor. Furthermore, the decoy receptors were shown to be involved in regulation of cell sensitivity/resistance to TRAIL effects in some cell lines. In our experiments, the increased level of DcR2 in FHC cells compared to HT-29 cells was demonstrated, suggesting that it could contribute to higher resistance of normal colon cells to TRAIL. The amount of DcR2 in FHC cells was also suppressed by combined treatment with TRAIL and CHX.

The FLICE-like inhibitory protein (cFLIP) may produce resistance at the level of the death-receptors by inhibition of caspase-8 activation. Furthermore, sensitization of tumor cells to TRAIL-induced apoptosis using an inducible pathway for degradation of FLIP protein was demonstrated [62]. Both cell lines expressed relatively high level of FLIP$_L$. The slight decrease of its expression after TRAIL treatment and its complete loss caused by co-treatment with CHX imply the possible role of FLIP$_L$ in the apoptotic response of colon cell lines.

Results of our experiments using the specific MEK1/2 inhibitor U0126 imply the protective role of the ERK1/2 pathway from the death receptor-mediated apoptosis, similarly as reported by others in human melanoma cell lines [63] or HeLa cells [64]. Combined treatment with TRAIL and U0126 significantly enhanced caspase-3 activity and potentiated PARP cleavage. U0126-mediated potentiation of TRAIL-induced apoptosis was more pronounced in cancer HT-29 cells compared to normal FHC cells.

Both HT-29 and FHC cells seem to be type II cells (intrinsic apoptotic pathway) because the Bid-dependent mitochondrial amplification loop was functional. The BH3-only protein Bid provides a link between the receptor-mediated and mitochondrial pathway. In our experiments, TRAIL-induced Bid cleavage was accompanied by increased APO2.7 protein expression, decrease of MMP, and activation of caspase-9. Moreover, the apoptosis induced by TRAIL was partially reversed by caspase-9 inhibition. Described responses were weaker in FHC cells compared to HT-29 cells and were potentiated by CHX in both cell lines. These results correspond with suggestion that in a cell line where DISC-generated initiator caspase activity is insufficient to fully activate enough caspase-3 to kill the cell, the cleavage of Bid followed by further events at the level of mitochondria (involving interplay between the Bcl-2 protein family members) is necessary for full caspase activation and apoptosis [6].

In colon cancer, the important role of especially Bax and Bak proteins was suggested [6, 65]. Hague et al. [12] reported no correlation between the expression of Bcl-2 family members and differential sensitivity of adenoma and carcinoma colon cell lines and no evidence supporting the concept of Bax inactivation as a critical mechanism to evade TRAIL-induced apoptosis *in vivo* in MSI-H tumors was reported [59]. Similarly, in our experiments the expression of neither Bax nor Bak was changed after any type of treatment in both HT-29 and FHC cells.

Based on the results of our study we can conclude that both HT-29 cancer and FHC non-cancer cells did not respond to TNF-α in the concentration used, while anti-Fas antibody and TRAIL induced more apparent response in HT-29 cells. Despite of the fact that a lot of other regulatory molecules could be considered [13] our results demonstrated that the regulation of apoptotic response of studied colon cell lines to TRAIL is accompanied by the significant changes of specific protein level in cells, particularly DR5, DcR2, FLIP, and Bid and showed the importance of mitochondrial pathway. Potentiation of several apoptotic parameters by U0126 implies an important role of ERK1/2 activation. The different level of the response of HT-29 and FHC cells to TRAIL-induced apoptosis could be attributed to the lower activity of mitochondrial pathway components followed by decreased caspase-9 and -3 activities and PARP cleavage.

DIFFERENT EFFECTS OF FATTY ACIDS ON NORMAL AND CANCER COLON CELLS AND THEIR INTERACTION

PUFAs

Our results clearly demonstrated higher sensitivity of normal FHC cells to both AA (n-6) and DHA (n-3). Even if relatively low concentration (50 μM) was used, AA and particularly DHA decreased FHC cell growth and viability, affected adhesion (increasing % of floating cells) and induced apoptosis of these cells. On the other hand, no effects on these parameters were observed in HT-29 cells. Previously, when using different experimental design (treatment of non-confluent 24 hours culture of HT-29 cells) and higher DHA concentration (100 μM), only slightly decreased total cell number and increased per cent of floating cells and no effects on apoptosis and viability were observed [35]. Some differences in the response of the same cell types to PUFAs may be explained by distinct response of cells growing in low or high cell density [66].

There are many reports presenting cytotoxic and/or antiproliferative effects of both n-6 and n-3 PUFAs. The tumor cell-specific effects and lower sensitivity of normal cells have been documented [67] [68]. On the contrary, there are few studies showing higher PUFA sensitivity of normal cells than in cancer or transformed cell of the same type [69] [66]. Unfortunately, many investigations did not compare neoplastic and non-neoplastic cells from the same species and tissue type. Our model compared normal and cancer colon cells presenting very similar growth characteristics. Although FHC cells were derived from normal colon tissue, their fetal character resembles that of HT-29 cells, which are well-differentiated and non-invasive cancer cells. Both cell lines grow rapidly and are able to differentiate to enterocyte-like cells.

The epidemiological and *in vivo* studies reported promoting role of n-6 PUFAs and their metabolites in colon cancer development [16]. On the other hand, AA exogenously added to cultured cells was shown to inhibit proliferation and induce apoptosis [70]. Our experiments are in accordance with these findings. Moreover, the sensitivity may differ according to the cell type, confluency, time and concentration used. However, DHA seems to be always more effective in suppression of proliferation and induction of colonic cell apoptosis.

DHA represents the longest and most unsaturated acyl chain (six double bonds) commonly found in cell membranes and it was shown to modify membrane properties by enhancing membrane lipid phase separation into lipid rafts [71] [72]. It was reported that n-3 PUFA feeding altered colonic caveolae microenvironment compared with n-6 PUFAs [73]. Changes of membrane fluidity and remodeling of lipid rafts and caveolae may further influence receptor-ligand interactions, signal transduction properties and various aspects of membrane-mediated cellular functions [74]. Our published results showed that pretreatment of HT-29 cells with either AA or DHA modulated cell cycle parameters and significantly potentiated apoptosis induced by TNF family molecules like anti-Fas and TRAIL [34] [35].

It was reported that accumulation of LDs may occur during apoptosis or differentiation [26]. LDs generated in response to elevated fatty acids content may serve as reservoir of signal molecules important for regulation of these processes. The connection between LDs and specific proteins known to have role in plasma membrane structure and function like caveolines and fatty acid transporter protein was also reported [25]. Our results showing significantly increased Nile red fluorescence (LD accumulation) after treatment of cells with both AA and DHA suggest a possible role of the amount of LDs in cell cytoplasm in final cell response. While percent of cell with decreased MMP and induction of ROS were comparable after treatment with both PUFAs, DHA, which was more effective, caused higher LD accumulation.

The higher effectiveness of DHA may also be connected with the fact that AA and DHA are not incorporated to HT-29 cell phospholipids (PLs) in the same manner. In the colonic cells, AA was incorporated in total cell PLs to a great extent, while DHA was preferentially found in cardiolipin (CL). CL is one of the essential components of the inner-mitochondrial membranes and belongs to the most diet-responsive and changeable fatty acid compounds among PLs [75]. The increase of CL unsaturation is associated with changes of mitochondria structural integrity, electron transport chain, membrane potential, and ROS production. Mitochondrial lipid peroxidation resulted in marked loss of CL content, cytochrome c oxidase activity, and release of cytochrome c, which is involved in the process of apoptosis [38] [76].

We and others have shown that AA or DHA supplemented to the culture medium are incorporated into cellular lipids, increasing the level of appropriate PUFA and dose-dependently stimulating ROS production and lipid peroxidation of HT-29 cells [35] [44] [30]. ROS serve as common messengers downstream of various stimulus-specific pathways including some TNF family members or PUFAs and are involved in regulation of immune responses, proliferation and apoptosis [77]. The ability of mitochondria to generate ROS may regulate mitochondrial-dependent death pathways [78]. Although ROS stimulation may not be required per se for apoptosis, a variety of stimuli (including TNF family members and probably PUFAs) may facilitate the process of apoptosis. Consequently the scavenging of ROS (e.g. by antioxidants) can delay or prevent apoptosis as we previously showed by addition of Trolox together with PUFAs [44]. Furthermore, the lower ROS production in tumour HT-29 may reflect different ability of PUFAs to be incorporated into cell membranes or their altered oxidative metabolism and antioxidative defense compared to normal cells. This is also supported by lower accumulation of LDs in cytoplasm and lower amount of cells with decreased MMP compared to FHC cells.

It may be supposed that alteration of lipid transport and metabolism, changes of expression of specific genes and regulating molecules and malfunction of certain signaling pathways affect sensitivity to PUFAs in transformed cells and their final response [23]. Our published studies suggest that differences in lipid metabolism of tumour and normal tissues may be responsible for observed distinct response of HT-29 and FHC cells to commercial parenteral lipid emulsions. Compared to the controls, supplementation with lipid emulsions containing mostly n-6 linoleic acid resulted in a multiple increase of linoleic and linolenic acids in total cell lipids, but the content of AA, DHA, and eicosapentaenoic acid decreased particularly in HT-29 cells. No effects of any type of emulsions were detected on cell proliferation and viability. However, the concentration of emulsions which did not affected HT-29 cells increased the percentage of floating and subG0/G1 FHC cells probably due to observed higher ROS production and lipid peroxidation. Emulsion with fish oil appeared to be most effective, probably due to high induced oxidative stress (our unpublished results). Co-treatment of cells with antioxidant Trolox reduced effects of lipid emulsion.

NaBt

NaBt in concentration of 3 mM significantly affected most of the parameters reflecting proliferation, differentiation and apoptosis of FHC and HT-29 cells in the similar way. It strongly suppressed cell growth, increasing the amount of cells in G0/G1 phase and decreasing percent of cells in S-phase and in G2/M phase in FHC and HT-29 cells, respectively. It was reported that NaBt exerts dose-dependent antiproliferative effects on colon cancer cell lines *in vitro* due to G0/G1 arrest caused mainly by modulation of cell cycle-related proteins [79]. It was confirmed by our previous results showing significant increase of the expression of cyclin-dependent kinase inhibitors p27[Kip1] and p21[Cip1/WAF1] after treatment of HT-29 cells with 5 mM of NaBt.

A variety of colon carcinoma cell lines has been shown to differentiate after NaBt treatment. This differentiation was mostly detected by an increase in cellular ALP and was accompanied by decreased growth rate [80]. In our experiments, FHC and HT-29 cells significantly differed in their ability to differentiate. Untreated FHC cells presented higher basic level of expression of CEA, which additionally increased after treatment with NaBt. CEA as well as ALP activity induced in HT-29 cells was only about one half of values observed in FHC cells. It was reported that during butyrate-induced differentiation of HT-29 cells alteration of lipid composition, namely increased esterification of neutral lipids, occurs

[81]. In our experiments, NaBt significantly increased accumulation of LDs and production of ROS, and a marked difference between FHC and HT-29 cell lines was observed. Thus, increased accumulation of lipids within the FHC cells could be connected to their better ability to differentiate.

In addition to differentiation, NaBt significantly increased percent of floating cells and decreased viability after 48 hours, which was followed by apoptosis of about 15% of cell population after 72 hours. In agreement with other authors, our results demonstrated that NaBt induced dissipation of MMP in colonic cells. About 25-28% FHC and HT-29 cell population with dissipated MMP after NaBt treatment confirmed the findings of other authors (Heerdt et al.) and imply the important role of this phenomenon in regulation of cytokinetic of colonic cells. The relationship between apoptosis, differentiation and associated events like LDs accumulation, changes of MMP and oxidative metabolism needs further investigation.

Our results with FHC cells are in contrast with those of Comalada et al. [82] who showed that in normal FHC cells or in differentiated cultures as well as in *in vivo* studies, the normal proliferation and regeneration of damaged epithelium is not affected by butyrate. However, luminal butyrate appears to have complex effects on colonic epithelial proliferation and recent evidence indicates that the response of proliferating cells to butyrate depends on the concentration, the physiological conditions during the study, the cell phenotype, differentiation status, and other factors present [41] [83].

We previously reported that despite similar response of both HT-29 and FHC cell lines to NaBt (G0/G1 arrest, decrease of growth rate and increase of apoptosis) they differ in the level and dynamics of measured parameters [84]. This is associated with declined telomerase activity and the level of mRNA for its catalytic subunit (hTERT). While both cell lines show similar kinetics of hTERT transcriptional silencing, the down-regulation of telomerase activity is faster in FHC cells. Our further experiments investigating the changes in histone acetylation and methylation pattern after NaBt and trichostatin treatment in HT-29 and FHC cells showed similar dynamic reorganization of chromatin in parallel with changes in its epigenetic modifications [85].

COMBINATION OF NaBt WITH PUFAs

The most important findings of our recent research represent possibilities of modulation of NaBt effects on colonic cells with other factors. We previously demonstrated that TNF-α potentiated apoptosis induced by NaBt, but suppressed NaBt-induced differentiation of HT-29 cells [54]. Moreover, we proved that the effects of TNF-α, inhibitors of AA metabolism as well as their combination depend on the cell differentiation status induced by NaBt [32]. Next, pretreatment of HT-29 cells with AA or DHA sensitized them to apoptotic effects of 5 mM NaBt [44].

In this report we demonstrated strong potentiation of apoptosis after co-treatment of colonic cells with NaBt together with AA or DHA in normal FHC cells. In HT-29 cells, lower effect was detected only in combination of NaBt with DHA. We detected primarily increased percentage of floating cells, which were primed for apoptosis as demonstrated by increased amount of cells in sub G0/G1 population and enhanced per cent of apoptotic cells after 72 hours. This mechanism may be analogous to "anoikis" (induction of apoptosis in response to loss of cell contact), typical of exfoliating epithelial cells. Our important findings include those documenting interaction of PUFAs and NaBt on the level of mitochondria where significantly higher percent of FHC population with decreased MMP was detected. Heerdt et al. [47] reported that intact MMP is required for the initiation and maintenance of the early p-53 independent p21$^{Cip1/WAF1}$ induction and growth arrest in G0/G1, while dissipation of the MMP is essential for the initiation and activation of downstream events that culminate in apoptosis of colon cancer cells after NaBt treatment.

In contrast to the effects of apoptosis, combined treatment of cells with NaBt and PUFAs did not result in additional suppression of cell growth. Surprisingly, attenuation of G0/G1 arrest and decreased S-phase induced by NaBt were detected after co-treatment with DHA. In our previous paper this effect was also demonstrated in pre-treatment regimen of HT-29 cells with PUFAs. We showed that in spite of reported activation of p27^{Kip1} and p21$^{Cip1/WAF1}$ by DHA [86], the NaBt-induced expression of p27^{Kip1} was markedly attenuated in cells pre-treated with both PUFAs used, which was probably reflected by increased S phase values in these cells. In this chapter, the strong potentiation of MMP decrease (45%) and resulting high percent (80%) of apoptotic cells with no effects on proliferation after combined treatment of FHC cells with NaBt and DHA suggest the role of MMP level in control of the balance between growth, differentiation and apoptosis.

Further, our interesting findings include the strong potentiation of LD accumulation in cytoplasm of both FHC and HT-29 cells after combined treatment of cells with NaBt and AA and especially with DHA. This finding corresponded with an increase of other parameters reflecting apoptosis and implies the role of this event in colonic cell apoptotic response. Next, we suppose that changes of membrane lipids as well as lipid content and metabolism may play an important role in the enhanced apoptotic effects of PUFAs, especially DHA together with NaBt. Several reports demonstrated effects of NaBt on the fatty acid composition of colon cancer cells [87] and suggest an interaction between butyrate and n-6 and n-3 PUFAs in the effects on differentiation and PUFA metabolism of human colon cancer cell lines [88]. They also showed different response of colon cancer cell lines to PUFAs and butyrate in growth, composition of membranes, and the relationship between membrane fatty acid composition and prostaglandin synthesis [89].

Associated with lipid changes, the oxidative metabolism may influence the cell response. The role of fatty acid oxidation and signaling in apoptosis has been well documented [90] and increased production of ROS by NaBt was reported to contribute to sensitization of HT-29 cells to apoptosis induced by TNF or Fas [91]. The results of other authors showed that fish oil or n-3 PUFAs primed mouse colonocytes *in vitro* as well as *in vivo* for elevated butyrate-induced apoptosis by enhancing the unsaturation of mitochondrial lipids, especially CL, resulting in ROS production and the dissipation of MMP [92] [39]. Similarly, Hossain et al [93] reported enhancing effects of phospholipids containing DHA on butyrate-induced growth inhibition, differentiation and apoptosis of colon cancer cells.

In contrast to other authors, who observed significant apoptotic effects *in vitro* or *in vivo* only with the n-3 type of PUFAs or fish oil, we demonstrated similar significant potentiating effects also with n-6 AA. This is reasonable with regard to evidence that AA can cause cell death through mitochondrial permeability transition [94]. However, diverse mechanisms of action of AA and DHA can be supposed. Reported different incorporation of these PUFAs within the cell (cell vs. mitochondrial membranes, various types of PLs) and production of different types of ROS (peroxide, superoxide, hydroxyl) as well as nitric radicals [95] may be the reason for different levels of lipid peroxidation, and cell cycle changes. While for AA more direct oxidative effects may be supposed, after DHA treatment other more complex mechanisms should be considered. This includes activation of specific intracellular receptors, altered expression of transcription factors, cell cycle and apoptosis regulating proteins as well as inactivation of prostaglandin family genes and lipoxygenases.

CONCLUSION

It may be concluded that although normal fetal colon FHC and adenocarcinoma HT-29 cell lines showed some similar characteristics (morphology, extensive proliferation, not full differentiation), they differ in their ability to respond to endogenous inductors of apoptosis like TRAIL as well as to exogenous dietary factors like PUFAs or NaBt produced from fiber.

Although both FHC and HT-29 cells did not respond to TNF-α in the concentration used, anti-Fas antibody and TRAIL induced a more apparent response in HT-29 cells. Despite the fact that a lot of other regulatory molecules could be considered, our results demonstrated that the regulation of apoptotic response of colon cell lines studied to TRAIL is accompanied by the significant changes of specific protein levels in cells, particularly DR5, DcR2, FLIP, and Bid and showed the importance of mitochondrial pathway. Potentiation of several apoptotic parameters by U0126 implies an important role of ERK1/2 activation. The different level of the response of HT-29 and FHC cells to TRAIL-induced apoptosis could be attributed to the lower activity of mitochondrial pathway components followed by decreased caspase-9 and -3 activities and PARP cleavage.

Lipid compounds, PUFAs and NaBt working individually or together initiated higher apoptotic response in FHC than in HT-29 cells. Moreover, NaBt also induced higher differentiation response in FHC cells, although the proliferation pattern of both cell types was similarly affected. The relationship between these effects and different lipid and oxidative metabolism of normal and cancer cells (changes of lipids in plasma as well as mitochondrial membrane and in cell cytoplasm associated with changes of MMP and oxidative metabolism) is supposed. The changes in cell lipids (composition of plasma and mitochondrial membranes) associated with induction of oxidative metabolism can consequently affect cell proliferation, differentiation and apoptosis. For better elucidation of these effects and relationship of individual processes the more detail studies of mechanism and their dynamics are necessary.

Generally, our research contributes to the knowledge of the role of endogenous apoptotic regulators as well as exogenous dietary factors and their interaction in colon carcinogenesis. From our results it follows that in speculation of chemopreventive effects of these compounds, it is necessary to take into consideration specific circumstances (healthy or afflicted colonic tissue), different response of normal and cancer epithelial cells and possibilities of mutual modulation of the effects of the factors operating in the lumen.

REFERENCES

[1]   Lifshitz, S., Lamprecht, S. A., Benharroch, D., Prinsloo, I., Polak-Charcon, S., and Schwartz, B. (2001). Apoptosis (programmed cell death) in colonic cells: from normal to transformed stage. *Cancer Lett* 163, 229-38.

[2]   Walczak, H., and Krammer, P. H. (2000). The CD95 (APO-1/Fas) and the TRAIL (APO-2L) apoptosis systems. *Exp Cell Res* 256, 58-66.

[3]   Manos, E. J., and Jones, D. A. (2001). Assessment of tumor necrosis factor receptor and Fas signaling pathways by transcriptional profiling. *Cancer Res* 61, 433-8.

[4]   MacFarlane, M. (2003). TRAIL-induced signalling and apoptosis. *Toxicol Lett* 139, 89-97.

[5]   Ozoren, N., and El-Deiry, W. S. (2002). Defining characteristics of Types I and II apoptotic cells in response to TRAIL. *Neoplasia* 4, 551-7.

[6]   Sprick, M. R., and Walczak, H. (2004). The interplay between the Bcl-2 family and death receptor-mediated apoptosis. *Biochim Biophys Acta* 1644, 125-32.

[7]   Jaattela, M. (1991). Biologic Activities and Mechanisms of Action of TNF-alpha/Cachektin. *Biology of Disease* 64, 724-742.

[8]   Decaudin, D., Beurdeley-Thomas, A., Nemati, F., Miccoli, L., Pouillart, P., Bourgeois, Y., Goncalves, R. B., Rouillard, D., and Poupon, M. F. (2001). Distinct experimental efficacy of anti-Fas/APO-1/CD95 receptor antibody in human tumors. *Exp Cell Res* 268, 162-8.

[9]   Krammer, P. H., Galle, P. R., Moller, P., and Debatin, K. M. (1998). CD95(APO-1/Fas)-mediated apoptosis in normal and malignant liver, colon, and hematopoietic cells. *Adv Cancer Res* 75, 251-73.

[10]   von Reyher, U., Strater, J., Kittstein, W., Gschwendt, M., Krammer, P. H., and Moller, P. (1998). Colon carcinoma cells use different mechanisms to escape CD95-mediated apoptosis. *Cancer Res* 58, 526-34.

[11]   Kim, K., Fisher, M. J., Xu, S. Q., and el-Deiry, W. S. (2000). Molecular determinants of response to TRAIL in killing of normal and cancer cells. *Clin Cancer Res* 6, 335-46.

[12]   Hague, A., Hicks, D. J., Hasan, F., Smartt, H., Cohen, G. M., Paraskeva, C., and MacFarlane, M. (2005). Increased sensitivity to TRAIL-induced apoptosis occurs during the adenoma to carcinoma transition of colorectal carcinogenesis. *Br J Cancer* 92, 736-42.

[13]   Van Geelen, C. M., de Vries, E. G., and de Jong, S. (2004). Lessons from TRAIL-resistance mechanisms in colorectal cancer cells: paving the road to patient-tailored therapy. *Drug Resist Updat* 7, 345-58.

[14]   Strater, J., and Moller, P. (2003). CD95 (Fas/APO-1)/CD95L in the gastrointestinal tract: fictions and facts. *Virchows Arch* 442, 218-25.

[15]   Bartsch, H., Nair, J., and Owen, R. W. (1999). Dietary polyunsaturated fatty acids and cancers of the breast and colorectum: emerging evidence for their role as risk modifiers. *Carcinogenesis* 20, 2209-18.

[16]   McEntee, M. F., and Whelan, J. (2002). Dietary polyunsaturated fatty acids and colorectal neoplasia. *Biomed Pharmacother* 56, 380-7.

[17] Moyad, M. A. (2005). An introduction to dietary/supplemental omega-3 fatty acids for general health and prevention: part II. *Urol Oncol* 23, 36-48.

[18] Larsson, S. C., Kumlin, M., Ingelman-Sundberg, M., and Wolk, A. (2004). Dietary long-chain n-3 fatty acids for the prevention of cancer: a review of potential mechanisms. *Am J Clin Nutr* 79, 935-45.

[19] Siddiqui, R. A., Shaikh, S. R., Sech, L. A., Yount, H. R., Stillwell, W., and Zaloga, G. P. (2004). Omega 3-fatty acids: health benefits and cellular mechanisms of action. *Mini Rev Med Chem* 4, 859-71.

[20] Kinsella, J. E. (1990). Lipids, membrane receptors, and enzymes: effects of dietary fatty acids. *JPEN J Parenter Enteral Nutr* 14, 200S-217S.

[21] DiMarzo, V. (1995). Arachidonic acid and eikosanoids as targets and effectors in second messenger interactions. *Prostaglandins, Leukotrienes and Essential Fatty Acids* 53, 239-254.

[22] Kliewer, S. A., Sundseth, S. S., Jones, S. A., Brown, P. J., Wisely, G. B., Koble, C. S., Devchand, P., Wahli, W., Willson, T. M., Lenhard, J. M., and Lehmann, J. M. (1997). Fatty acids and eicosanoids regulate gene expression through direct interactions with peroxisome proliferator-activated receptors alpha and gamma. *Proc Natl Acad Sci U S A* 94, 4318-23.

[23] Sampath, H., and Ntambi, J. M. (2005). Polyunsaturated fatty acid regulation of genes of lipid metabolism. *Annu Rev Nutr* 25, 317-40.

[24] Mills, S. C., Windsor, A. C., and Knight, S. C. (2005). The potential interactions between polyunsaturated fatty acids and colonic inflammatory processes. *Clin Exp Immunol* 142, 216-28.

[25] Martin, S., and Parton, R. G. (2006). Lipid droplets: a unified view of a dynamic organelle. *Nat Rev Mol Cell Biol* 7, 373-8.

[26] Finstad, H. S., Drevon, C. A., Kulseth, M. A., Synstad, A. V., Knudsen, E., and Kolset, S. O. (1998). Cell proliferation, apoptosis and accumulation of lipid droplets in U937-1 cells incubated with eicosapentaenoic acid. *Biochem J* 336 (Pt 2), 451-9.

[27] Inazawa, Y., Nakatsu, M., Yasugi, E., Saeki, K., and Yuo, A. (2003). Lipid droplet formation in human myeloid NB4 cells stimulated by all trans retinoic acid and granulocyte colony-stimulating factor: possible involvement of peroxisome proliferator-activated receptor gamma. *Cell Struct Funct* 28, 487-93.

[28] Al-Saffar, N. M., Titley, J. C., Robertson, D., Clarke, P. A., Jackson, L. E., Leach, M. O., and Ronen, S. M. (2002). Apoptosis is associated with triacylglycerol accumulation in Jurkat T-cells. *Br J Cancer* 86, 963-70.

[29] Maziere, C., Conte, M. A., Degonville, J., Ali, D., and Maziere, J. C. (1999). Cellular enrichment with polyunsaturated fatty acids induces an oxidative stress and activates the transcription factors AP1 and NFkappaB. *Biochem Biophys Res Commun* 265, 116-22.

[30] Muzio, G., Salvo, R. A., Trombetta, A., Autelli, R., Maggiora, M., Terreno, M., Dianzani, M. U., and Canuto, R. A. (1999). Dose-dependent inhibition of cell proliferation induced by lipid peroxidation products in rat hepatoma cells after enrichment with arachidonic acid. *Lipids* 34, 705-11.

[31] Zhang, H., and Sun, X. F. (2002). Overexpression of cyclooxygenase-2 correlates with advanced stages of colorectal cancer. *Am J Gastroenterol* 97, 1037-41.

[32] Kovarikova, M., Hofmanova, J., Soucek, K., and Kozubik, A. (2004). The effects of TNF-alpha and inhibitors of arachidonic acid metabolism on human colon HT-29 cells depend on differentiation status. *Differentiation* 72, 23-31.

[33] Vondracek, J., Stika, J., Soucek, K., Minksova, K., Blaha, L., Hofmanova, J., and Kozubik, A. (2001). Inhibitors of arachidonic acid metabolism potentiate tumour necrosis factor-alpha-induced apoptosis in HL-60 cells. *Eur J Pharmacol* 424, 1-11.

[34] Hofmanova, J., Vaculova, A., and Kozubik, A. (2005). Polyunsaturated fatty acids sensitize human colon adenocarcinoma HT-29 cells to death receptor-mediated apoptosis. *Cancer Lett* 218, 33-41.

[35] Vaculova, A., Hofmanova, J., Andera, L., and Kozubik, A. (2005). TRAIL and docosahexaenoic acid cooperate to induce HT-29 colon cancer cell death. *Cancer Lett* 229, 43-8.

[36] Penzo, D., Tagliapietra, C., Colonna, R., Petronilli, V., and Bernardi, P. (2002). Effects of fatty acids on mitochondria: implications for cell death. *Biochim Biophys Acta* 1555, 160-5.

[37] Henry-Mowatt, J., Dive, C., Martinou, J. C., and James, D. (2004). Role of mitochondrial membrane permeabilization in apoptosis and cancer. *Oncogene* 23, 2850-60.

[38] Chapkin, R. S., Hong, M. Y., Fan, Y. Y., Davidson, L. A., Sanders, L. M., Henderson, C. E., Barhoumi, R., Burghardt, R. C., Turner, N. D., and Lupton, J. R. (2002). Dietary n-3 PUFA alter colonocyte mitochondrial membrane composition and function. *Lipids* 37, 193-9.

[39] Ng, Y., Barhoumi, R., Tjalkens, R. B., Fan, Y. Y., Kolar, S., Wang, N., Lupton, J. R., and Chapkin, R. S. (2005). The role of docosahexaenoic acid in mediating mitochondrial membrane lipid oxidation and apoptosis in colonocytes. *Carcinogenesis* 26, 1914-21.

[40] Malis, C. D., Weber, P. C., Leaf, A., and Bonventre, J. V. (1990). Incorporation of marine lipids into mitochondrial membranes increases susceptibility to damage by calcium and reactive oxygen species: evidence for enhanced activation of phospholipase A2 in mitochondria enriched with n-3 fatty acids. *Proc Natl Acad Sci U S A* 87, 8845-9.

[41] Sengupta, S., Muir, J. G., and Gibson, P. R. (2006). Does butyrate protect from colorectal cancer? *J Gastroenterol Hepatol* 21, 209-18.

[42] Andoh, A., Tsujikawa, T., and Fujiyama, Y. (2003). Role of dietary fiber and short-chain fatty acids in the colon. *Curr Pharm Des* 9, 347-58.

[43] Pool-Zobel, B. L., Selvaraju, V., Sauer, J., Kautenburger, T., Kiefer, J., Richter, K. K., Soom, M., and Wolfl, S. (2005). Butyrate may enhance toxicological defence in primary, adenoma and tumor human colon cells by favourably modulating expression of glutathione S-transferases genes, an approach in nutrigenomics. *Carcinogenesis* 26, 1064-76.

[44] Hofmanova, J., Vaculova, A., Lojek, A., and Kozubik, A. (2005). Interaction of polyunsaturated fatty acids and sodium butyrate during apoptosis in HT-29 human colon adenocarcinoma cells. *Eur J Nutr* 44, 40-51.

[45] Heerdt, B. G., Houston, M. A., and Augenlicht, L. H. (1997). Short-chain fatty acid-initiated cell cycle arrest and apoptosis of colonic epithelial cells is linked to mitochondrial function. *Cell Growth Differ* 8, 523-32.

[46] Heerdt, B. G., Houston, M. A., Anthony, G. M., and Augenlicht, L. H. (1998). Mitochondrial membrane potential (delta psi(mt)) in the coordination of p53-independent proliferation and apoptosis pathways in human colonic carcinoma cells. *Cancer Res* 58, 2869-75.

[47] Heerdt, B. G., Houston, M. A., Wilson, A. J., and Augenlicht, L. H. (2003). The intrinsic mitochondrial membrane potential (Deltapsim) is associated with steady-state mitochondrial activity and the extent to which colonic epithelial cells undergo butyrate-mediated growth arrest and apoptosis. *Cancer Res* 63, 6311-9.

[48] Mei, S., Ho, A. D., and Mahlknecht, U. (2004). Role of histone deacetylase inhibitors in the treatment of cancer (Review). *Int J Oncol* 25, 1509-19.

[49] Place, R. F., Noonan, E. J., and Giardina, C. (2005). HDAC inhibition prevents NF-kappa B activation by suppressing proteasome activity: down-regulation of proteasome subunit expression stabilizes I kappa B alpha. *Biochem Pharmacol* 70, 394-406.

[50] Sasahara, Y., Mutoh, M., Takahashi, M., Fukuda, K., Tanaka, N., Sugimura, T., and Wakabayashi, K. (2002). Suppression of promoter-dependent transcriptional activity of inducible nitric oxide synthase by sodium butyrate in colon cancer cells. *Cancer Lett* 177, 155-61.

[51] Hofmanova, J., Machala, M., and Kozubik, A. (2000). Epigenetic mechanisms of the carcinogenic effects of xenobiotics and in vitro methods of their detection. *Folia Biol (Praha)* 46, 165-73.

[52] Vondracek, J., Soucek, K., Sheard, M. A., Chramostova, K., Andrysik, Z., Hofmanova, J., and Kozubik, A. (2006). Dimethyl sulfoxide potentiates death receptor-mediated apoptosis in the human myeloid leukemia U937 cell line through enhancement of mitochondrial membrane depolarization. *Leuk Res* 30, 81-9.

[53] Chopin, V., Slomianny, C., Hondermarck, H., and Le Bourhis, X. (2004). Synergistic induction of apoptosis in breast cancer cells by cotreatment with butyrate and TNF-alpha, TRAIL, or anti-Fas agonist antibody involves enhancement of death receptors' signaling and requires P21(waf1). *Exp Cell Res* 298, 560-73.

[54] Kovarikova, M., Pachernik, J., Hofmanova, J., Zadak, Z., and Kozubik, A. (2000). TNF-alpha modulates the differentiation induced by butyrate in the HT-29 human colon adenocarcinoma cell line. *Eur J Cancer* 36, 1844-52.

[55] Vaculova, A., Hofmanova, J., Soucek, K., Kovarikova, M., and Kozubik, A. (2002). Tumor necrosis factor-alpha induces apoptosis associated with poly(ADP-ribose) polymerase cleavage in HT-29 colon cancer cells. *Anticancer Res* 22, 1635-9.

[56] Teitelbaum, J. E., and Allan Walker, W. (2001). Review: the role of omega 3 fatty acids in intestinal inflammation. *J Nutr Biochem* 12, 21-32.

[57] Wilson, C. A., and Browning, J. L. (2002). Death of HT29 adenocarcinoma cells induced by TNF family receptor activation is caspase-independent and displays features of both apoptosis and necrosis. *Cell Death Differ* 9, 1321-33.

[58] Zhang, X. D., Nguyen, T., Thomas, W. D., Sanders, J. E., and Hersey, P. (2000). Mechanisms of resistance of normal cells to TRAIL induced apoptosis vary between different cell types. *FEBS Lett* 482, 193-9.

[59] Koornstra, J. J., Jalving, M., Rijcken, F. E., Westra, J., Zwart, N., Hollema, H., de Vries, E. G., Hofstra, R. W., Plukker, J. T., de Jong, S., and Kleibeuker, J. H. (2005). Expression of tumour necrosis factor-related apoptosis-inducing ligand death receptors in sporadic and hereditary colorectal tumours: potential targets for apoptosis induction. *Eur J Cancer* 41, 1195-202.

[60] Jin, Z., McDonald, E. R., 3rd, Dicker, D. T., and El-Deiry, W. S. (2004). Deficient tumor necrosis factor-related apoptosis-inducing ligand (TRAIL) death receptor transport to the cell surface in human colon cancer cells selected for resistance to TRAIL-induced apoptosis. *J Biol Chem* 279, 35829-39.

[61] Muppidi, J. R., Tschopp, J., and Siegel, R. M. (2004). Life and death decisions: secondary complexes and lipid rafts in TNF receptor family signal transduction. *Immunity* 21, 461-5.

[62] Kim, Y., Suh, N., Sporn, M., and Reed, J. C. (2002). An inducible pathway for degradation of FLIP protein sensitizes tumor cells to TRAIL-induced apoptosis. *J Biol Chem* 277, 22320-9.

[63] Zhang, X. D., Borrow, J. M., Zhang, X. Y., Nguyen, T., and Hersey, P. (2003). Activation of ERK1/2 protects melanoma cells from TRAIL-induced apoptosis by inhibiting Smac/DIABLO release from mitochondria. *Oncogene* 22, 2869-81.

[64] Tran, S. E., Holmstrom, T. H., Ahonen, M., Kahari, V. M., and Eriksson, J. E. (2001). MAPK/ERK overrides the apoptotic signaling from Fas, TNF, and TRAIL receptors. *J Biol Chem* 276, 16484-90.

[65] LeBlanc, H., Lawrence, D., Varfolomeev, E., Totpal, K., Morlan, J., Schow, P., Fong, S., Schwall, R., Sinicropi, D., and Ashkenazi, A. (2002). Tumor-cell resistance to death receptor--induced apoptosis through mutational inactivation of the proapoptotic Bcl-2 homolog Bax. *Nat Med* 8, 274-81.

[66] Diggle, C. P. (2002). In vitro studies on the relationship between polyunsaturated fatty acids and cancer: tumour or tissue specific effects? *Prog Lipid Res* 41, 240-53.

[67] Begin, M. E., Ells, G., and Horrobin, D. F. (1988). Polyunsaturated fatty acid-induced cytotoxicity against tumor cells and its relationship to lipid peroxidation. *J Natl Cancer Inst* 80, 188-94.

[68] Nwankwo, J. O. (2001). Repression of cellular anaplerosis as the hypothesized mechanism of gamma-linolenic acid-induced toxicity to tumor cells. *Med Hypotheses* 56, 582-8.

[69] Griffiths, G., Jones, H. E., Eaton, C. L., and Stobart, A. K. (1997). Effect of n-6 polyunsaturated fatty acids on growth and lipid composition of neoplastic and non-neoplastic canine prostate epithelial cell cultures. *Prostate* 31, 29-36.

[70] Cao, Y., Pearman, A. T., Zimmerman, G. A., McIntyre, T. M., and Prescott, S. M. (2000). Intracellular unesterified arachidonic acid signals apoptosis. *Proc Natl Acad Sci U S A* 97, 11280-5.

[71] Stillwell, W., and Wassall, S. R. (2003). Docosahexaenoic acid: membrane properties of a unique fatty acid. *Chem Phys Lipids* 126, 1-27.

[72] Shaikh, S. R., Dumaual, A. C., Castillo, A., LoCascio, D., Siddiqui, R. A., Stillwell, W., and Wassall, S. R. (2004). Oleic and docosahexaenoic acid differentially phase separate from lipid raft molecules: a comparative NMR, DSC, AFM, and detergent extraction study. *Biophys J* 87, 1752-66.

[73] Ma, D. W., Seo, J., Davidson, L. A., Callaway, E. S., Fan, Y. Y., Lupton, J. R., and Chapkin, R. S. (2004). n-3 PUFA alter caveolae lipid composition and resident protein localization in mouse colon. *Faseb J* 18, 1040-2.

[74] Ma, D. W., Seo, J., Switzer, K. C., Fan, Y. Y., McMurray, D. N., Lupton, J. R., and Chapkin, R. S. (2004). n-3 PUFA and membrane microdomains: a new frontier in bioactive lipid research. *J Nutr Biochem* 15, 700-6.

[75] Watkins, S. M., Carter, L. C., and German, J. B. (1998). Docosahexaenoic acid accumulates in cardiolipin and enhances HT-29 cell oxidant production. *J Lipid Res* 39, 1583-8.

[76] Cristea, I. M., and Degli Esposti, M. (2004). Membrane lipids and cell death: an overview. *Chem Phys Lipids* 129, 133-60.

[77] Aw, T. Y. (1999). Molecular and cellular responses to oxidative stress and changes in oxidation-reduction imbalance in the intestine. *Am J Clin Nutr* 70, 557-65.

[78] Fernandez-Checa, J. C. (2003). Redox regulation and signaling lipids in mitochondrial apoptosis. *Biochem Biophys Res Commun* 304, 471-9.

[79] Coradini, D., Pellizzaro, C., Marimpietri, D., Abolafio, G., and Daidone, M. G. (2000). Sodium butyrate modulates cell cycle-related proteins in HT29 human colonic adenocarcinoma cells. *Cell Prolif* 33, 139-46.

[80] Lupton, J. R. (1995). Butyrate and colonic cytokinetics: differences between in vitro and in vivo studies. *Eur J Cancer Prev* 4, 373-8.

[81] Madesh, M., Benard, O., and Balasubramaian, K. A. (1996). Butyrate-induced alteration in lipid composition of human colon cell line HT-29. *Biochem Mol Biol Int* 38, 659-64.

[82] Comalada, M., Bailon, E., de Haro, O., Lara-Villoslada, F., Xaus, J., Zarzuelo, A., and Galvez, J. (2006). The effects of short-chain fatty acids on colon epithelial proliferation and survival depend on the cellular phenotype. *J Cancer Res Clin Oncol* 132, 487-97.

[83] Mariadason, J. M., Velcich, A., Wilson, A. J., Augenlicht, L. H., and Gibson, P. R. (2001). Resistance to butyrate-induced cell differentiation and apoptosis during spontaneous Caco-2 cell differentiation. *Gastroenterology* 120, 889-99.

[84] Fajkus, J., Borsky, M., Kunicka, Z., Kovarikova, M., Dvorakova, D., Hofmanova, J., and Kozubik, A. (2003). Changes in telomerase activity, expression and splicing in response to differentiation of normal and carcinoma colon cells. *Anticancer Res* 23, 1605-12.

[85] Bartova, E., Pachernik, J., Harnicarova, A., Kovarik, A., Kovarikova, M., Hofmanova, J., Skalnikova, M., Kozubek, M., and Kozubek, S. (2005). Nuclear levels and patterns

of histone H3 modification and HP1 proteins after inhibition of histone deacetylases. *J Cell Sci* 118, 5035-46.

[86]　Narayanan, B. A., Narayanan, N. K., and Reddy, B. S. (2001). Docosahexaenoic acid regulated genes and transcription factors inducing apoptosis in human colon cancer cells. *Int J Oncol* 19, 1255-62.

[87]　Horvath, P. J., Awad, A. B., and Andersen, M. (1993). Differential effect of butyrate on lipids of human colon cancer cells. *Nutr Cancer* 20, 283-91.

[88]　Awad, A. B., Horvath, P. J., and Andersen, M. S. (1991). Influence of butyrate on lipid metabolism, survival, and differentiation of colon cancer cells. *Nutr Cancer* 16, 125-33.

[89]　Awad, A. B., Kamei, A., Horvath, P. J., and Fink, C. S. (1995). Prostaglandin synthesis in human cancer cells: influence of fatty acids and butyrate. *Prostaglandins Leukot Essent Fatty Acids* 53, 87-93.

[90]　Tang, D. G., La, E., Kern, J., and Kehrer, J. P. (2002). Fatty acid oxidation and signaling in apoptosis. *Biol Chem* 383, 425-42.

[91]　Giardina, C., Boulares, H., and Inan, M. S. (1999). NSAIDs and butyrate sensitize a human colorectal cancer cell line to TNF-alpha and Fas ligation: the role of reactive oxygen species. *Biochim Biophys Acta* 1448, 425-38.

[92]　Hong, M. Y., Chapkin, R. S., Barhoumi, R., Burghardt, R. C., Turner, N. D., Henderson, C. E., Sanders, L. M., Fan, Y. Y., Davidson, L. A., Murphy, M. E., Spinka, C. M., Carroll, R. J., and Lupton, J. R. (2002). Fish oil increases mitochondrial phospholipid unsaturation, upregulating reactive oxygen species and apoptosis in rat colonocytes. *Carcinogenesis* 23, 1919-25.

[93]　Hossain, Z., Konishi, M., Hosokawa, M., and Takahashi, K. (2006). Effect of polyunsaturated fatty acid-enriched phosphatidylcholine and phosphatidylserine on butyrate-induced growth inhibition, differentiation and apoptosis in Caco-2 cells. *Cell Biochem Funct* 24, 159-65.

[94]　Scorrano, L., Penzo, D., Petronilli, V., Pagano, F., and Bernardi, P. (2001). Arachidonic acid causes cell death through the mitochondrial permeability transition. Implications for tumor necrosis factor-alpha aopototic signaling. *J Biol Chem* 276, 12035-40.

[95]　Narayanan, B. A., Narayanan, N. K., Simi, B., and Reddy, B. S. (2003). Modulation of inducible nitric oxide synthase and related proinflammatory genes by the omega-3 fatty acid docosahexaenoic acid in human colon cancer cells. *Cancer Res* 63, 972-9.

In: Cell Apoptosis Research Trends
Editor: Charles V. Zhang, pp. 207-230

ISBN: 1-60021-424-X
© 2007 Nova Science Publishers, Inc

Chapter VII

Targeting Cell Cycle in a Coordinated Strategy for Anticancer Treatment and Neuroprotection

Vilen A. Movsesyan and Alexander G. Yakovlev

Department of Neuroscience, Georgetown University, Washington, D. C. 20007

ABSTRACT

Developing brain is highly vulnerable to injury caused by radiation and chemotherapy leading to DNA damage. These stimuli activate the intrinsic apoptotic pathway that results in extensive neuronal loss over prolonged periods of time. High susceptibility of the developing neurons to apoptosis represents the major problem in treatment of childhood brain tumors by radiation and chemotherapy that cause significant loss of healthy neurons leading to development of severe cognitive and psychological deficits. Therefore, development of anti-cancer treatment strategies aimed at preservation of the developing brain is highly significant.

In this study, we made the first step in development of such a strategy using *in vitro* models of rat primary neurons and human tumor cell lines. Our results demonstrate that an inhibitor of cyclin-dependent kinases, flavopiridol potentiates death of tumor cells in cultures, while protecting cerebellar and cortical neurons from apoptosis after irradiation and treatment with the DNA damaging drug, etoposide. Taken together, our experimental data suggest that the neuroprotective effect of flavopiridol may be mediated by repression of specific BH3-only genes through the cyclin D1/CDK4/Rb/E2F-dependend pathway.

Future development and use of specific synthetic inhibitors of cyclin-dependent kinases in the clinic may have a significant impact on the improvement of the existing strategies to treat childhood brain tumors.

ABBREVIATIONS

AMC	Aminomethylcoumarin
Smac/Diablo	Second mitochondria-derived activator of caspase
IAP	Inhibitor of apoptosis
AIF	Apoptosis-inducing factor
Bcl2	B-cell leukemia 2 oncogene
Bak	Bcl2 antagonist killer
Bax	Bcl2-associated X protein BH3
Bcl2	Homology region 3
PARP-1	Poly (ADP-ribose) polymerase 1
MAPK	Mitogen-activated protein kinase
TRAIL	Tumor necrosis factor-related apoptosis-inducing ligand
ERK	Extracellular signal-regulated kinase
FLIP	FLICE (caspase-8) inhibitory protein
iNOS	Inducible nitric oxide synthase

INTRODUCTION

Apoptosis plays a critical role in normal development and control of cancer. During brain development nearly half of neural cells die as a result of activation of the intrinsic apoptotic program [1, 2], and aberrations in the pathways of physiological neuronal death often lead to alterations in brain development [3-5]. On the other hand, mutations in essential apoptotic genes may result in tumor formation, which are not rare in the developing brain [6].

Most common brain tumors in children are localized in the cerebellum. Medulloblastoma (MB) is one of them that usually occurs in children in the first decade of life and represents 15-20% of intracranial neoplasms. Surgery is the primary choice of MB treatment; however, in many cases complete removal of the tumor is not possible, which increases a possibility of MB spread to other areas of the brain, spinal cord, and bone marrow. Therefore, after surgery MB patients are often treated with radiation and/or chemotherapy. While this treatment strategy has markedly improved the survival rate, in young children it causes significant loss of healthy tissues and leads to development of serious cognitive and psychological deficits, which are less profound in adult patients.

It was common believe in the past that brain injury in children is less harmful compared to adults. However, more recent reports have provided strong experimental evidence that the developing brain is significantly more vulnerable to damage, which explains the fact that radiation and chemotherapy cause more severe damage in children. Rapidly accumulating knowledge suggests that such difference is based on different age-dependent susceptibility to neuronal apoptosis [7]. Thus, Nakaya and coauthors reported that age-related differences in apoptotic sensitivity contribute to the increased vulnerability of the young mouse brain to radiation [8]. These authors found significantly greater number of apoptotic neurons in the cortex of irradiated 1-day-old animals relative to older groups. Furthermore, Li and Wong demonstrated similar age-dependent differences in the extent of radiation-induced apoptosis in the rat spinal cord [9]. It has also been shown that cerebellar granule neurons (CGNs) lose their competence to die *in vitro* in response to radiation and a DNA damaging agent camptothecin as a function of age in culture [10].

Loss of developing neurons after radiation and chemotherapy results from activation of the intrinsic apoptotic pathway. In this pathway, DNA damage caused by radiation and common anticancer drugs triggers cell death through alterations at the level of mitochondria and the release of mitochondrial cytochrome *c* (cyto-C), Smac/Diablo, HtrA2/Omi, AIF, and endonuclease G to cell cytosol [11-18]. Each of these proteins may contribute to cell death. Thus, in the presence of ATP or dATP, cyto-C binds to cytosolic Apaf-1 leading to the assembly of the apoptosome and activation of the initiator caspase-9, which, in turn, activates executioner caspases [17]. Initiation of cyto-C release is primarily regulated by members of the Bcl-2 protein family [19]; however, most recent reports indicate that active caspases 3 and 7 significantly enhance apoptotic events in mitochondria [20]. Smac/Diablo and HtrA2/Omi are involved in activation of caspases *via* binding and inhibiting members of the IAP protein family, although they can also induce cell death independently of IAPs or caspases [11-15]. In healthy cells, mitochondrial AIF plays an important physiological role by regulating oxidative processes and neuronal survival [21, 22]. However, in response to apoptotic stimuli, it is able to enter cell nuclei and cause condensation of chromatin, large-scale DNA fragmentation, and cell death [16, 23]. These effects are independent of caspases and the oxidoreductase activity of AIF.

Apoptotic alterations at the mitochondrial level largely depend on oligomerization of multidomain Bak and Bax on mitochondrial membranes [24]. Prosurvival Bcl-2 proteins inhibit the release of cyto-C from mitochondria and reduce cell death [19]. BH3-only proteins are believed to bind to and neutralize Bcl-2-like family members [25], and their induction or activation promotes oligomerization of proapoptotic Bak and Bax [26]. A critical role in the initiation of this pathway has been attributed to the p53 tumor suppressor protein [27]. In response to DNA damage, this transcription factor induces *BAX* and BH3-only genes, such as *NOXA* and *PUMA*, resulting in the release of cyto-C and caspase activation [28-31]. However, p53 is also responsible for activation of caspase-independent pathways of cell death induced *via* Bax-dependent AIF translocation to cell nuclei [16, 32] or p53-dependent induction of AMID (AIF-homologous mitochondrion-associated inducer of death) also known as PRG3 [33, 34].

Moreover, in addition to activation of the proapoptotic Bcl-2 family members, p53 induces expression of cyclin D1, cyclin-dependent protein kinases (CDKs), and other proteins involved in the cell cycle and DNA repair [35-37]. In proliferating cells, cyclin D1 is expressed in the G_1 phase and forms an active complex with CDK4/6 that phosphorylate retinoblastoma (Rb) protein [38]. Phosphorylation of Rb, in turn, inhibits its function as an inhibitor of E2F transcription factors [39]. An important role of this pathway in neuronal survival has been demonstrated in many instances. Thus, Clarke and collaborators [40] reported that disruption of the *RB* gene in mice results in widespread neuronal death. Remarkably, Rb-deficient neurons can be rescued by E2F1 deficiency [41]. On the other hand, overexpression of E2F1 can induce neuronal apoptosis under certain conditions [42]. Activation of CDK4/6 is also essential for the execution of the death program, since specific inhibition of these CDKs by p16ink4 or their dominant negative mutants prevents neuronal death [43].

Hence, the increasing evidence suggests that the cyclin D1/CDK4/Rb/E2F pathway essentially contributes to neuronal death. Our knowledge about contribution of DNA damage and cell cycle to neuronal death is largely based on the systematic studies reported by David Park and Lloyd Greene, who demonstrated that p53 and CDKs can be activated independently [44]. Their data clearly indicate that the inhibition of either pathway alone inhibits cyto-C release, caspase-3 activation, and improves neuronal survival [44]. On the other hand, inhibition of the cyclin D1-dependent kinases leads to proliferation arrest and suppression of MB [45].

A role for DNA damage in neurodegeration has been strongly supported by the observations that mice deficient in DNA damage repair pathways display aberrant neuronal loss and impaired neurodevelopment [46-48]. Thus, mice carrying a *p53* null mutation are more resistant to neuronal injury [49, 50], and p53 overexpression is sufficient to cause neuronal death [48, 51]. It has also been demonstrated that synthetic inhibitors of p53 are neuroprotective in models of acute brain injury [52, 53]. Consistent with these reports, we have shown that caspase-dependent apoptosis can be inhibited in neuroblastoma cells by overexpression of the p53 dominant negative form [54].

An alternative mechanism that does not depend on p53 and proapoptotic Bcl-2 proteins can also contribute to cyto-C release and caspase activation after DNA damage, although this mechanism operates in late stages of cell death [55]. Activation of PARP-1 is a key event in the initiation of this pathway. PARP-1 is a well examined nuclear enzyme activated by DNA damage, which participates in DNA repair, replication, cell differentiation, and death [56]; however, a specific mechanism, by which PARP-1 signals to mitochondria, remains to be investigated.

Notably, mutations in *p53* and *PARP-1* genes lead to development of MB in mice. Thus, Wetmore and colleagues [57] have demonstrated that *p53* deletion markedly increases the number of MB in *PATCHED1* (*PTCH*) heterozygous animals. *PTCH* mutations have been found in a number of sporadic MB, suggesting that this gene or other members of the Hedgehog (Hh) pathway play a role in MB development. These data suggest that inactivation of *p53* and other genes controlling cell cycle, DNA repair, and apoptosis promotes MB formation through the increase in genetic instability [58]. This hypothesis is supported by the observation of Tong and colleagues [59] who demonstrated that increased genetic instability caused by abrogation of p53 and PARP-1 function results in activation of the Hh pathway (without underlying *PTCH* mutations) and growth of MB.

Hence, it becomes increasingly clear that the progression of MB tumors employs the same molecular mechanisms that control survival of normal neural tissue in the developing brain. Therefore, activation of such proapoptotic mechanisms in cancer cells by radiation and chemotherapy leads to simultaneous induction of apoptosis in normal brain neurons as well as cognitive and psychological deficits later in development. Thus, development of anti-cancer treatment strategies aimed on preservation of the developing brain is an important goal. Analysis of the recent progress in the fields of anticancer therapy and neuroprotection suggests that inhibition of the cell cycle using CDK inhibitors may provide a key for the development of such a strategy.

In this report, we demonstrate that an inhibitor of CDKs, flavopiridol potentiates death of tumor cells *in vitro*, while protecting primary cerebellar and cortical neurons from apoptosis after irradiation and treatment with the DNA damaging agent etoposide. We also found that the neuroprotective action of flavopiridol targets induction of specific *BH3-only* genes, the release of proapoptotic factors from mitochondria and can be achieved as late as 2 h postinjury.

MATERIALS AND METHODS

Daoy and SH-SY5Y Cell Cultures—Daoy human MB cells [60] were obtained from the American Tissue Culture Collection (Manassas, VA). Cells were grown in Minimum Essential Medium (Eagle) with 2 mM L-glutamine and Earle's BSS containing 1.5 g/L sodium bicarbonate, 0.1 mM non-essential amino acids, 1.0 mM sodium pyruvate, and 10% of fetal bovine serum in a humidified atmosphere of 5% CO_2 in air at 37°C. SH-SY5Y human neuroblastoma cells were obtained from the Cancer Cell Line Repository at Georgetown University Lombardi Cancer Center (Washington, D. C.). Cells were grown in Dulbecco's modified Eagle's medium supplemented with 10% fetal bovine serum, 100 units/ml penicillin, and 100 mg/ml streptomycin. The cultures were maintained in a logarithmic phase by passage every 2–3 days.

Rat cortical neuronal cultures—Rat cortical neurons (RCNs) were derived from embryonic cortices (Taconic, Germantown, NY) as described previously [61]. Briefly, cortices from 17 to 18-day-old SpragueDawley rat embryos were cleaned from meninges and blood vessels in Krebs-Ringers bicarbonate buffer containing 0.3% bovine serum albumin (BSA), (Invitrogen Corp., Carlsbad, CA). Cortices were minced and dissociated in the same buffer with 1800 U/ml trypsin (Sigma, St. Louis, MO) at 37°C for 20 min. Following the addition of 200 U/ml DNase I (Sigma) and 3600 U/ml soybean trypsin inhibitor (Sigma) to the suspension, cells were triturated through a 5 ml pipette. After the tissue was allowed to settle for 5-10 min, the supernatant was collected, and the remaining tissue pellet was retriturated. The combined supernatants were centrifuged through a 4% BSA layer and the cell pellet was resuspended in neuronal seeding medium (NSM), which consisted of Neurobasal Medium (Invitrogen) supplemented with 1.1% 100 × antibiotic/antimycotic solution (Invitrogen), 25 µM Na-glutamate, 0.5 mM L-glutamine, and 2% B27 supplement (Invitrogen). Cells were seeded at a density of 5×10^5 cells/ml onto 96-well tissue culture plates or 10 cm Petri dishes (Corning, Corning, NY, USA) precoated with poly-D-lysine (70-150 kDa, Sigma). This technique allows to obtain cultures containing 91 ± 5% of neurons as estimated by immunostaining with antibodies to the neuronal specific enolase and the neuronal nuclear antigen [7].

Cerebellar granule neurons—Briefly, the meninges and blood vessels were removed from cerebella of 7-day-old Sprague-Dawley rat pups (Taconic) while in Krebs-Ringerr's bicarbonate buffer containing 0.3% bovine serum albumin (Invitrogen). The cerebella were dissociated at 37°C for 20 min in the same buffer with the addition of 1800 U/ml trypsin (Sigma). The dissociated cells were triturated through a 5 ml pipette following the addition of 200 U/ml DNase I (Sigma) and 3600 U/ml soybean trypsin inhibitor (Sigma). The supernatant was collected after the tissue settled for 10 min; the pellet was retriturated as noted above. The supernatants were combined and centrifuged through a 4% BSA solution. The pellet was resuspended in seeding medium containing Neurobasal Medium (Invitrogen), 25 mM KCl, 0.5 mM L-glutamine, 50 µg/ml gentamicin, and 2% B27 supplement (Invitrogen). Cells were seeded at 5×10^5 cells/ml onto culture plates (Corning) precoated with poly-D-lysine (Sigma).

Assessment of Cell Viability—Cell viability was measured by retention and deesterification of calcein AM [62] or by the LDH release assay [63]. In brief, media in culture plates was replaced with 5 µM calcein AM (Molecular Probes, Eugene, OR) in Locke's buffer containing 154 mM NaCl, 5.6 mM KCl, 3.6 mM $NaHCO_3$, 2.3 mM $CaCl_2$, 1.2 mM $MgCl_2$, 5.6 mM glucose, and 5 mM Hepes, pH 7.4. After incubation at 37°C for 30 min, fluorescence was measured using a CytoFluor 4000 fluorometer (PerSeptive Biosystems Inc., Framingham, MA) at 485 nm excitation and 560 nm emission wavelengths.

LDH release was measured using CytoTox-96 non-radioactive cytotoxicity assay kit (Promega, Madison, WI) according to the manufacturer's protocol. Relative absorbance was measured at 490 nm using a Multiskan Ascent microplate reader (Labsystems Inc., Helsinki, Finland).

Immunoblotting—Cells were harvested and washed with ice-cold phosphate-buffered saline. To analyze a total protein content, cells were lysed on ice in a solution containing 50 mM Tris-HCl, pH 7.5, 150 mM NaCl, 1 mM EGTA, 1 mM phenylmethylsulfonyl fluoride, 0.5% Nonidet P-40, 0.25% SDS, 5 µg/ml leupeptin, and 5 µg/ml aprotinin. To obtain cytosolic fractions, cells were lysed for 5 min on ice in a solution consisting of 20 mM Hepes, pH 7.2, 80 mM KCl, 1 mM EDTA, 1 mM EGTA, 1 mM dithiothreitol, 1 mM 4-(2-aminoethyl)benzenesulfonyl fluoride, 10 µg/ml aprotinin, 10 µg/ml leupeptin, 1 µM pepstatin, 250 mM sucrose, and 200 µg/ml digitonin. After removal of cell debris by centrifugation at 20,000 x *g* for 20 min, a portion of the lysate was then fractionated by SDS-PAGE, and the separated proteins were transferred to a nitrocellulose filter. The filter was stained with Ponceau S to confirm equal loading and transfer of samples and was then probed with specific antibodies. Immune complexes were detected with appropriate secondary antibodies and chemiluminescence reagents (Pierce Biotechnology Inc., Rockford, IL). A polyclonal rabbit antibody to active caspase-3 was obtained from *Cell Signaling Technology* (New England Biolabs, #9661); a monoclonal mouse antibody to caspase-9 was from Medical and Biological Laboratories (MBL, clone 5B4); a monoclonal mouse antibody to β-actin was from Sigma (A5441); polyclonal rabbit antibodies to procaspase-3 were from Santa Cruz Biotechnology (sc-7148); a mouse monoclonal antibody to cyto-C was from BD Biosciences (556433); and a mouse monoclonal antibody to p53 was from Oncogene Research Products (OP43).

Reverse Transcription-PCR—Abundance of mRNAs encoding proapoptotic members of the Bcl-2 family was examined by reverse transcription (RT)-PCR. Total cellular RNA was isolated by acidic phenol extraction [64] and treated with DNase I. RT was performed using 10 µg of total RNA in a 20 µl Moloney murine leukemia virus reverse transcription reaction (*Invitrogen*). One-tenth of the resulting cDNA was amplified by PCR. PCR primers were as follows (sense and antisense, respectively): *BAX*, 5'- TGCAGAGGATGATTGCTGAC -3' and 5'- GATCAGCTCGGGCACTTTAG -3'; *HRK*, 5'- TCTGGAAGACACCCTCTGCT -3' and 5'- CCAGGCTCAGAAAGCAAAAC -3'; *NOXA*, 5'- GGAGTGCACCGGACATAACT -3' and 5'- TGTCTCCAATTCTCCGCAGT -3'; *PUMA*, 5'- ATACTGGACTGCCAGCCTTG -3' and 5'- CTCGGTCACCATGAGTCCTT -3'; *SIVA*, 5'- CAAGCAGCTCCTTTTCCAAG -3' and 5'- TCTCACGCAGGATGAACAAG -3'; and rat *Histone 4G* RNA, 5'-ATGTCTGGACGAGGGAAAGGCGGC-3' and 5'-CCGTGACCGTCTTGCGCTTGGC GTG-3'. Amplification profiles included denaturation for 30 s at 94°C, annealing for 15 s at 55°C, and primer extension for 60 s at 72°C. Numbers of cycles were estimated to be optimal to provide a linear relationship between the amount of input template and the amount of PCR product generated over a wide concentration range from 1 to 20 µg of total RNA as described in detail previously [65]. The amplification products were analyzed by electrophoresis in 2% agarose gels in the presence of 0.5 µg/ml ethidium bromide. After being stained, UV light gel images were captured and analyzed using the *Scion Image* software. Levels of individual mRNA were expressed in arbitrary units as the proportion of individual PCR product mean optical density (inverted video image) to a control (*Histone 4G*) product mean optical density obtained from the same RNA sample.

Assay for caspase activity— Caspase-3-like activity was measures as described by us previously [66]. Cells were grown in 96-well plates. Culture media in wells was replaced with 50 µl of caspase assay buffer (10 mM Hepes/KOH, pH 7.4, 2 mM EDTA, 0.1% Chaps, 5 mM dithiothreitol) containing 20 µM Ac-DEVD-AMC (Biomol International, Plymouth Meeting, PA). Accumulation free AMC that resultes from cleavage of the aspartate–AMC bond was monitored continuously in each well using a CytoFluor 4000 fluorometer at 360-nm excitation and 460-nm emission wavelengths. The emission from each well was plotted against time. Linear regression analysis of the initial velocity (slope) of each curve yielded an activity for each sample. A standard curve was established using known concentrations of free AMC and the results were normalized based on the standard curve and expressed as picomoles (pmol) AMC per minute. The interpretation of the in-plate assay is based on the fact that all wells were plated with and contain the same number of cells.

Results

Flavopiridol promotes apoptosis induced in MB and neuroblastoma cells by radiation and etoposide. Radiation and chemotherapy are commonly used to eliminate tumor cells remaining after surgical removal of solid tumors. Although shown to negatively affect neurological function in children, this strategy is still widely applicable for treatment of childhood brain tumors including MB. Besides, recent reports indicate that cells derived from a variety of solid tumors can be effectively treated by an experimental drug flavopiridol, which on the other hand, has been shown to protect neurons in various models of neuronal degeneration including DNA damage [67-71]. Considering that radiation and many commonly used chemotherapeutic drugs induce cell death through DNA damage-dependent mechanisms, we assumed that treatment with flavopiridol may provide a way to protect normal neurons during a course of DNA damaging therapy. In order to test this hypothesis, we examined effects of flavopiridol on apoptosis and survival of primary neurons and tumor cells *in vitro*.

To our knowledge, an effect of flavopiridol on survival of MB cells has not been investigated. Therefore, we first examined an ability of this compound alone and in combination with radiation and etoposide to induce apoptosis in Daoy cells derived from human MB [60]. Measurements of DEVD-AMC cleavage in cytosolic extracts of Daoy cells made 12 h after treatment showed ~40-fold induction of caspase-3-like activity after etoposide treatment, ~20-fold induction after irradiation, and ~12-fold increase after treatment with flavopiridol (Fig.1A). Any combination of the treatments demonstrated an additive effect with a higher DEVDase activity in the presence of flavopiridol. The highest caspase activation (~62-fold) was observed after simultaneous treatment with flavopiridol, radiation, and etoposide. Overall viability of cells measured 24 h after treatment correlated inversely with levels of DEVD-AMC cleavage demonstrating the lowest viability in cells treated with the combination of flavopiridol, radiation, and etoposide (Fig.1B). Notably, survival of cells after irradiation or treatment with flavopiridol was significantly higher compared to treatment with etoposide, suggesting that, in addition to caspase-dependent apoptosis, irradiation and flavopiridol activate caspase-independent death of Daoy cells.

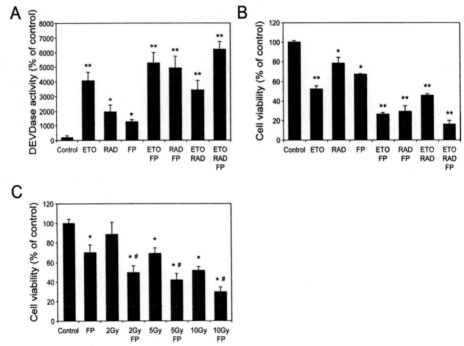

Figure 1. Flavopiridol promotes apoptosis of human tumor cells induced by radiation and/or etoposide in vitro. A, Daoy cells were treated with etoposide (ETO) and/or irradiated (RAD) with 10 Gy in the presence or absence of 1 μM flavopiridol (FP). Untreated cells served as a control. Caspase-3-like activity was assayed fluorometrically 12 h after treatment and expressed as a percentage of control ± S.D. (n = 5). *, p < 0.05; **, p < 0.01 compared to control cells. B, Daoy cell viability was measured using calcein AM assay 24 h after treatment. Data are expressed as a percentage of the value for irradiated cells in the absence of FP ± S.D. *, p < 0.05; **, p < 0.01 compared to control cells. C, SH-SY5Y cells were treated with 1 μM FP or irradiated with indicated doses in the absence or presence of 1 μM FP. Cell viability was measured using calcein AM assay 24 h after treatment. Data are expressed as a percentage of the value for control cells ± S.D. *, p < 0.01; compared to control cells. #, p < 0.01; compared to the value of cells irradiated with the same dose in the absence of FP.

A similar additive proapoptotic effect of radiation and flavopiridol was observed in the SH-SY5Y human neuroblastoma cell line (Fig.1C). In these cells, radiation resulted in dose-dependent decrease of cell viability 24 h after treatment from ~90% to ~50% at 2 and 10 Gy, correspondingly. Viability of SH-SY5Y cells treated with flavopiridol alone decreased to approximately 70%. However, cell loss after irradiation in the presence of flavopiridol was significantly higher compare to a single treatment. Thus, the rate of cell death in SH-SY5Y cultured irradiated with 2 Gy in the presence of 1 μM flavopiridol was similar to that of 10 Gy-irradiated cells in the absence of flavopiridol (Fig.1C).

Age-dependent susceptibility of cerebellar and cortical neurons to apoptosis induced by DNA damage. Radiation and chemotherapy lead to DNA damage that kills normal neurons through activation of the intrinsic apoptotic pathway. It has been convincingly demonstrated that both radiation and chemotherapy most harmfully affect the developing nervous system. This effect may be mediated by a high apoptotic potential in developing neurons. Therefore, in order to test neuroprotective ability of flavopiridol against radiation and chemotherapy, we utilized *in vitro* models of vulnerable primary cortical and cerebellar neurons at early stages of their development.

Previously we have shown that the apoptotic potential of cortical neurons markedly declines between days 7 and 14 of postnatal development *in vivo* and *in vitro* [7]. In this study, we demonstrate that a similar age-dependent regulation of the apoptotic ability also takes place the rat cerebellum. This finding is based on measurements of caspase-3 activity and cell viability of primary CGNs cultured for 1, 7, or 14 days *in vitro* (DIV) prior to treatment with etoposide (Fig.2A,B). It is essential to note that 1 DIV CGNs correspond to a day 8 of postnatal development *in vivo*, as 7 days old pups were used for cell cultures. Results show that incubation of 1 DIV CGNs in the presence of 50 μM etoposide leads to ~6-fold activation of caspase-3 (Fig. 2A) and death of approximately 55% CGNs (Fig.2B). In contrast, changes in caspase-3 activity in 7DIV and 14 DIV cells were not detected for at least 24 h after the treatment (Fig.2A,B). Differential activation of caspase-3 in 1 DIV and 14 DIV CGNs was confirmed by Western immoblotting using specific antibodies that recognize the active form of this caspase Fig.2C).

In parallel, we have analyzed AIF release from mitochondria in 1 DIV and 14 DIV CGNs during a time course after etoposide treatment. In contrast to caspase-3 activity, no difference in AIF release was observed between cultures of different ages (Fig.2C), suggesting independence of this event from caspase-3 activity. Furthermore, using antibodies that recognize procaspase-3, we demonstrated that the age-dependent decline in the ability of CGNs to undergo caspase-dependent apoptosis positively correlates with the expression levels of caspase-3 protein in 1, 7, and 14 DIV CGNs, as well as in the rat cerebellum isolated at postnatal days 1, 7, and 14 (Fig.2D).

Taken together these results determined our choice of using 1DIV RCNs and 1 DIV CGNs as a model of radiation- and chemotherapy-induced neuronal death.

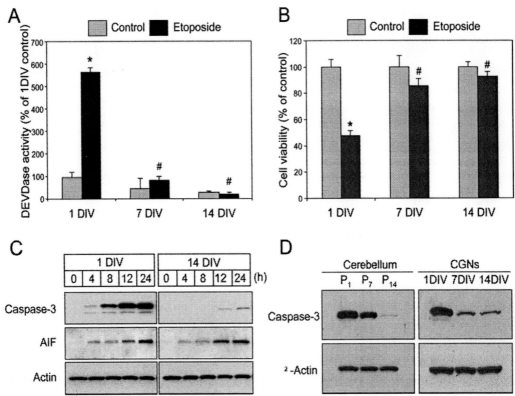

Figure 2. Age-depence of apoptotic susceptibility of rat cerebellar ganule neurons. A, one, 7, or 14 DIV CGNs were treated with 50 μM etoposide for 6 hr. Untreated cultures (Control) served as negative controls. Caspase-3-like activity in cytosolic extracts from treated or control cells was assayed fluorometrically. Protease activity is expressed in as a percentage of 1 DIV Control ± SD (n = 6). *, p < 0.001, compared to caspase-3 activity in control 1DIV cells, by ANOVA, followed by the Student-Newman-Keuls test. #, p < 0.001, compared to caspase-3 activity in etoposide-treated 1 DIV neurons. B, One, 7, or 14 DIV primary neurons were treated with 50 μM etoposide for 24 hr, and cell viability was analyzed by measurement of calcein AM fluorescence. Data are expressed as a percentage of the value for control cells not exposed to etoposide ± SD (n = 6). *, p < 0.01, compared to caspase-3 activity in control 1DIV cells. #, p < 0.05, compared to caspase-3 activity in etoposide-treated 1 DIV neurons. C, twenty micrograms of cytosolic proteins from 1 or 14 DIV CGNs were separated in 12% SDS-PAGE followed by staining with an anti-active caspase-3, anti-AIF, and anti-β-Actin antibodies. D, Western blot analysis of the abundance of procaspase-3 in the protein extracts isolated from rat cerebellum on the indicated days of postnatal development. Twenty microgram aliquots of cytosolic protein extracts were subjected to 12% SDS-PAGE, transferred to a membrane, and probed with corresponding antibodies. Beta-actin protein abundance was used as a control for gel loading and transfer.

*Flavopiridol protects immature cortical and cerebellar neurons from apoptosis induced by radiation and etoposide.*It has been shown previously that flavopiridol protects primary neurons against cell death induced by a variety of proapoptotic stimuli including potassium withdrawal, kainic acid, or *ara-c* [67-71]. More recently, it was demonstrated that flavopiridol also protects RCNs against etoposide-induced cell death [72]. Here we confirmed these findings using a wide range of flavopiridol concentrations. Caspase-3-like activity was measured using a fluorogenic assay, and cell death was assessed by measuring LDH release to the culture medium. Results clearly demonstrate a dose-dependent neuroprotective effect of flavopiridol that markedly inhibited both caspase activity and cell death induced by etoposide (Fig.3A,B).

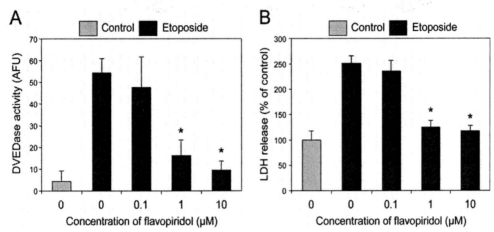

Figure 3. Flavopiridol protects rat cortical neurons from etoposide-induced apoptosis. A, rat primary cortical 1 DIV neurons were treated with 50 μM etoposide for 6 h in the absence or presence of the increasing concentrations of flavopiridol. Untreated cultures (Control) served as negative controls. Caspase-3-like (DEVDase) activity in cytosolic extracts from treated or control cells was assayed fluorometrically. Protease activity is expressed in arbitrary fluorescence units (AFU) ± S.D. (n = 6). *, p < 0.01 compared with DEVDase activity in etoposide-treated cells in the absence of flavopiridol by ANOVA, followed by the Student-Newman-Keuls test. B, the cultures were treated with etoposide for 24 h, and cell viability was analyzed by measurement of calcein AM fluorescence. Data are expressed as a percentage of the value for control cultures not exposed to etoposide ± S.D. (n = 6). *, p < 0.05 compared with viability of etoposide-treated cells in the absence of flavopiridol.

Moreover, we have also found that flavopiridol potently inhibits caspase-3 activation induced by radiation and protects 1DIV RCNs from radiation-induced death (Fig.4A,B). This effect of flavopiridol was observed at 5×10^3 times lower concentration (0.1 μM) compared to that of another CDKs inhibitor, olomoucine (0.5mM). Similarly, flavopiridol demonstrated neuroprotection in CGNs model of radiation-induced death, while olomoucine was much less potent (Fig.4C).

We next examined whether neuroprotection by flavopiridol could be achieved when this compound was applied at different time after irradiation of RCNs. Results shown in Figure 5 indicate that flavopiridol remains potently neuroprotective when administered as late as 2 h after irradiation (Fig.5).

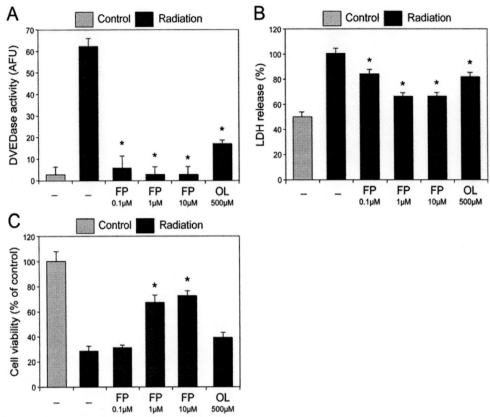

Figure 4. Flavopiridol protects neurons from radiation-induced apoptosis. A, rat primary cortical 1 DIV neurons received 10 Gy of radiation in the absence or presence of indicated concentrations of flavopiridol (FP) or olomoucine (OL). Untreated nonirradiated cells served as a control. Caspase-3-like activity was assayed 12 h after irradiation fluorometrically and expressed in arbitrary fluorescence units (AFU) $\pm$ S.D. (n = 6). *, $p < 0.01$ compared to irradiated cells in the absence of CDK inhibitors. B, cell viability was assessed by measuring LDH release 24 h after treatment. Data are expressed as a percentage of the value for irradiated cells in the absence of CDK inhibitors $\pm$ S.D. *, $p < 0.05$. C, rat primary cerebellar granule 1 DIV neurons received 10 Gy of radiation in the absence or presence of indicated concentrations of flavopiridol (FP), or olomoucine (OL). Cell viability was measured using calcein AM assay 24 h after treatment. Data are expressed as a percentage of the value for irradiated cells in the absence of CDK inhibitors $\pm$ S.D. *, $p < 0.05$.

Flavopiridol inhibits the neuronal intrinsic apoptotic pathway through repression of proapoptotic BH3-only genes. Previous studies have demonstrated that DNA damage activates intrinsic mechanisms of neuronal death induced in parallel by p53 and CDK activity, and that the inhibition of either pathway markedly improves neuronal survival [44]. Nevertheless, molecular targets common for p53- and CDK-dependent pathways responsible for mitochondrial damage, cyto-C release, and caspase activation, have not been identified.

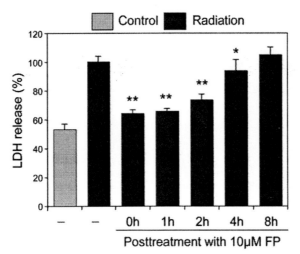

Figure 5. A time window of neuroprotection by flavopiridol. A, 1 DIV RCNs received 10 Gy of radiation in the presence of 1 μM flavopiridol (FP) or were subsequently treated with FL at indicated time after irradiation. Untreated nonirradiated cells served as an additional control. Cell viability was assessed by measuring LDH release 24 h after treatment. Data are expressed as a percentage of the value for irradiated cells in the absence of FL ± S.D. *, $p < 0.05$; **, $p < 0.01$ compared to irradiated cells in the absence of CDK inhibitors.

Using Western blotting analysis of control 1 DIV RCN cultures and RCNs treated with etoposide in the absence or presence of flavopiridol, we confirmed results of preceding reports [44] showing that CDK inhibition does not affect accumulation of p53 after DNA damage, but blocks the release of cyto-C and activation of caspases 3 and 9 (Fig.6). We also found that, in addition to cyto-C, flavopiridol is able to inhibit AIF release from mitochrondria suggesting its potential ability to inhibit caspase-independent neuronal death.

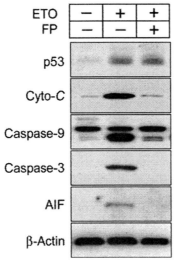

Figure 6. Flavopiridol inhibits neuronal apoptosis at the level of mutochondria. A, 1 DIV rat CGNs were treated with etoposide (ETO) in the presence or absence of 1μM flavopiridol (FP). Six h after treatment cytosol from control and treated neurons was isolated and subjected to Western blotting using the antibodies against p53, Cyto-c, procaspase-9, active caspase-3, AIF, and β-actin.

The release of cyto-C, AIF, and other proapoptotic mitochondrial factor can be mediated by the members of Bcl-2 family [73]. A number of genes encoding these proteins, such as *BAX, NOXA, HRK, PUMA,* and *SIVA,* contain p53 response elements, and their regulation has been shown to depend on p53 activity [29, 30, 74-76]. On the other hand, it has also been demonstrated that the expression of these genes is directly regulated by E2F transcription factors, and that inhibition of E2F1-mediated induction of *NOXA* or *PUMA* inhibits apoptosis in mouse fibroblasts and SAOS2 cells [77]. Given these reports, we examined an effect of flavopiridol on expression of these proapoptotic genes in CGNs after etoposide treatment using a semiquantitative RT-PCR approach.

Gene-specific PCR primers for rat *BAX, HRK, PUMA,* and *SIVA* were design according to mRNA sequences in the GenBank; however, at the time of this study, rat *NOXA* mRNA or gene sequences were not available. Therefore, in order to get an informative rat *NOXA* nucleotide sequence, we performed BLAST search of the rat genome using the mouse *NOXA* cDNA probe. Homology analysis followed by the computer-based prediction of a gene structure revealed that the *Rn18_2036* supercontig located on *Rattus norvegicus* chromosome 18 includes the *NOXA* gene, which is 92% identical to its mouse ortholog. Using cDNA from the embryonic rat brain and *Rn18_2036* sequence, we amplified a complete coding sequence or the rat *NOXA* cDNA, sequenced it and submitted to the GenBank under the accession number *AY788892.*

Further analysis of the identified part of *Rn18_2036* using *Genomatix Gene2Promoter* software resulted in the prediction of a promoter region of the rat gene. Remarkably, similar to the mouse *NOXA* promoter [77], the rat predicted promoter contains both p53 and E2F consensus binding sites at similar positions relative to the predicted transcription start (Fig. 7).

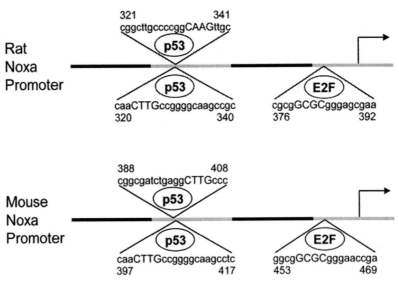

Figure 7. Schematic representation of the rat and mouse NOXA promoters. p53 and E2F binding sites are indicated with corresponding nucleotide sequences relative to theanscription start sites (square arrows) of the rat and mouse NOXA promoters.

Results of semiquantitative RT-PCR revealed that treatment of rat CGNs with etoposide results in marked induction of *NOXA* mRNA expression. *PUMA* mRNA levels appeared to be only slightly increased, and expression of *BAX, Harakiri* (*HRK*), and *SIVA* was not affected by etoposide in CGNs (Fig.8A,B). Importantly, pretreatment with flavopiridol potently inhibited expression of *NOXA*, *PUMA*, and *SIVA*, but did not affect expression of *BAX* and *HRK*.

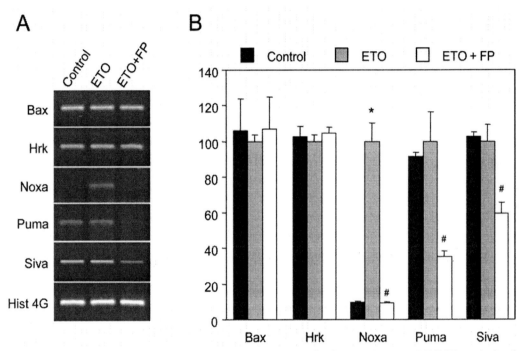

Figure 8. mRNA expression profiles of proapoptotic BCL-2 family members. A, RT-PCR analysis of the abundance of mRNAs encoding specified members of the BCL-2 family in 1 DIV rat CGNs. Cell cultures were treated with etoposide (ETO) with or without flavopiridol (FL). Total cellular RNA was subjected to RT and PCR with gene-specific or Histone 4G (Hist 4G) primers. PCR products were analyzed by agarose gel electrophoresis and staining with ethidium bromide. B, results of semiquantitative RT-PCR analysis of selected mRNA demonstrate that flavopiridol potently inhibits expression of NOXA, PUMA, and SIVA genes. Data present mean values of three independent experiments and are expressed as a percentage of the values for ETO treatment ± S.D. *, p < 0.001 compared to untreated control. #, p < 0.01 compared to etoposide-treated samples.

CONCLUSION

High sensitivity of the developing brain to programmed cell death represents the major obstacle in treatment of childhood brain tumors using radiation or chemotherapy. Our research has been aimed at enhancing the destruction of tumor cells without injuring normal neural tissue. This study represents an initial step in developing of such a strategy.

Our approach is based on the previous findings demonstrating that one of the well-characterized CDK inhibitors, flavopiridol provides evident long-term neuroprotection in various models of neuronal injury [42, 43, 67-70, 72, 78-81]. On the other hand, it has been shown that this compound demonstrates remarkable anticancer properties, at least *in vitro*, by killing several types of tumor cells alone or in combination with other chemotherapeutic drugs [82].

Although flavopiridol has been primarily characterized as a potent inhibitor of cell cycle-related CDKs, a mechanism of its anticancer action remains to be elucidated. Recent reports indicate that one of the mechanisms, by which flavopiridol induces apoptosis is mediated by the release of cyto-C and AIF from mitochondria [83], and that cell death induced by flavopiridol associates with activation of p38 MAPK, the decline in ERK activity, and can be suppressed by inhibition of caspase-9 [84]. Moreover, in addition to inhibition of CDKs, flavopiridol can activate expression of TRAIL receptors, inhibits expression of survivin, FLIP, Bcl-xL, XIAP, p21WAF1/CIP1, iNOS, and NO production [85-88], phosphorylation of RNA polymerase II, and reduces RNA synthesis [89].

In this study, we show that flavopiridol efficiently kills human MB-derived Daoy and SH-SY5Y neuroblastoma cells demonstrating an additive effect when combined with radiation and/or etoposide treatment. Furthermore, we confirmed previous reports and provided new data showing that at similar doses radiation and etoposide induce caspase-dependent cell death in 1 DIV RCN and CGN cultures, and that flavopiridol is able to prevent caspase activation in both types of neurons and protects them from death in a dose dependent fashion. Moreover, we have found that neuroprotection by flavopiridol can also be achieved when this compound is administered as late as 2 h after injury.

To address a mechanism of neuroprotection mediated by flavopiridol, we have investigated the major events in activation of the intrinsic apoptotic pathway and found that flavopiridol blocked the release of both cyto-C and AIF, activation-specific proteolysis of caspase-9, and activation of caspase-3, while it did not affect DNA damage-induced accumulation of p53.

Previously we have shown that, in human neuroblastoma cells, the release of apoptogenic factors from mitochondria depends of the p53-dependent induction of the *NOXA* gene [54]. Therefore, in this study, we examined an effect of flavopiridol on expression of this and other proapoptotic genes of BCL-2 family, known to be regulated by p53. In order to obtain an informative nucleotide sequence for the RT-PCR analysis, we cloned and sequenced the complete coding region of rat *NOXA* cDNA. Besides information for PCR experiments, this cDNA allowed us to identify a potential promoter region of the rat *NOXA* gene and to analyze its structure in comparison to that of previously identified mouse *NOXA* promoter. Results of *in silico* analysis demonstrate that both mouse and rat promoters contain very similar binding sites for p53 and E2F transcription factors at almost identical positions relative to each other and to the transcription start sites.

Results of RT-PCR analysis revealed that, in rat neurons, flavopiridol blocked expression of *NOXA*, *PUMA*, and *SIVA* genes, but did not affect expression of *BAX* and *HRK*. Importantly, promoters of all three genes repressed by flavopiridol has been described to contain both p53 and E2F regulatory elements that control transcriptional activity of these genes [74, 77].

Taken together, our results suggest that DNA damage and cell cycle re-entry cooperate in the induction of neuronal apoptosis through transcriptional activation of specific BH3-only genes, such as *NOXA*, *PUMA*, or *SIVA,* and the neuroprotective effect of flavopiridol may be mediated by repression of these genes through the cyclin D1/CDK4/Rb/E2F-dependend pathway. Induction of these *BH3*-only gene requires activation of both p53 and cyclin D1/CDK4/Rb-dependent E2F transcription factors. Activity of either factor alone is not sufficient to trigger gene induction and cyto-C release, caspase activation, and cell death (Fig.9). Hence, inhibition of E2F activity by flavopiridol may represent a central point in the mechanism of neuroprotection. Future studies will address this hypothesis and mechanisms by which flavopiridol enhances cell death in dividing cancer cells, while development and use of specific synthetic inhibitors of cyclin-dependent kinases in the clinic may have a significant impact on the improvement of the existing strategies to treat childhood brain tumors.

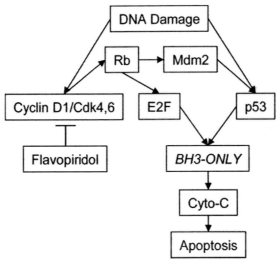

Figure 9. Schematic presentation of the hypothetical intrinsic p53- and cyclin D1-dependent apoptotic pathway.

ACKNOWLEDGEMENTS

This work was supported by NINDS, National Institutes of Health Grants R21 R21NS048974 and R21 NS051216 to A.G.Y.

Flavopiridol was provided to V.A.M. by the Developmental Therapeutics Program, Division of Cancer Treatment and Diagnosis, National Cancer Institute, National Institutes of Heath.

REFERENCES

[1]   Haydar, TF; Kuan, CY; Flavell, RA and Rakic, P. (1999). The role of cell death in regulating the size and shape of the mammalian forebrain. *Cereb Cortex, 9*, 621-626.

[2]   Becker, EB and Bonni, A. (2004). Cell cycle regulation of neuronal apoptosis in development and disease. *Prog Neurobiol, 72*, 1-25

[3]   Kuida, K; Zheng, TS; Na, S; Kuan, C; Yang, D; Karasuyama, H; Rakic, P and Flavell, RA. (1996). Decreased apoptosis in the brain and premature lethality in CPP32-deficient mice. *Nature, 384*, 368-372

[4]   Kuida, K; Haydar, TF; Kuan, CY; Gu, Y; Taya, C; Karasuyama, H; Su, MS; Rakic, P and Flavell, RA. (1998). Reduced apoptosis and cytochrome c-mediated caspase activation in mice lacking caspase 9. *Cell, 94*, 325-337

[5]   Honarpour, N; Gilbert, SL; Lahn, BT; Wang, X and Herz, J. (2001). Apaf-1 deficiency and neural tube closure defects are found in fog mice. *Proc Natl Acad Sci U S A, 98*, 9683-9687.

[6]   Kalifa, C and Grill, J. (2005). The therapy of infantile malignant brain tumors: current status? *J Neurooncol, 75*, 279-285

[7]   Yakovlev, AG; Ota, K; Wang, G; Movsesyan, V; Bao, WL; Yoshihara, K and Faden, AI. (2001). Differential expression of apoptotic protease-activating factor-1 and caspase-3 genes and susceptibility to apoptosis during brain development and after traumatic brain injury. *J Neurosci, 21*, 7439-7446.

[8]   Nakaya, K; Hasegawa, T; Flickinger, JC; Kondziolka, DS; Fellows-Mayle, W and Gobbel, GT. (2005). Sensitivity to radiation-induced apoptosis and neuron loss declines rapidly in the postnatal mouse neocortex. *Int J Radiat Biol, 81*, 545-554

[9]   Li, YQ and Wong, CS. (2000). Radiation-induced apoptosis in the neonatal and adult rat spinal cord. *Radiat Res, 154*, 268-276

[10]   Romero, AA; Gross, SR; Cheng, KY; Goldsmith, NK and Geller, HM. (2003). An age-related increase in resistance to DNA damage-induced apoptotic cell death is associated with development of DNA repair mechanisms. *J Neurochem, 84*, 1275-1287

[11]   Liu, Z; Sun, C; Olejniczak, ET; Meadows, RP; Betz, SF; Oost, T; Herrmann, J; Wu, JC and Fesik, SW. (2000). Structural basis for binding of Smac/DIABLO to the XIAP BIR3 domain. *Nature, 408*, 1004-1008.

[12]   Roberts, DL; Merrison, W; MacFarlane, M and Cohen, GM. (2001). The inhibitor of apoptosis protein-binding domain of Smac is not essential for its proapoptotic activity. *J Cell Biol, 153*, 221-228.

[13]   Martins, LM; Iaccarino, I; Tenev, T; Gschmeissner, S; Totty, NF; Lemoine, NR; Savopoulos, J; Gray, CW; Creasy, CL; Dingwall, C and Downward, J. (2002). The serine protease Omi/HtrA2 regulates apoptosis by binding XIAP through a reaper-like motif. *J Biol Chem, 277*, 439-444

[14]   Ravagnan, L; Roumier, T and Kroemer, G. (2002). Mitochondria, the killer organelles and their weapons. *J Cell Physiol, 192*, 131-137.

[15]   van Loo, G; Saelens, X; van Gurp, M; MacFarlane, M; Martin, SJ and Vandenabeele, P. (2002). The role of mitochondrial factors in apoptosis: a Russian roulette with more than one bullet. *Cell Death Differ, 9*, 1031-1042.

[16]   Cregan, SP; Fortin, A; MacLaurin, JG; Callaghan, SM; Cecconi, F; Yu, SW; Dawson, TM; Dawson, VL; Park, DS; Kroemer, G and Slack, RS. (2002). Apoptosis-inducing factor is involved in the regulation of caspase- independent neuronal cell death. *J Cell Biol, 158*, 507-517.

[17] Li, P; Nijhawan, D; Budihardjo, I; Srinivasula, SM; Ahmad, M; Alnemri, ES and Wang, X. (1997). Cytochrome c and dATP-dependent formation of Apaf-1/caspase-9 complex initiates an apoptotic protease cascade. *Cell, 91,* 479-489

[18] Li, LY; Luo, X and Wang, X. (2001). Endonuclease G is an apoptotic DNase when released from mitochondria. *Nature, 412,* 95-99.

[19] Cory, S and Adams, JM. (2002). The Bcl2 family: regulators of the cellular life-or-death switch. *Nat Rev Cancer, 2,* 647-656.

[20] Lakhani, SA; Masud, A; Kuida, K; Porter, GA, Jr.; Booth, CJ; Mehal, WZ; Inayat, I and Flavell, RA. (2006). Caspases 3 and 7: key mediators of mitochondrial events of apoptosis. *Science, 311,* 847-851

[21] Miramar, MD; Costantini, P; Ravagnan, L; Saraiva, LM; Haouzi, D; Brothers, G; Penninger, JM; Peleato, ML; Kroemer, G and Susin, SA. (2001). NADH oxidase activity of mitochondrial apoptosis-inducing factor. *J Biol Chem, 276,* 16391-16398.

[22] Klein, JA; Longo-Guess, CM; Rossmann, MP; Seburn, KL; Hurd, RE; Frankel, WN; Bronson, RT and Ackerman, SL. (2002). The harlequin mouse mutation downregulates apoptosis-inducing factor. *Nature, 419,* 367-374.

[23] Loeffler, M; Daugas, E; Susin, SA; Zamzami, N; Metivier, D; Nieminen, AL; Brothers, G; Penninger, JM and Kroemer, G. (2001). Dominant cell death induction by extramitochondrially targeted apoptosis-inducing factor. *Faseb J, 15,* 758-767.

[24] Wei, MC; Zong, WX; Cheng, EH; Lindsten, T; Panoutsakopoulou, V; Ross, AJ; Roth, KA; MacGregor, GR; Thompson, CB and Korsmeyer, SJ. (2001). Proapoptotic BAX and BAK: a requisite gateway to mitochondrial dysfunction and death. *Science, 292,* 727-730

[25] Puthalakath, H and Strasser, A. (2002). Keeping killers on a tight leash: transcriptional and post-translational control of the pro-apoptotic activity of BH3-only proteins. *Cell Death Differ, 9,* 505-512

[26] Huang, DC and Strasser, A. (2000). BH3-Only proteins-essential initiators of apoptotic cell death. *Cell, 103,* 839-842

[27] Schuler, M and Green, DR. (2001). Mechanisms of p53-dependent apoptosis. *Biochem Soc Trans, 29,* 684-688.

[28] Pearson, AS; Spitz, FR; Swisher, SG; Kataoka, M; Sarkiss, MG; Meyn, RE; McDonnell, TJ; Cristiano, RJ and Roth, JA. (2000). Up-regulation of the proapoptotic mediators Bax and Bak after adenovirus-mediated p53 gene transfer in lung cancer cells. *Clin Cancer Res, 6,* 887-890

[29] Igata, E; Inoue, T; Ohtani-Fujita, N; Sowa, Y; Tsujimoto, Y and Sakai, T. (1999). Molecular cloning and functional analysis of the murine bax gene promoter. *Gene, 238,* 407-415.

[30] Oda, E; Ohki, R; Murasawa, H; Nemoto, J; Shibue, T; Yamashita, T; Tokino, T; Taniguchi, T and Tanaka, N. (2000). Noxa, a BH3-only member of the Bcl-2 family and candidate mediator of p53-induced apoptosis. *Science, 288,* 1053-1058.

[31] Yu, J; Zhang, L; Hwang, PM; Kinzler, KW and Vogelstein, B. (2001). PUMA induces the rapid apoptosis of colorectal cancer cells. *Mol Cell, 7,* 673-682.

[32] Arnoult, D; Parone, P; Martinou, JC; Antonsson, B; Estaquier, J and Ameisen, JC. (2002). Mitochondrial release of apoptosis-inducing factor occurs downstream of

cytochrome c release in response to several proapoptotic stimuli. *J Cell Biol, 159*, 923-929

[33] Wu, M; Xu, LG; Li, X; Zhai, Z and Shu, HB. (2002). AMID, an apoptosis-inducing factor-homologous mitochondrion-associated protein, induces caspase-independent apoptosis. *J Biol Chem, 277*, 25617-25623

[34] Ohiro, Y; Garkavtsev, I; Kobayashi, S; Sreekumar, KR; Nantz, R; Higashikubo, BT; Duffy, SL; Higashikubo, R; Usheva, A; Gius, D; Kley, N and Horikoshi, N. (2002). A novel p53-inducible apoptogenic gene, PRG3, encodes a homologue of the apoptosis-inducing factor (AIF). *FEBS Lett, 524*, 163-171

[35] Evan, G and Littlewood, T. (1998). A matter of life and cell death. *Science, 281*, 1317-1322

[36] Smith, ML; Kontny, HU; Bortnick, R and Fornace, AJ, Jr. (1997). The p53-regulated cyclin G gene promotes cell growth: p53 downstream effectors cyclin G and Gadd45 exert different effects on cisplatin chemosensitivity. *Exp Cell Res, 230*, 61-68

[37] Del Sal, G; Murphy, M; Ruaro, E; Lazarevic, D; Levine, AJ and Schneider, C. (1996). Cyclin D1 and p21/waf1 are both involved in p53 growth suppression. *Oncogene, 12*, 177-185

[38] Dyson, N. (1998). The regulation of E2F by pRB-family proteins. *Genes Dev, 12*, 2245-2262

[39] Chellappan, SP; Hiebert, S; Mudryj, M; Horowitz, JM and Nevins, JR. (1991). The E2F transcription factor is a cellular target for the RB protein. *Cell, 65*, 1053-1061

[40] Clarke, AR; Maandag, ER; van Roon, M; van der Lugt, NM; van der Valk, M; Hooper, ML; Berns, A and te Riele, H. (1992). Requirement for a functional Rb-1 gene in murine development. *Nature, 359*, 328-330

[41] Tsai, KY; Hu, Y; Macleod, KF; Crowley, D; Yamasaki, L and Jacks, T. (1998). Mutation of E2f-1 suppresses apoptosis and inappropriate S phase entry and extends survival of Rb-deficient mouse embryos. *Mol Cell, 2*, 293-304

[42] O'Hare, MJ; Hou, ST; Morris, EJ; Cregan, SP; Xu, Q; Slack, RS and Park, DS. (2000). Induction and modulation of cerebellar granule neuron death by E2F-1. *J Biol Chem, 275*, 25358-25364

[43] Park, DS; Levine, B; Ferrari, G and Greene, LA. (1997). Cyclin dependent kinase inhibitors and dominant negative cyclin dependent kinase 4 and 6 promote survival of NGF-deprived sympathetic neurons. *J Neurosci, 17*, 8975-8983

[44] Morris, EJ; Keramaris, E; Rideout, HJ; Slack, RS; Dyson, NJ; Stefanis, L and Park, DS. (2001). Cyclin-dependent kinases and P53 pathways are activated independently and mediate Bax activation in neurons after DNA damage. *J Neurosci, 21*, 5017-5026

[45] Uziel, T; Zindy, F; Sherr, CJ and Roussel, MF. (2006). The CDK Inhibitor p18(Ink4c) is a Tumor Suppressor in Medulloblastoma. *Cell Cycle, 5*

[46] Frank, KM; Sharpless, NE; Gao, Y; Sekiguchi, JM; Ferguson, DO; Zhu, C; Manis, JP; Horner, J; DePinho, RA and Alt, FW. (2000). DNA ligase IV deficiency in mice leads to defective neurogenesis and embryonic lethality via the p53 pathway. *Mol Cell, 5*, 993-1002

[47] Eliasson, MJ; Sampei, K; Mandir, AS; Hurn, PD; Traystman, RJ; Bao, J; Pieper, A; Wang, ZQ; Dawson, TM; Snyder, SH and Dawson, VL. (1997). Poly(ADP-ribose)

polymerase gene disruption renders mice resistant to cerebral ischemia. *Nat Med, 3*, 1089-1095

[48]   Xiang, H; Hochman, DW; Saya, H; Fujiwara, T; Schwartzkroin, PA and Morrison, RS. (1996). Evidence for p53-mediated modulation of neuronal viability. *J Neurosci, 16*, 6753-6765

[49]   Crumrine, RC; Thomas, AL and Morgan, PF. (1994). Attenuation of p53 expression protects against focal ischemic damage in transgenic mice. *J Cereb Blood Flow Metab, 14*, 887-891

[50]   Martin, LJ; Kaiser, A; Yu, JW; Natale, JE and Al-Abdulla, NA. (2001). Injury-induced apoptosis of neurons in adult brain is mediated by p53-dependent and p53-independent pathways and requires Bax. *J Comp Neurol, 433*, 299-311

[51]   Slack, RS; Belliveau, DJ; Rosenberg, M; Atwal, J; Lochmuller, H; Aloyz, R; Haghighi, A; Lach, B; Seth, P; Cooper, E and Miller, FD. (1996). Adenovirus-mediated gene transfer of the tumor suppressor, p53, induces apoptosis in postmitotic neurons. *J Cell Biol, 135*, 1085-1096

[52]   Culmsee, C; Zhu, X; Yu, QS; Chan, SL; Camandola, S; Guo, Z; Greig, NH and Mattson, MP. (2001). A synthetic inhibitor of p53 protects neurons against death induced by ischemic and excitotoxic insults, and amyloid beta-peptide. *J Neurochem, 77*, 220-228

[53]   Greig, NH; Mattson, MP; Perry, T; Chan, SL; Giordano, T; Sambamurti, K; Rogers, JT; Ovadia, H and Lahiri, DK. (2004). New Therapeutic Strategies and Drug Candidates for Neurodegenerative Diseases: p53 and TNF-{alpha} Inhibitors, and GLP-1 Receptor Agonists. *Ann N Y Acad Sci, 1035*, 290-315

[54]   Yakovlev, AG; Di Giovanni, S; Wang, G; Liu, W; Stoica, B and Faden, AI. (2004). Bok and Noxa are essential mediators of p53-dependent apoptosis. *J Biol Chem,*

[55]   Yu, SW; Wang, H; Poitras, MF; Coombs, C; Bowers, WJ; Federoff, HJ; Poirier, GG; Dawson, TM and Dawson, VL. (2002). Mediation of poly(ADP-ribose) polymerase-1-dependent cell death by apoptosis-inducing factor. *Science, 297*, 259-263

[56]   Smulson, ME; Simbulan-Rosenthal, CM; Boulares, AH; Yakovlev, A; Stoica, B; Iyer, S; Luo, R; Haddad, B; Wang, ZQ; Pang, T; Jung, M; Dritschilo, A and Rosenthal, DS. (2000). Roles of poly(ADP-ribosyl)ation and PARP in apoptosis, DNA repair, genomic stability and functions of p53 and E2F-1. *Adv Enzyme Regul, 40*, 183-215

[57]   Wetmore, C; Eberhart, DE and Curran, T. (2001). Loss of p53 but not ARF accelerates medulloblastoma in mice heterozygous for patched. *Cancer Res, 61*, 513-516

[58]   Eberhart, CG. (2003). Medulloblastoma in mice lacking p53 and PARP: all roads lead to Gli. *Am J Pathol, 162*, 7-10

[59]   Tong, WM; Ohgaki, H; Huang, H; Granier, C; Kleihues, P and Wang, ZQ. (2003). Null mutation of DNA strand break-binding molecule poly(ADP-ribose) polymerase causes medulloblastomas in p53(-/-) mice. *Am J Pathol, 162*, 343-352

[60]   He, XM; Ostrowski, LE; von Wronski, MA; Friedman, HS; Wikstrand, CJ; Bigner, SH; Rasheed, A; Batra, SK; Mitra, S; Brent, TP and et al. (1992). Expression of O6-methylguanine-DNA methyltransferase in six human medulloblastoma cell lines. *Cancer Res, 52*, 1144-1148

[61] Movsesyan, VA; Stoica, BA and Faden, AI. (2004). MGLuR5 activation reduces beta-amyloid-induced cell death in primary neuronal cultures and attenuates translocation of cytochrome c and apoptosis-inducing factor. *J Neurochem, 89*, 1528-1536

[62] Eldadah, BA; Yakovlev, AG and Faden, AI. (1997). The role of CED-3-related cysteine proteases in apoptosis of cerebellar granule cells. *J Neurosci, 17*, 6105-6113

[63] Sinensky, MC; Leiser, AL and Babich, H. (1995). Oxidative stress aspects of the cytotoxicity of carbamide peroxide: in vitro studies. *Toxicol Lett, 75*, 101-109

[64] Chomczynski, P and Sacchi, N. (1987). Single-step method of RNA isolation by acid guanidinium thiocyanate- phenol-chloroform extraction. *Anal Biochem, 162*, 156-159

[65] Yakovlev, AG and Faden, AI. (1994). Sequential expression of c-fos protooncogene, TNF-alpha, and dynorphin genes in spinal cord following experimental traumatic injury. *Mol Chem Neuropathol, 23*, 179-190.

[66] Stoica, BA; Movsesyan, VA; Lea, PMt and Faden, AI. (2003). Ceramide-induced neuronal apoptosis is associated with dephosphorylation of Akt, BAD, FKHR, GSK-3beta, and induction of the mitochondrial-dependent intrinsic caspase pathway. *Mol Cell Neurosci, 22*, 365-382

[67] Padmanabhan, J; Park, DS; Greene, LA and Shelanski, ML. (1999). Role of cell cycle regulatory proteins in cerebellar granule neuron apoptosis. *J Neurosci, 19*, 8747-8756

[68] Verdaguer, E; Jorda, EG; Alvira, D; Jimenez, A; Canudas, AM; Folch, J; Rimbau, V; Pallas, M and Camins, A. (2005). Inhibition of multiple pathways accounts for the antiapoptotic effects of flavopiridol on potassium withdrawal-induced apoptosis in neurons. *J Mol Neurosci, 26*, 71-84

[69] Verdaguer, E; Jorda, EG; Stranges, A; Canudas, AM; Jimenez, A; Sureda, FX; Pallas, M and Camins, A. (2003). Inhibition of CDKs: a strategy for preventing kainic acid-induced apoptosis in neurons. *Ann N Y Acad Sci, 1010*, 671-674

[70] Jorda, EG; Verdaguer, E; Canudas, AM; Jimenez, A; Bruna, A; Caelles, C; Bravo, R; Escubedo, E; Pubill, D; Camarasa, J; Pallas, M and Camins, A. (2003). Neuroprotective action of flavopiridol, a cyclin-dependent kinase inhibitor, in colchicine-induced apoptosis. *Neuropharmacology, 45*, 672-683

[71] Courtney, MJ and Coffey, ET. (1999). The mechanism of Ara-C-induced apoptosis of differentiating cerebellar granule neurons. *Eur J Neurosci, 11*, 1073-1084

[72] Di Giovanni, S; Movsesyan, V; Ahmed, F; Cernak, I; Schinelli, S; Stoica, B and Faden, AI. (2005). Cell cycle inhibition provides neuroprotection and reduces glial proliferation and scar formation after traumatic brain injury. *Proc Natl Acad Sci U S A, 102*, 8333-8338

[73] Festjens, N; van Gurp, M; van Loo, G; Saelens, X and Vandenabeele, P. (2004). Bcl-2 family members as sentinels of cellular integrity and role of mitochondrial intermembrane space proteins in apoptotic cell death. *Acta Haematol, 111*, 7-27

[74] Fortin, A; MacLaurin, JG; Arbour, N; Cregan, SP; Kushwaha, N; Callaghan, SM; Park, DS; Albert, PR and Slack, RS. (2004). The proapoptotic gene SIVA is a direct transcriptional target for the tumor suppressors p53 and E2F1. *J Biol Chem, 279*, 28706-28714

[75] Nakano, K and Vousden, KH. (2001). PUMA, a novel proapoptotic gene, is induced by p53. *Mol Cell, 7*, 683-694

[76]   Miyashita, T and Reed, JC. (1995). Tumor suppressor p53 is a direct transcriptional activator of the human bax gene. *Cell, 80,* 293-299

[77]   Hershko, T and Ginsberg, D. (2004). Up-regulation of Bcl-2 homology 3 (BH3)-only proteins by E2F1 mediates apoptosis. *J Biol Chem, 279,* 8627-8634

[78]   Stefanis, L; Park, DS; Friedman, WJ and Greene, LA. (1999). Caspase-dependent and - independent death of camptothecin-treated embryonic cortical neurons. *J Neurosci, 19,* 6235-6247

[79]   Giovanni, A; Keramaris, E; Morris, EJ; Hou, ST; O'Hare, M; Dyson, N; Robertson, GS; Slack, RS and Park, DS. (2000). E2F1 mediates death of B-amyloid-treated cortical neurons in a manner independent of p53 and dependent on Bax and caspase 3. *J Biol Chem, 275,* 11553-11560

[80]   Park, DS; Farinelli, SE and Greene, LA. (1996). Inhibitors of cyclin-dependent kinases promote survival of post-mitotic neuronally differentiated PC12 cells and sympathetic neurons. *J Biol Chem, 271,* 8161-8169

[81]   Osuga, H; Osuga, S; Wang, F; Fetni, R; Hogan, MJ; Slack, RS; Hakim, AM; Ikeda, JE and Park, DS. (2000). Cyclin-dependent kinases as a therapeutic target for stroke. *Proc Natl Acad Sci U S A, 97,* 10254-10259

[82]   Shapiro, GI. (2006). Cyclin-dependent kinase pathways as targets for cancer treatment. *J Clin Oncol, 24,* 1770-1783

[83]   Newcomb, EW; Tamasdan, C; Entzminger, Y; Alonso, J; Friedlander, D; Crisan, D; Miller, DC and Zagzag, D. (2003). Flavopiridol induces mitochondrial-mediated apoptosis in murine glioma GL261 cells via release of cytochrome c and apoptosis inducing factor. *Cell Cycle, 2,* 243-250

[84]   Pepper, C; Thomas, A; Fegan, C; Hoy, T and Bentley, P. (2003). Flavopiridol induces apoptosis in B-cell chronic lymphocytic leukaemia cells through a p38 and ERK MAP kinase-dependent mechanism. *Leuk Lymphoma, 44,* 337-342

[85]   Rosato, RR; Almenara, JA; Cartee, L; Betts, V; Chellappan, SP and Grant, S. (2002). The cyclin-dependent kinase inhibitor flavopiridol disrupts sodium butyrate-induced p21WAF1/CIP1 expression and maturation while reciprocally potentiating apoptosis in human leukemia cells. *Mol Cancer Ther, 1,* 253-266

[86]   Quiney, C; Dauzonne, D; Kern, C; Fourneron, JD; Izard, JC; Mohammad, RM; Kolb, JP and Billard, C. (2004). Flavones and polyphenols inhibit the NO pathway during apoptosis of leukemia B-cells. *Leuk Res, 28,* 851-861

[87]   Miyashita, K; Shiraki, K; Fuke, H; Inoue, T; Yamanaka, Y; Yamaguchi, Y; Yamamoto, N; Ito, K; Sugimoto, K and Nakano, T. (2006). The cyclin-dependent kinase inhibitor flavopiridol sensitizes human hepatocellular carcinoma cells to TRAIL-induced apoptosis. *Int J Mol Med, 18,* 249-256

[88]   Rosato, RR; Dai, Y; Almenara, JA; Maggio, SC and Grant, S. (2004). Potent antileukemic interactions between flavopiridol and TRAIL/Apo2L involve flavopiridol-mediated XIAP downregulation. *Leukemia, 18,* 1780-1788

[89]   Chen, R; Keating, MJ; Gandhi, V and Plunkett, W. (2005). Transcription inhibition by flavopiridol: mechanism of chronic lymphocytic leukemia cell death. *Blood, 106,* 2513-2519

In: Cell Apoptosis Research Trends
Editor: Charles V. Zhang, pp. 231-244

ISBN: 1-60021-424-X
© 2007 Nova Science Publishers, Inc

Chapter VIII

Modulating Apoptosis Thresholds in Cancer Therapy

Christoph Herold and Matthias Ocker[7]

Department of Medicine 1, University Hospital Erlangen, Ulmenweg 18, D-91054
Erlangen, Germany

Abstract

The ability to induce and execute apoptosis is essential for the maintenance and regulation of tissue and organ homeostasis. A common feature of malignant cells is their capability of evading apoptosis, which contributes to the high rate of resistance to radio- or chemotherapy of tumor cells. Although some progress has been made in inhibiting tumor growth by developing receptor-tyrosine-kinase antagonists that inhibit survival and growth promoting pathways, resistance against these and other compounds is mainly due to defects in apoptosis regulation. In contrast, low apoptotic thresholds may sensitize tumors to well tolerated and clinical established treatment regimens. Therefore, downstream targets of the apoptosis machinery, e.g. survivin, caspases or members of the bcl-2 family, seem to represent interesting new targets for cancer therapy. Modulating apoptotic thresholds can be achieved by reconstituting genetically lost or epigenetically silenced pro-apoptotic factors, e.g. by viral gene transfer or inducing re-expression with DNA methyltransferase or histone deacetylase inhibitors, or by inhibiting the activity or expression of anti-apoptotic factors. Here, RNAi-based approaches, e.g. against anti-apoptotic bcl-2, seem to hold great promise to overcome treatment resistance of a variety of malignant tumor types. Overall, restoring apoptosis sensitivity or lowering apoptotic thresholds will lead to better tolerated and more potent treatment strategies.

---

[7] Correspondence to: Dr. Matthias Ocker, Assistant Professor of Experimental Medicine, Department of Medicine 1, University Hospital Erlangen, Ulmenweg 18, D-91054 Erlangen, Germany, Phone: ++49-9131-8535057, Fax: ++49-9131-8535058, Matthias.Ocker@med1.imed.uni-erlangen.de

INTRODUCTION

The control of cell number, tissue integrity and homeostasis is crucial for the survival of multicellular organisms. Besides controlling proliferation of cells, the coordinated removal and killing of unwanted or unnecessary cells is a highly conserved mechanism to achieve this end. This process has been termed "apoptosis" in 1972 in analogy to the Greek for "falling off", resembling the falling of leaves from trees in autumn [1]. Apoptosis plays several critical roles in various physiologic processes ranging from embryonic development, immune maturation and response and tissue homeostasis in adult organisms. The dysregulation of apoptosis has been linked to the pathogenesis of human cancer diseases and especially to the high resistance of some cancers to conventional radio- and/or chemotherapy [2]. Recently, several genes and proteins have been identified that contribute to this dysregulation and the better understanding of how these processes themselves are regulated has lead to the development of novel therapeutic approaches, esp. applying gene technology approaches, small molecule inhibitors or epigenetic modulators.

In this chapter, we will review the basic features of apoptosis in health and cancer diseases and point out to strategies to overcome or modulate apoptotic thresholds for cancer therapy.

BASIC FEATURES OF APOPTOSIS

Apoptosis was first recognized from a morphologic point of view by cell shrinkage, chromatin condensation and fragmentation (pyknosis and karyorrhexis), formation of apoptotic blebbing and apoptotic bodies and finally phagocytosis of apoptotic remnants by macrophages without signs of inflammation. Importantly, and in contrast to necrosis, membranes stay intact until the final stages of apoptosis execution [3]. These morphologic phenomena are caused by a series of distinct biochemical features that are executed in a coordinated fashion, e.g. the hydrolytic cleavage of cellular proteins by the caspase cascade [4], the distinct pattern of internucleosomal DNA degradation and changes in membrane architecture (i.e. expression of phosphatidylserine on the outer layers of the plasma membrane) that are necessary for the phagocytic recognition by macrophages [5, 6].

The process of programmed cell death can be divided into three phases: initiation, regulation and execution that follow each other rapidly. Initiation of apoptosis can be observed after stimulating the intrinsic or mitochondrial pathway by genotoxic damage or microtubule disruption (e.g. after radio- or chemotherapy for cancer treatment), hypoxia, hypoglycaemia, reactive oxygen species, withdrawal of growth factors or via the extrinsic pathway, i.e. stimulation of death receptors [7] (see figure 1). The phase of regulation involves the cellular decision to undergo the irreversible pathway to cell death. Here, a connection between the extrinsic and the intrinsic apoptosis induction pathways can further enhance and accelerate the final execution phase. The execution phase finally leads to the above described morphologic alterations. In the following sections, the different pathways of induction, regulation and execution of apoptosis will be highlighted in regard of their meaning for cancer development and therapy. Although the different pathways are described separately for better understanding, it is important to note that they are closely interconnected and maybe even interdependent on each other.

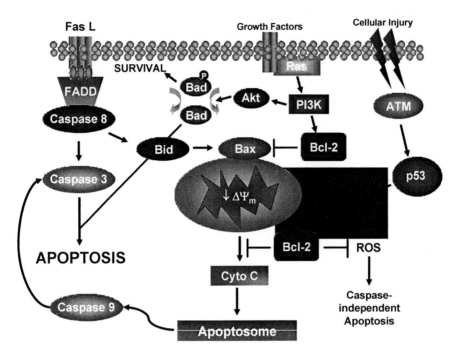

Figure 1. Schematic overview on extrinsic and intrinsic pathways of apoptosis induction. The mitochondrion represents the central decision making point for both the receptor mediated extrinsic as well as the physiologic intrinsic pathway. The release of Cytochrome *c* from the mitochondrial intermembrane space leads to the formation of the apoptosome that activates the initiator caspase 9. The receptor mediated pathway leads to the activation of intracellular death domains (FADD) that activate the initiator caspase 8. Both initiator caspases can activate the effector caspase 3 that is the final executioner of apoptosis. Activated caspase 8 can also disturb the mitochondrial integrity and lead to the loss of the mitochondrial membrane potential $\Delta\Psi_m$ with consecutive release of Cytochrome *c*. The regulation of the mitochondrial integrity is dependent on different pro- and anti-apoptotic members of the bcl-2 protein family that are differentially regulated by growth factor withdrawal or cellular stress.

EXTRINSIC (RECEPTOR MEDIATED) INDUCTION OF APOPTOSIS

Death receptors represent a growing family of transmembrane receptors present on most cell types. After binding of apoptosis inducing ligands (e.g. FAS ligand, TNF-α, TNF-β, TRAIL) a cytoplasmic domain involved in protein-protein interactions called death domain (DD) is activated and recruits several adapter proteins. DD in turn binds to several inactive molecules of pro-caspase 8. Although pro-caspases are thought to represent inactive pro-forms of these proteases, they seem to posses a low intrinsic activity that allows the pro-caspase to activate other pro-caspases when brought to close proximity, e.g. after DD binding [4, 8, 9]. The thus activated initator caspase 8 then triggers a cascade of caspase activation by cleaving and activating other downstream caspases and factors leading to a destabilization of the mitochondrial membrane, e.g. cleavage of bid to t-bid [10]. The DD mediated activation of pro-caspases can be inhibited by the protein FLIP, which is commonly overexpressed in cancer cells to protect the cell from ligand induced cell death [11, 12]. Inhibition of FLIP might therefore represent a novel target for restoring apoptosis sensitivity in some tumors.

INTRINSIC (MITOCHONDRIAL) INDUCTION OF APOPTOSIS

Different cellular stress factors (e.g. irradiation, growth factor withdrawal, mitotic catastrophe etc.) lead to increased permeability of the mitochondrial membrane with subsequent release of Cytochrome c. In the cytosol, Cytochrome c then binds to the apoptosis activating factor-1 (Apaf-1) and this complex then activates the initiator caspase 9 [13]. Futhermore, antagonists of endogenous caspase inhibitors known as IAPs like Smac/Diablo are also released from mitochondria and enhance apoptotic activity [14]. The further processes resemble those described for the extrinsic pathway: activation of effector caspases and execution of cell death.

The mitochondrial pathway is strongly involved in the apoptotic response to radio- or chemotherapy. Normally, the integrity of the mitochondrial membrane is controlled by the family of bcl-2 proteins, which consists of anti-apoptotic (e.g. bcl-2, bcl-X_L, Mcl-1, Bfl-1, A1, etc.) and pro-apoptotic members (e.g. bax, bad, bak, bik, bcl-X_S, etc.) [14-17]. According to the rheostat model, the balance between pro- and anti-apoptotic factors decides about the execution of cell death [18]. Here, the ratio of bcl-2 and bax seems to be most important. Therefore, overexpression of anti-apoptotic bcl-2 or loss of pro-apoptotic bax has been shown in various human tumor diseases [19, 20].

Although the exact mechanisms of how pro- and anti-apoptotic factors decide about cell fate still remains elusive, several models have been proposed so far [21]. Pro- and anti-apoptotic bcl-2 family members share conserved stretches of amino-acid sequences that were termed bcl-2 homology (BH) domains. Among these, the BH3-containing or BH3-only proteins possess pro-apoptotic properties, e.g. bax or bak. Normally, these pro-apoptotic molecules are located in the cytoplasm in a folded configuration that hides the C-terminal mitochondrial localizing domain and renders the protein inactive. Upon a death stimulus, bax can unfold, localize to the outer mitochondrial membrane [22, 23], form multimers [24, 25], and induce mitochondrial permeabilization with subsequent loss of $\Delta\Psi_m$ and release of Cytochrome c. A death stimulus can e.g. be the caspase-mediated cleavage of bid to t-bid upon activation of the receptor-mediated pathway that activates the BH3-domain of t-bid [21, 26]. Furthermore, the BH3-domain can also antagonize the effect of anti-apoptotic bcl-2 family members [27, 28]. Activated bax-multimers can induce the mitochondrial permeability transition (MPT) pore, where the voltage-dependent anion channel (VDAC) seems to be an integral part [29]. According to this model, opening of VDAC then disrupts the electron gradient and leads to loss of $\Delta\Psi_m$. Another controversially debated model indicates that multimerized bax molecules can themselves form pores in the mitochondrial membrane that are large enough for releasing Cytochrome c [21].

Irrespective of the precise mechanism that leads to bax-mediated disruption of the mitochondrial integrity, a variety of factors are involved in upstream signaling or regulation of these molecules and several of them have been shown to be dysregulated in cancer diseases.

REGULATION OF APOPTOSIS IN CANCER

As the efficacy of cancer therapy is largely dependent on the capability to induce programmed cell death, evasion of apoptosis has been termed as one of the hallmarks of cancer [30]. Normally, proliferation and apoptosis are tightly balanced to maintain tissue homeostasis and either enhanced proliferation or impaired apoptosis lead to a net gain in cell number, i.e. tumor formation. Therefore, defects in either apoptosis induction or apoptosis execution are central for the survival of successful tumor cells (see table 1). Here, the two basic principles of apoptosis evasion (failure of apoptosis induction and failure in apoptosis execution) will be highlighted.

Table 1: Role of apoptosis deficiency in cancer [98]

Feature	Mechanism
Net increase in cell number	Cell death < cell proliferation
Longevity	Accumulation of genetic and epigenetic lesions
Proto-oncogene tolerance	Nullify pro-apoptotic effects
Immune surveillance	Resistance to immune attack
Growth factor-independence	Survival without growth factors
Angiogenesis	Resistance to hypoxia, hypoglycemia
Metastasis	Survival without attachment, resistance to anoikis
Radio-/Chemoresistance	Increased apoptotic threshold

As discussed previously, induction of apoptosis can be mediated via an extrinsic or an intrinsic pathway. Loss of expression of DD-receptors like Fas (CD95) or members of the TRAIL-family are commonly observed in human cancers and contribute to impaired overall survival [31-34]. Furthermore, inactivating point mutations in either the ligand binding domain or in the intracellular DD itself have been described [35, 36]. Another commonly observed strategy is the overexpression of so called "decoy" receptors, i.e. members of the TNF-receptor superfamily that bind pro-apoptotic ligands but do not have an intracellular DD [34, 37, 38].

Other mechanisms of apoptosis resistance in cancer focus on the intrinsic pathway and the downstream effectors of apoptosis execution. As indicated, overexpression of anti-apoptotic bcl-2 family members like bcl-2 or bcl-X_L stabilizes mitochondrial membranes or inhibits the effect of pro-apoptotic molecules like bax or t-bid [19]. Vice versa, the loss of pro-apoptotic bcl-2 family members also leads to impaired apoptosis in cancer cells. As the execution of cell death by activated caspases is normally tightly controlled, several endogenous caspase inhibitors like livin, survivin or the IAP family or mitochondrial apoptosis inhibitors like SMAC have been identified [39-44]. Another commonly denoted phenomenon in cancer is the overexpression of growth factor receptors like erb-B1 (EGFR) or erb-B2 (Her2) which can modulate the threshold of mitochondria associated apoptotic factors, e.g. by phosphorylating and thus inactivating caspase 9 by the downstream kinase AKT [45]. The phosphorylation of these proteins can also be mediated by the mitogen-activated protein kinase (MAPK) pathway.

The regulation of these pathways seems in some cancers to be dependent on the functional state of p53, which itself is commonly mutated in various cancer diseases. Therefore, mutations of p53 have been associated with impaired apoptosis due to e.g. overexpression of survivin, bcl-2 or other anti-apoptotic molecules [46-48]. Additionally, p53 has also been shown to regulate the extrinsic apoptosis cascade by influencing the expression of TRAIL- or Fas-receptors [49, 50].

STRATEGIES TO MODULATE APOPTOTIC THRESHOLDS IN CANCER THERAPY

The increasing understanding of regulatory processes involved in apoptosis induction and execution has offered the prospect of novel therapeutic approaches. In fact, the selective induction of apoptosis via the extrinsic pathway has been pursued in different cancers using recombinant antibodies [51, 52]. Yet, the application of Fas- or TRAIL-receptor specific antibodies or other activators like recombinant human tumor necrosis factor (rhTNF) or soluble Fas ligands (FasL/CD95L) showed significant levels of toxicity in various organs up to sepsis like syndromes [53, 54]. Gene therapy approaches using adenoviral vectors for re-expressing Fas or FasL have not shown convincing results overall [51].

As indicated, the activity of apoptotic pathways is usually under control of pro-survival (e.g. growth factor signaling) pathways. Here, the development of novel small molecule inhibitors of different kinases involved in this signaling has lead to better treatment responses due to shifts in apoptotic thresholds, too. Yet, due to the novelty of these treatments, e.g. targeting EGFR, MAPK, AKT or PI3K pathways, no final conclusion on the specifity and efficacy can be drawn.

Different strategies have been followed to modulate the bax/bcl-2 ratio in tumor cells. Antisense oligonucleotides against bcl-X_L or bcl-2 have been successfully used in various cancer cell models *in vitro*. However, preclinical and clinical studies did not demonstrate convincing anti-cancer efficacies due to unspecific effects (e.g. complement activation, thrombocytopenia, cytokine release, hepatotoxicity) and the need to introduce chemical modifications (e.g. locked nucleic acids) to stabilize these molecules [55-64]. Recently, a novel and more specific approach to suppress the function of oncogenes has been described in *C. elegans* and mammals: RNA interference (RNAi) [65-67]. Briefly, short double-stranded RNA molecules (siRNA) with homology to the mRNA sequence of a target gene are processed enzymatically and lead to the cleavage and suppression of the target mRNA with high specificity [68, 69]. We applied this technology to silence the expression of anti-apoptotic bcl-2 in a pancreatic cancer model, which is otherwise highly resistant to chemotherapy. In this setting, the knockdown of bcl-2 by siRNA lead to the induction of apoptosis and reduced cell proliferation *in vitro* and in a xenograft model *in vivo* [70]. This effect was mediated by a loss of bcl-2 and an enhanced destabilization of the mitochondrial transmembrane potential with subsequent loss of $\Delta\Psi_m$ (Figure 2). We next applied the cytidine analogue Gemcitabine, which has become the gold standard in the treatment of advanced pancreatic cancer, after silencing bcl-2. In cells with silenced bcl-2, Gemcitabine was able to induce cell death even at concentrations 10-times below the effective threshold of Gemcitabine monotherapy (Figure 3). This approach seems to be highly specific, as control experiments did not show any efficacy of this low dose of Gemcitabine. Although the broad application of RNAi based therapies is still limited, these experiments have shown that modulating the apoptotic bax/bcl-2 threshold restores sensitivity to chemotherapy even at very low doses. This indicates that with this adjunctive treatment either an enhanced efficacy or less undesired side effects could be achieved in patients who may otherwise be untreatable at all. This approach has also been successfully applied in other model system [71-79].

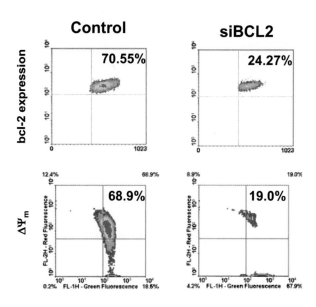

Figure 2. siRNA-mediated knockdown of anti-apoptotic bcl-2 [70]. Pancreatic cancer cells were transfected with siRNA against bcl-2 (siBCL2) and the expression of bcl-2 was determined by flow cytometry using a FITC-labeled anti-bcl-2 antibody. In untreated controls, 70.55% of cells expressed bcl-2, while in treated cells, only 24.27% showed a positive staining. This correlated with the loss of $\Delta\Psi_m$ as evidenced after JC-1 staining in flow cytometry. Here, 68.9% of untreated controls exhibited intact mitochondria, while transfection with siBCL2 reduced this level to 19%.

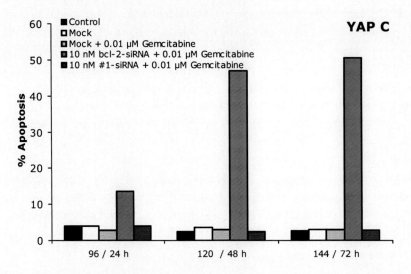

Figure 3. Loss of bcl-2 sensitizes human pancreatic cancer cells for Gemcitabine treatment. Cells were pre-transfected with siBCL2 and then incubated with Gemcitabine at an ineffective concentration. Apoptosis induction was determined by flow cytometry after propidium ioded staining. Only in cells with silenced bcl-2, significant levels of apoptosis were reached, while in controls (mock transfected or control siRNA #1), no changes compared to control were seen.

Recently, so called epigenetic events have attracted attention as novel cancer therapeutics [80, 81]. Among these, small molecule inhibitiors of histone deacetylases (HDAC), are currently undergoing clinical trials for different tumor diseases [82-84]. These compounds have been shown to modulate the bax/bcl-2 ratio of cancer cells as well as the expression of TRAIL- and other death receptors [85-90]. We and others have shown that HDAC inhibitors can restore sensitivity to radio- or chemotherapy in otherwise resistant tumor entities at low molecular concentrations [91-95]. These effects are explained by either a transcriptional regulation of apoptosis relevant target genes like bcl-2, a suppression of multidrug resistance proteins or a halt in cell cycle progression, e.g. by upregulation of $p21^{cip1/wafl}$, which allows conventional cytostatic or cytotoxic drugs to exert their growth-inhibitory effect for a longer time period [91].

Another principle of epigenetic regulation is the hypermethylation of DNA elements [81]. Here, novel compounds like Zebularine also show promising results in preclinical settings of human cancer diseases and the combination of classical cytotoxic agents, targeted therapies and epigenetic modulators can provide potent, specific and well tolerable treatment options [81, 96].

CONCLUSIONS

The increasing understanding of apoptosis pathways, its regulation and dysregulation in cancer leads to the generation of a variety of novel treatment approaches. Especially epigenetic events enter the focus of current research, as the discussion of the role of so-called cancer stem cells seems to lead to a paradigm shift in the understanding of cancer biology and treatment [97]. Targeted approaches, using RNAi based strategies or specific tyrosine kinase inhibitors, as well as the development of HDAC inhibitors will soon enter clinical oncology and an increased overall survival and clinical benefit is expected for otherwise untreatable patients.

REFERENCES

[1]   Kerr, J.F., A.H. Wyllie, and A.R. Currie, Apoptosis: a basic biological phenomenon with wide-ranging implications in tissue kinetics. *Br J Cancer*, 1972. 26(4): p. 239-57.

[2]   Reed, J.C., Dysregulation of apoptosis in cancer. *J Clin Oncol*, 1999. 17(9): p. 2941-53.

[3]   Kumar, V., A.K. Abbas, and N. Fausto, Cellular Adaptions, Cell Injury, and Cell Death, in *Pathologic Basis of Disease*, V. Kumar, A.K. Abbas, and N. Fausto, Editors. 2005, Elsevier Saunders: Philadelphia. p. 3-46.

[4]   Shi, Y., Caspase activation: revisiting the induced proximity model. *Cell*, 2004. 117(7): p. 855-8.

[5]   Schultz, D.R. and W.J. Harrington, Jr., Apoptosis: programmed cell death at a molecular level. *Semin Arthritis Rheum*, 2003. 32(6): p. 345-69.

[6]   Kasof, G., et al., Overview: A Matter of Life and Death, in *Signalling Pathways in Apoptosis*, D. Watters and M. Lavin, Editors. 1999, Harwood Academic Publishers: Amsterdam. p. 1-28.

[7]   Fennell, D.A., Apoptosis: molecular physiology and significance for cancer therapeutics, in *Introduction to the Cellular and Molecular Biology of Cancer*, M. Knowles and P. Selby, Editors. 2005, Oxford University Press: Oxford. p. 210-28.

[8]   Salvesen, G.S. and V.M. Dixit, Caspase activation: the induced-proximity model. *Proc Natl Acad Sci U S A*, 1999. 96(20): p. 10964-7.

[9]   Srinivasula, S.M., et al., Autoactivation of procaspase-9 by Apaf-1-mediated oligomerization. *Mol Cell*, 1998. 1(7): p. 949-57.

[10]   Esposti, M.D., The roles of Bid. *Apoptosis*, 2002. 7(5): p. 433-40.

[11]   Ryu, B.K., et al., Increased expression of cFLIP(L) in colonic adenocarcinoma. *J Pathol*, 2001. 194(1): p. 15-9.

[12]   Zhou, X.D., et al., Overexpression of cellular FLICE-inhibitory protein (FLIP) in gastric adenocarcinoma. *Clin Sci (Lond)*, 2004. 106(4): p. 397-405.

[13]   Schafer, Z.T. and S. Kornbluth, The apoptosome: physiological, developmental, and pathological modes of regulation. *Dev Cell*, 2006. 10(5): p. 549-61.

[14]   Brenner, C. and G. Kroemer, The Mitochondrion: Decisive for Cell Death Control?, in *Signalling Pathways in Apoptosis*, D. Watters and M. Lavin, Editors. 1999, Harwood Academic Publishers: Amsterdam. p. 207-26.

[15]   Thomadaki, H. and A. Scorilas, BCL2 family of apoptosis-related genes: functions and clinical implications in cancer. *Crit Rev Clin Lab Sci*, 2006. 43(1): p. 1-67.

[16]   Armstrong, J.S., Mitochondrial membrane permeabilization: the sine qua non for cell death. *Bioessays*, 2006. 28(3): p. 253-60.

[17]   Mohamad, N., et al., Mitochondrial apoptotic pathways. *Biocell*, 2005. 29(2): p. 149-61.

[18] Korsmeyer, S.J., et al., Bcl-2/Bax: a rheostat that regulates an anti-oxidant pathway and cell death. *Semin Cancer Biol,* 1993. 4(6): p. 327-32.

[19] Reed, J.C., Bcl-2 family proteins: regulators of apoptosis and chemoresistance in hematologic malignancies. *Semin Hematol,* 1997. 34(4 Suppl 5): p. 9-19.

[20] Reed, J.C., et al., BCL-2 family proteins: regulators of cell death involved in the pathogenesis of cancer and resistance to therapy. *J Cell Biochem,* 1996. 60(1): p. 23-32.

[21] Sharpe, J.C., D. Arnoult, and R.J. Youle, Control of mitochondrial permeability by Bcl-2 family members. *Biochim Biophys Acta,* 2004. 1644(2-3): p. 107-13.

[22] Wolter, K.G., et al., Movement of Bax from the cytosol to mitochondria during apoptosis. *J Cell Biol,* 1997. 139(5): p. 1281-92.

[23] Hsu, Y.T., K.G. Wolter, and R.J. Youle, Cytosol-to-membrane redistribution of Bax and Bcl-X(L) during apoptosis. *Proc Natl Acad Sci U S A,* 1997. 94(8): p. 3668-72.

[24] Tan, Y.J., W. Beerheide, and A.E. Ting, Biophysical characterization of the oligomeric state of Bax and its complex formation with Bcl-XL. *Biochem Biophys Res Commun,* 1999. 255(2): p. 334-9.

[25] Antonsson, B., et al., Bax is present as a high molecular weight oligomer/complex in the mitochondrial membrane of apoptotic cells. *J Biol Chem,* 2001. 276(15): p. 11615-23.

[26] Yin, X.M., Bid, a BH3-only multi-functional molecule, is at the cross road of life and death. *Gene,* 2006. 369: p. 7-19.

[27] Kelekar, A. and C.B. Thompson, Bcl-2-family proteins: the role of the BH3 domain in apoptosis. *Trends Cell Biol,* 1998. 8(8): p. 324-30.

[28] Huang, D.C. and A. Strasser, BH3-Only proteins-essential initiators of apoptotic cell death. *Cell,* 2000. 103(6): p. 839-42.

[29] Jacotot, E., et al., Mitochondrial membrane permeabilization during the apoptotic process. *Ann N Y Acad Sci,* 1999. 887: p. 18-30.

[30] Hanahan, D. and R.A. Weinberg, The hallmarks of cancer. *Cell,* 2000. 100(1): p. 57-70.

[31] Wajant, H., K. Pfizenmaier, and P. Scheurich, TNF-related apoptosis inducing ligand (TRAIL) and its receptors in tumor surveillance and cancer therapy. *Apoptosis,* 2002. 7(5): p. 449-59.

[32] Zhang, L. and B. Fang, Mechanisms of resistance to TRAIL-induced apoptosis in cancer. *Cancer Gene Ther,* 2005. 12(3): p. 228-37.

[33] Wajant, H., CD95L/FasL and TRAIL in tumour surveillance and cancer therapy. *Cancer Treat Res,* 2006. 130: p. 141-65.

[34] Bouralexis, S., D.M. Findlay, and A. Evdokiou, Death to the bad guys: targeting cancer via Apo2L/TRAIL. *Apoptosis,* 2005. 10(1): p. 35-51.

[35] Takahashi, H., et al., FAS death domain deletions and cellular FADD-like interleukin 1beta converting enzyme inhibitory protein (long) overexpression: alternative mechanisms for deregulating the extrinsic apoptotic pathway in diffuse large B-cell lymphoma subtypes. *Clin Cancer Res,* 2006. 12(11 Pt 1): p. 3265-71.

[36] Shen, L., et al., Frequent deletion of Fas gene sequences encoding death and transmembrane domains in nasal natural killer/T-cell lymphoma. *Am J Pathol,* 2002. 161(6): p. 2123-31.

[37] LeBlanc, H.N. and A. Ashkenazi, Apo2L/TRAIL and its death and decoy receptors. *Cell Death Differ*, 2003. 10(1): p. 66-75.

[38] Ozoren, N. and W.S. El-Deiry, Cell surface Death Receptor signaling in normal and cancer cells. *Semin Cancer Biol*, 2003. 13(2): p. 135-47.

[39] Deveraux, Q.L., et al., Endogenous inhibitors of caspases. *J Clin Immunol*, 1999. 19(6): p. 388-98.

[40] Deveraux, Q.L. and J.C. Reed, IAP family proteins--suppressors of apoptosis. *Genes Dev*, 1999. 13(3): p. 239-52.

[41] Roy, N., et al., The c-IAP-1 and c-IAP-2 proteins are direct inhibitors of specific caspases. *Embo J*, 1997. 16(23): p. 6914-25.

[42] Deveraux, Q.L., et al., X-linked IAP is a direct inhibitor of cell-death proteases. *Nature*, 1997. 388(6639): p. 300-4.

[43] LaCasse, E.C., et al., The inhibitors of apoptosis (IAPs) and their emerging role in cancer. *Oncogene*, 1998. 17(25): p. 3247-59.

[44] Ambrosini, G., C. Adida, and D.C. Altieri, A novel anti-apoptosis gene, survivin, expressed in cancer and lymphoma. *Nat Med*, 1997. 3(8): p. 917-21.

[45] Cardone, M.H., et al., Regulation of cell death protease caspase-9 by phosphorylation. *Science*, 1998. 282(5392): p. 1318-21.

[46] Vegran, F., et al., Association of p53 gene alterations with the expression of antiapoptotic survivin splice variants in breast cancer. *Oncogene*, 2006.

[47] Tsutsui, S., et al., Bcl-2 protein expression is associated with p27 and p53 protein expressions and MIB-1 counts in breast cancer. *BMC Cancer*, 2006. 6(1): p. 187.

[48] Yu, J. and L. Zhang, The transcriptional targets of p53 in apoptosis control. *Biochem Biophys Res Commun*, 2005. 331(3): p. 851-8.

[49] Sheikh, M.S., et al., p53-dependent and -independent regulation of the death receptor KILLER/DR5 gene expression in response to genotoxic stress and tumor necrosis factor alpha. *Cancer Res*, 1998. 58(8): p. 1593-8.

[50] Liu, X., et al., p53 upregulates death receptor 4 expression through an intronic p53 binding site. *Cancer Res*, 2004. 64(15): p. 5078-83.

[51] Timmer, T., E.G. de Vries, and S. de Jong, Fas receptor-mediated apoptosis: a clinical application? *J Pathol*, 2002. 196(2): p. 125-34.

[52] de Jong, S., et al., Death receptor ligands, in particular TRAIL, to overcome drug resistance. *Cancer Metastasis Rev*, 2001. 20(1-2): p. 51-6.

[53] Kakinuma, C., et al., Acute toxicity of an anti-Fas antibody in mice. *Toxicol Pathol*, 1999. 27(4): p. 412-20.

[54] Abe, R., et al., Toxic epidermal necrolysis and Stevens-Johnson syndrome are induced by soluble Fas ligand. *Am J Pathol*, 2003. 162(5): p. 1515-20.

[55] Waters, J.S., et al., Phase I clinical and pharmacokinetic study of bcl-2 antisense oligonucleotide therapy in patients with non-Hodgkin's lymphoma. *J Clin Oncol*, 2000. 18(9): p. 1812-23.

[56] Cotter, F.E., J. Waters, and D. Cunningham, Human Bcl-2 antisense therapy for lymphomas. *Biochim Biophys Acta*, 1999. 1489(1): p. 97-106.

[57] Pepper, C., et al., Antisense-mediated suppression of Bcl-2 highlights its pivotal role in failed apoptosis in B-cell chronic lymphocytic leukaemia. *Br J Haematol*, 1999. 107(3): p. 611-5.

[58] Chi, K.C., et al., Effects of Bcl-2 modulation with G3139 antisense oligonucleotide on human breast cancer cells are independent of inherent Bcl-2 protein expression. *Breast Cancer Res Treat*, 2000. 63(3): p. 199-212.

[59] Wacheck, V., et al., Bcl-2 antisense oligonucleotides chemosensitize human gastric cancer in a SCID mouse xenotransplantation model. *J Mol Med*, 2001. 79(10): p. 587-93.

[60] Jansen, B., et al., Chemosensitisation of malignant melanoma by BCL2 antisense therapy. *Lancet*, 2000. 356(9243): p. 1728-33.

[61] Lebedeva, I. and C.A. Stein, Antisense oligonucleotides: promise and reality. *Annu Rev Pharmacol Toxicol*, 2001. 41: p. 403-19.

[62] Gautschi, O., et al., Activity of a novel bcl-2/bcl-xL-bispecific antisense oligonucleotide against tumors of diverse histologic origins. *J Natl Cancer Inst*, 2001. 93(6): p. 463-71.

[63] Hopkins-Donaldson, S., et al., Induction of apoptosis and chemosensitization of mesothelioma cells by Bcl-2 and Bcl-xL antisense treatment. *Int J Cancer*, 2003. 106(2): p. 160-6.

[64] Stahel, R.A. and U. Zangemeister-Wittke, Antisense oligonucleotides for cancer therapy-an overview. *Lung Cancer*, 2003. 41 Suppl 1: p. S81-8.

[65] Fire, A., et al., Potent and specific genetic interference by double-stranded RNA in Caenorhabditis elegans. *Nature*, 1998. 391(6669): p. 806-11.

[66] Elbashir, S.M., et al., Duplexes of 21-nucleotide RNAs mediate RNA interference in cultured mammalian cells. *Nature*, 2001. 411(6836): p. 494-8.

[67] Elbashir, S.M., W. Lendeckel, and T. Tuschl, RNA interference is mediated by 21- and 22-nucleotide RNAs. *Genes Dev*, 2001. 15(2): p. 188-200.

[68] Meister, G. and T. Tuschl, Mechanisms of gene silencing by double-stranded RNA. *Nature*, 2004. 431(7006): p. 343-9.

[69] Tuschl, T., RNA interference and small interfering RNAs. *Chembiochem*, 2001. 2(4): p. 239-45.

[70] Ocker, M., et al., Variants of bcl-2 specific siRNA for silencing antiapoptotic bcl-2 in pancreatic cancer. *Gut*, 2005. 54(9): p. 1298-308.

[71] Rödel, F., et al., Survivin as a radioresistance factor, and prognostic and therapeutic target for radiotherapy in rectal cancer. *Cancer Res*, 2005. 65(11): p. 4881-7.

[72] Ling, X. and F. Li, Silencing of antiapoptotic survivin gene by multiple approaches of RNA interference technology. *Biotechniques*, 2004. 36(3): p. 450-4, 456-60.

[73] Chawla-Sarkar, M., et al., Downregulation of Bcl-2, FLIP or IAPs (XIAP and survivin) by siRNAs sensitizes resistant melanoma cells to Apo2L/TRAIL-induced apoptosis. *Cell Death Differ*, 2004. 11(8): p. 915-23.

[74] Collis, S.J., et al., Enhanced radiation and chemotherapy-mediated cell killing of human cancer cells by small inhibitory RNA silencing of DNA repair factors. *Cancer Res,* 2003. 63(7): p. 1550-4.

[75]   Kosciolek, B.A., et al., Inhibition of telomerase activity in human cancer cells by RNA interference. *Mol Cancer Ther*, 2003. 2(3): p. 209-16.

[76]   Salvi, A., et al., Small interfering RNA urokinase silencing inhibits invasion and migration of human hepatocellular carcinoma cells. *Mol Cancer Ther*, 2004. 3(6): p. 671-8.

[77]   Schmitz, J.C., T.M. Chen, and E. Chu, Small interfering double-stranded RNAs as therapeutic molecules to restore chemosensitivity to thymidylate synthase inhibitor compounds. *Cancer Res*, 2004. 64(4): p. 1431-5.

[78]   Takei, Y., et al., A small interfering RNA targeting vascular endothelial growth factor as cancer therapeutics. *Cancer Res*, 2004. 64(10): p. 3365-70.

[79]   Yano, J., et al., Antitumor Activity of Small Interfering RNA/Cationic Liposome Complex in Mouse Models of Cancer. *Clin Cancer Res*, 2004. 10(22): p. 7721-6.

[80]   Liu, T., et al., Histone deacetylase inhibitors: multifunctional anticancer agents. *Cancer Treat Rev*, 2006. 32(3): p. 157-65.

[81]   Esteller, M., CpG island methylation and histone modifications: biology and clinical significance. *Ernst Schering Res Found Workshop*, 2006(57): p. 115-26.

[82]   Kuendgen, A., et al., The histone deacetylase (HDAC) inhibitor valproic acid as monotherapy or in combination with all-trans retinoic acid in patients with acute myeloid leukemia. *Cancer*, 2006. 106(1): p. 112-9.

[83]   Kelly, W.K., et al., Phase I study of an oral histone deacetylase inhibitor, suberoylanilide hydroxamic acid, in patients with advanced cancer. *J Clin Oncol*, 2005. 23(17): p. 3923-31.

[84]   Ryan, Q.C., et al., Phase I and pharmacokinetic study of MS-275, a histone deacetylase inhibitor, in patients with advanced and refractory solid tumors or lymphoma. *J Clin Oncol*, 2005. 23(17): p. 3912-22.

[85]   Wang, S., et al., Activation of Mitochondrial Pathway is Crucial for Tumor Selective Induction of Apoptosis by LAQ824. *Cell Cycle*, 2006. 5(15).

[86]   Ziauddin, M.F., et al., Valproic Acid, an Antiepileptic Drug with Histone Deacetylase Inhibitory Activity, Potentiates the Cytotoxic Effect of Apo2L/TRAIL on Cultured Thoracic Cancer Cells through Mitochondria-Dependent Caspase Activation. *Neoplasia*, 2006. 8(6): p. 446-57.

[87]   Pathil, A., et al., HDAC inhibitor treatment of hepatoma cells induces both TRAIL-independent apoptosis and restoration of sensitivity to TRAIL. *Hepatology*, 2006. 43(3): p. 425-34.

[88]   Schuchmann, M., et al., Histone deacetylase inhibition by valproic acid down-regulates c-FLIP/CASH and sensitizes hepatoma cells towards CD95- and TRAIL receptor-mediated apoptosis and chemotherapy. *Oncol Rep*, 2006. 15(1): p. 227-30.

[89]   Ganslmayer, M., et al., A quadruple therapy synergistically blocks proliferation and promotes apoptosis of hepatoma cells. *Oncol Rep*, 2004. 11(5): p. 943-50.

[90]   Herold, C., et al., The histone-deacetylase inhibitor Trichostatin A blocks proliferation and triggers apoptotic programs in hepatoma cells. *J Hepatol*, 2002. 36(2): p. 233-40.

[91]   Ocker, M., et al., The histone-deacetylase inhibitor SAHA potentiates proapoptotic effects of 5-fluorouracil and irinotecan in hepatoma cells. *J Cancer Res Clin Oncol*, 2005. 131(6): p. 385-94.

[92] Chinnaiyan, P., et al., Modulation of radiation response by histone deacetylase inhibition. *Int J Radiat Oncol Biol Phys*, 2005. 62(1): p. 223-9.

[93] Zhang, Y., et al., Enhancement of radiation sensitivity of human squamous carcinoma cells by histone deacetylase inhibitors. *Radiat Res*, 2004. 161(6): p. 667-74.

[94] Maiso, P., et al., The histone deacetylase inhibitor LBH589 is a potent antimyeloma agent that overcomes drug resistance. *Cancer Res*, 2006. 66(11): p. 5781-9.

[95] Piacentini, P., et al., Trichostatin A enhances the response of chemotherapeutic agents in inhibiting pancreatic cancer cell proliferation. *Virchows Arch*, 2006. 448(6): p. 797-804.

[96] Neureiter, D., et al., Apoptosis, proliferation and differentiation patterns are influenced by Zebularine and SAHA in pancreatic cancer models. *Scand J Gastroenterol*, 2006. in press.

[97] Neureiter, D., C. Herold, and M. Ocker, Gastrointestinal cancer - only a deregulation of stem cell differentiation? *Int J Mol Med*, 2006. 17(3): p. 483-9.

[98] Reed, J.C., Apoptosis and Cancer, in *Cancer Medicine* 7, D.W. Kufe, et al., Editors. 2006, BC Decker Inc.: Hamilton. p. 41-52.

In: Cell Apoptosis Research Trends
Editor: Charles V. Zhang, pp. 245-261

ISBN: 1-60021-424-X
© 2007 Nova Science Publishers, Inc

Chapter IX

Effects of Pegfilgrastim Administration on Neutrophil Apoptosis in Breast Cancer Patients Treated with Dose-Dense Chemotherapy Regimens

Rosangela Invernizzi[8], Donatella Grasso, Erica Travaglino, Chiara Benatti, Elena Collovà, Mariangela Manzoni, Marco Danova, and Alberto Riccardi

Internal Medicine and Medical Oncology, University of Pavia and IRCCS Policlinico S. Matteo, Pavia, Italy

ABSTRACT

Recent clinical studies show that dose-dense chemotherapy regimens offer the potential benefits of improved both response rate and survival for breast cancer patients. Pegfilgrastim is a covalent conjugate of filgrastim and polyethylene glycol with an increased elimination half-life due to decreased serum clearance. Few data are available on biological effects of pegylated granulocyte colony-stimulating factor. At our center two clinical trials are currently ongoing that will further evaluate pegfilgrastim utilized in dose-dense regimens for breast cancer patients both as neoadjuvant and adjuvant approach. Twenty patients were enrolled in two different dose-dense multicenter clinical trials. Four patients received four courses of concomitant Anthracycline and Taxane chemotherapy as primary systemic treatment and sixteen patients received eight courses of Anthracycline and Taxane on sequential scheme, both every 2 weeks (dose dense schedule), with pegfilgrastim administered from 24 to 72 hours after each chemotherapy course. On peripheral blood buffy coat smears obtained before starting treatment and immediately before each chemotherapy course we analyzed the following parameters in neutrophils: apoptosis by TUNEL technique, actin polymerization using phalloidin

---

[8] Correspondence: R. Invernizzi, M.D., Clinica Medica III, Policlinico S. Matteo, Piazzale Golgi, 27100 Pavia (Italy) Tel. 0039 0382 502160 Fax 0039 0382 526223 e-mail: r.invernizzi@smatteo.pv.it

labeled with FITC, and alkaline phosphatase activity by cytochemistry. Our aim was to evaluate the influence of pegfilgrastim on these biological features in patients exposed to chemotherapy. After stimulation with pegfilgrastim in all patients we observed: stability of the absolute neutrophil count for the whole duration of treatment and no infectious events; a reduction of neutrophil apoptosis rate in comparison with that observed in control patients treated with standard chemotherapy courses without filgrastim support; persistent abnormalities of actin assembly in neutrophils, indicative of changes in cytoskeleton organization; a significant increase of the leukocyte alkaline phosphatase activity, that is a sensitive marker of myeloid differentiation. In conclusion, these results suggest that pegfilgrastim may improve the neutrophil function in patients with cancer exposed to chemotherapy by inhibiting the accelerated apoptosis and prolonging survival. This effect may, at least in part, be dependent on the influence of pegfilgrastim on actin cytoskeleton organization.

INTRODUCTION

Neutrophils, that form the first line of defense against bacterial and fungal infections, are terminally differentiated cells produced in the bone marrow from myeloid stem cells. They have a very short life in the circulation (8-20h), but their life can increase once they enter infected or inflamed tissues. Aged neutrophils undergo spontaneous apoptosis (programmed cell death) in the absence of cytokines or other proinflammatory agents prior to their removal by macrophages. This phagocytosis of intact apoptotic neutrophils prevents them from causing tissue damage via the release of toxic reactive oxygen species and granule enzymes. In acute inflammation, neutrophil numbers within tissues can be high not only because of influx from the circulation with concomitant expansion of the peripheral pool due to an enhanced neutrophil production and maturation in the bone marrow, but also because their constitutive apoptotic pathway is delayed by the action of local inflammatory mediators (Weinmann et al, 2003). Several reports suggest that the dysregulation of neutrophil apoptosis may contribute to the expansion or the reduction of the peripheral pool under pathologic conditions and that, at least under pathophysiologic conditions, apoptosis is critical for the modulation of the peripheral neutrophil count and critically contributes to the homeostasis of the functional neutrophil pool in the circulation (Weinmann et al, 2003; Wesche et al, 2005).

Apoptotic neutrophils display morphological and biochemical modifications and show molecular alterations on their cell surface; they are non-functional and lose the ability to move by chemotaxis, generate a respiratory burst or degranulate.

The biologic mechanisms that regulate neutrophil apoptosis are not fully understood (Akgul et al, 2001). It is well known that several factors can accelerate or suppress neutrophil apoptosis in vitro (Colott et al, 1992). Inflammatory cytokines including IL-1beta, IL-2, IL-5, granulocyte colony-stimulating factor (G-CSF) and granulocyte macrophage colony-stimulating factor (GM-CSF), macrophage inhibitory factor (MIF) as well as microbial substances, like lipopolysaccharide (LPS) can prolong neutrophil survival by inhibiting apoptosis (Baumann et al, 2003; Bruno et al, 2005). Also glucocorticoids such as dexamethasone delay neutrophil apoptosis. The proinflammatory mediator tumor necrosis factor alpha (TNF-alpha) reduces the neutrophil life span by accelerating apoptosis (Goepel et al, 2004). Neutrophils express significant levels of Fas and are susceptible to Fas-induced apoptosis (Watson et al, 1999); cycloheximide and actinomycin (inhibitors of translation and transcription, respectively) as well as chemotherapeutic drugs accelerate neutrophil apoptosis in vitro.

The intracellular mechanism that controls the apoptotic machinery in neutrophils subsequent to cytokine stimulation involves the Bcl-2 family of apoptosis associated genes, including antiapoptotic (Bcl-2, Bcl-X_L, Mcl-1, A1) and proapoptotic members (Bax, Bak, Bad, Bcl-X_S, Bik) (Dibbert, 1999; Moulding et al, 2001). The shift of the balance between the proapoptotic factor Bax and the antiapoptotic factor Bcl-X_L toward the proapoptotic gene product is known to promote apoptosis of human neutrophils in vivo. In particular, the high rate of constitutive apoptosis that is a special feature of human neutrophils may be explained by the constitutive expression of proapoptotic proteins of the Bcl-2 family (Villunger et al, 2000) and is associated with translocation of Bid and Bax to the mitochondria and truncation of Bid, with subsequent release of cytochrome c, Omi/HtrA2 and Smac/DIABLO into the cytosol and increased enzymatic activity of caspase-8, -9, and -3 (Maianski et al, 2004). Neutrophil extended survival that is mediated by cytokines or other exogenous factors appears to involve the inducible expression of the survival proteins A1 and Mcl-1, via transcriptional signalling pathways regulated by NFkB and MAPKs (Akgul et al, 2001; Moulding et al, 2001).

Among mediators that prolong neutrophil life span, G-CSF is one of the most clinically relevant and widely used in the treatment of various conditions associated with neutropenia (Morstyn et al, 1988). Besides stimulation of neutrophil myeloid precursors in bone marrow, this growth factor has a clear antiapoptotic effect on mature neutrophils (Adachi et al, 1994; Molloy et al, 2005). Recently, the mechanisms of G-CSF-mediated prosurvival effects with respect to the mitochondrial death pathway have been studied. It has been demonstrated that the antiapoptotic properties of G-CSF are mediated through inhibition of Bid/Bax-dependent mitochondrial dysfunction and caspase activation (Maianski et al, 2002, 2004). Also an increased expression of the IAP2 protein due to upregulation by G-CSF through activation of the JAK2-STAT3 pathway may contribute to G-CSF-mediated antiapoptosis (Hasegawa et al, 2003). On the other hand, neutrophil hematopoietic growth factors could reactivate the gene of the antiapoptotic protein survivin in mature neutrophils under inflammatory conditions in vivo in a cell cycle-independent manner (Altznauer et al, 2004).

Chemotherapeutic agents exert their cytotoxic effects on tumor cells mainly by inducing apoptosis (Kerr et al, 1994; Tsangaris et al, 1996; Hannun, 1997; Kamesaki, 1998; Schmitt et al, 1999; Kaufmann and Earnshaw, 2000). For example, doxorubicin exerts its antineoplastic effect by activating apoptotic pathways involving Fas receptor and caspase-3 modulations (Ciobotaro et al, 2003). The lack of cytotoxic selectivity of these drugs results in dose-limiting effects, particularly myelosuppression, manifested by leukopenia and neutropenia associated with life-threatening infections. However, only a few reports regarding their effect on normal peripheral blood leukocytes have been published. Some authors found that myeloablative therapy caused a time-dependent increase in in vivo granulocyte apoptosis and suggested that peripheral blood leukocyte apoptosis is a major determinant of chemotherapy-induced leukopenia (Chucklovin, 2000; Kimhi et al, 2004). In cancer patients exposed to chemotherapy also neutrophil defective bactericidal activities and other functional abnormalities have been reported.

Since the two myeloid colony stimulating factors G-CSF and GM-CSF increase mature neutrophil survival, both are extensively used as an adjunct to chemotherapy-induced neutropenia to accelerate the recovery of neutrophil counts and their quantitative effects have been evaluated (Vose and Armitage, 1995). The qualitative effects of these growth factors on functionally defective neutrophils have also been investigated (Leavey et al, 1998; Lejeune et al, 1998, 1999): GM-CSF corrective effect on neutrophil microbicidal activity after chemotherapy proved to be consistently superior in comparison with that of G-CSF as well as its effect on neutrophil accelerated apoptosis (Lejeune et al, 1999). Moreover, differences between the effect of G-CSF (filgrastim) and its glycosylated form (lenograstim) on neutrophil functions have been reported by some authors (Azzarà et al, 2001; Mattii et al, 2005) that may be explained by a different interaction of the two G-CSFs with the specific receptor on the cell membrane.

Pegfilgrastim is a covalent conjugate of filgrastim and polyethylene glycol with an increased elimination half-life due to decreased serum clearance, that has been recently introduced for the treatment of chemotherapy-induced neutropenia (Molineux et al, 1999; Green et al, 2003). Its long half-life allows for administration in a single dose per chemotherapy cycle (Holmes et al, 2002). Pegfilgrastim appears to be self-regulating: it remains in the blood while a patient is neutropenic and is cleared rapidly when the neutrophil count increases. Few data are available on biological effects of pegylated G-CSF. At our center two clinical trials are currently ongoing that will further clinically evaluate pegfilgrastim utilized in dose-dense regimens for breast cancer patients both as neoadjuvant and adjuvant approach. We analyzed neutrophil apoptosis and other biological features in these groups of patients after a different number of treatment courses. Our aim was to assess the effect of pegfilgrastim on the neutrophil chemotherapy-induced accelerated apoptosis as well as its influence on cytoskeleton organization.

PATIENTS AND METHODS

Patient Characteristics

Twenty consecutive patients with breast cancer were enrolled in two different dose-dense multicenter clinical trials. Their median age was 50.5 years (range 42-63). Four patients received four courses of concomitant Anthracycline (Epirubicin 75 mg/m^2) and Taxane (Docetaxel 80 mg/m^2) at a two-week interval as primary systemic treatment (neoadjuvant therapy) and sixteen patients received eight courses of chemotherapy (four courses of concomitant Epirubicin 90 mg/m^2 and Cyclophosphamide 600 mg/m^2 or concomitant Epirubicin 90 mg/m^2, Cyclophosphamide 600 mg/m^2 and 5-Fluorouracil 600 mg/m^2, followed by 4 courses of Docetaxel 175 mg/m^2) on sequential scheme, every 2 weeks (adjuvant therapy). Pegfilgrastim (6 mg subcutaneously) was administered from 24 to 72 hours after each chemotherapy course. No steroids were administered as antiemetics. Control patients were five breast cancer patients treated every four weeks with standard chemotherapy courses (CEF) without growth factor support.

Peripheral blood samples for blood cell count and biological studies were obtained before starting treatment and immediately before each chemotherapy course.

Samples from 9 healthy female volunteers, age-matched, who presented normal blood count and no evidence of inflammatory or infectious disease, were run in parallel with the patient samples.

Informed consent was obtained from patients and controls, and the study was in accordance with the ethical standards of the institutional committee on human experimentation and with the Helsinki Declaration.

Analysis of Apoptosis

Apoptosis was measured on peripheral blood buffy coat smears using a terminal-deoxynucleotidyltransferase (TdT)-mediated deoxyuridine triphosphate (dUTP) nick end labeling (TUNEL) technique (Negoescu et al, 1996). The "In situ Cell Death Detection Kit, AP" (Roche, Mannheim, Germany), which containes calf thymus TdT, fluorescein-dUTP and an alkaline phosphatase anti-fluorescein sheep Fab fragment, was used according to the manufacturer's instructions, improved by the fixation of smears with cold acetone for 7 min and by the omission of cell permeabilization. A negative control was performed by omitting TdT from the labeling mixture. The apoptotic index (AI) was expressed as the percentage of positive nuclei for at least 500 counted neutrophils at a magnification of x1250.

Actin Polymerization

Distribution of F-actin was evaluated as described elsewhere with some modifications (Mattii, 2004, 2005). Peripheral blood buffy coat smears were air-dried and fixed with a solution of acetone at -20 C° for 15 min. After six washes with PBS, cells were then incubated with 10 µg/ml FITC-phalloidin (Molecular Probes, Eugene, OR) for 60 min. Then, samples were washed three times in PBS, and, after counterstaining with a Hoechst 33258 pentahydrate (Molecular Probes) 100 ng/ml solution, they were mounted with *p*-phenylendiamine-glycerol solution. Slides were examined at magnification x400 or x1000 by a Zeiss fluorescence microscope. F-actin cell distribution of 100 cells for sample was determined.

Leukocyte Alkaline Phosphatase (LAP) Cytochemical Assay

Neutrophil LAP was determined on peripheral blood smears by the cytochemical naphthol AS-MX phosphate/ Fast Garnet GBC staining and scored as previously described (Hayhoe and Quaglino, 1958), with individual cells being scored from 0 to 4. One hundred neutrophils were examined consecutively in each smear. The result was then calculated as the sum of the partial scores. The normal range of LAP score in our laboratory was 40 to 140.

Statistical Analysis

Student's paired t-test and analysis of variance (ANOVA) were employed in analysis of differences between cohorts. Correlations were assessed according to Pearson's correlation coefficient. All tests were 2-sided. A P value of <0.05 was considered statistically significant. Stata 7.0 (Stata Corporation, College Station, TX) was used for computation.

RESULTS

Neutrophil Count and Clinical Outcome

After stimulation with pegfilgrastim in 6 patients we observed stability of the absolute neutrophil counts within the range of normal levels for the whole duration of treatment, whereas in 14 cases neutrophil counts increased significantly at the end of the first chemotherapy course (P=0.03) and remained stable thereafter (Fig. 1A). In all patiens but one, differential counts showed a significant increase of immature cells (myelocytes, metamyelocytes, band neutrophils) (P<0.0001) (Fig. 1B). No infectious events occurred. No specific adverse effects were attributed to use of pegfilgrastim.

In control patients we found a non significant drug-induced decrease in neutrophil counts two weeks after each course of treatment and a return to normal values within two other weeks with no significant increase of immature cells .

Apoptosis

In basal conditions the neutrophil AI was similar in breast cancer patients (median 2.3%, range 0.4-27%) and in healthy volunteers (median 3%, range 1-14%) (P=0.15) (Fig. 2 and 3A).

Chemotherapy induced a significant in vivo apoptosis of peripheral neutrophils from 17 patients treated according the dose-dense trials with pegfilgrastim support from the end of the second course forward (median 12 %, range 1-77%) (P=0.0002). The highest apoptosis levels were observed after Docetaxel courses (Fig. 3B). In 3 cases, however, the AI remained stable within the normal range for the whole duration of treatment.

In patients treated with standard chemotherapy without growth factor support the neutrophil AI increased significantly after the first course of treatment (median 41.5%, range 5-78%). The apoptotic effect was demonstrated in all cases and no correlation was found between the chemotherapy course number and apoptosis. The apoptosis levels observed after chemotherapy in these cases were significantly higher than those found in patients treated with dose-dense chemotherapy plus pegfilgastrim (P=0.019) (Fig. 3A).

A)

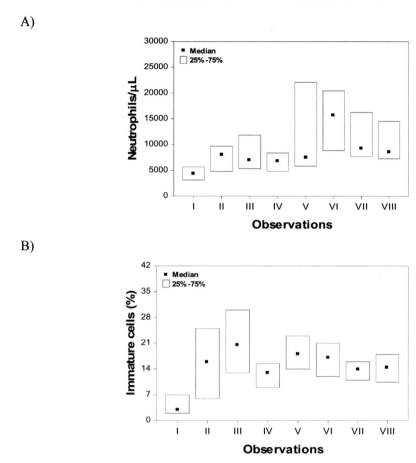

B)

Figure 1. Sequential evaluation of absolute neutrophil count (A) and percentage of circulating immature cells (B) in patients treated with dose-dense chemotherapy and pegfilgrastim. Data are shown as median values plus interquartile range.

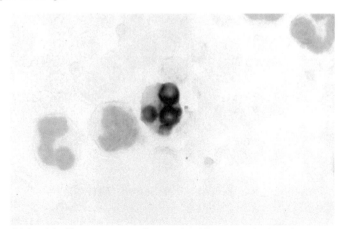

Figure 2. Peripheral blood buffy coat smear from a healthy control, showing an apoptotic neutrophil. TUNEL technique, x1250.

A)

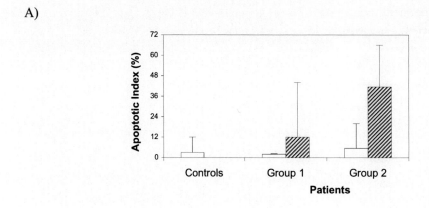

B)

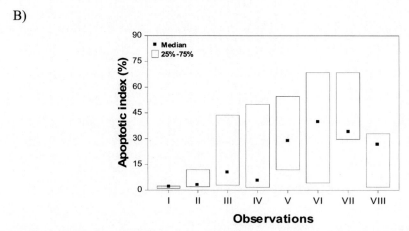

Figure 3. Apoptosis in peripheral blood neutrophils.(A) Apoptotic index in healthy volunteers and in breast cancer patients before and during (striped columns) chemotherapy. Group 1 and group 2 include respectively patients treated with dose-dense chemotherapy and pegfilgrastim, and patients treated with standard chemotherapy. Data are shown as median values plus upper quartile. (B) Sequential evaluation of the apoptotic index in patients treated with dose-dense chemotherapy and pegfilgrastim. Data are shown as median values plus interquartile range.

Actin Organization

It was evaluated by using Phalloidin, which binds the polymeric form of actin (Wulf, 1979). Neutrophil movement is largely based on the ability of actin to change rapidly between its monomeric (G) and filamentous/polymerized (F) forms. An altered cytoskeletal structure could represent the cause of functional alterations of neutrophils (Hall, 1998).

In our patients, observations were performed before therapy, at the third course and at the end of the treatment.

As shown in Fig. 4A and 5A, neutrophils from healthy volunteers and from untreated patients showed predominantly F-actin spread throughout the cell cytoplasm with low intensity (median 85%, range 73-89%, and 85%, range 16-99% respectively, P=0.8). In contrast, many neutrophils from patients treated with dose-dense chemotherapy and pegfilgrastim exhibited an increased actin polymerization distributed principally at the cell membrane and frequently polarized in focal areas (median 39%, range 5-98%) (Fig. 4B, 5A and B). The difference versus basal condition was statistically significant (P=0.0006). In samples from standard chemotherapy-treated patients a median of 20% (range 19-44%) of neutrophils with abnormal actin organization was detected. The difference between the two patient populations was significant (P=0.01) (Fig. 5A).

A)

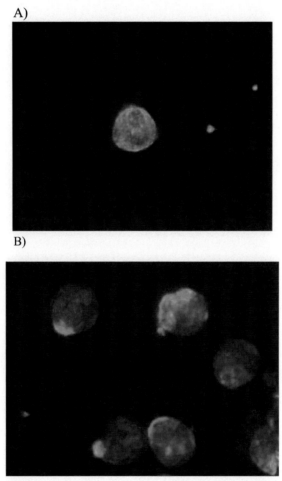

B)

Fig. 4. FITC-Phalloidin allows to reveal F-actin spread throughout the cell cytoplasm in a peripheral blood neutrophil from a healthy control (A) and an enhanced amount of F-actin polarized in focal areas in neutrophils from a patient treated with dose-dense chemotherapy and pegfilgrastim (B). x1250.

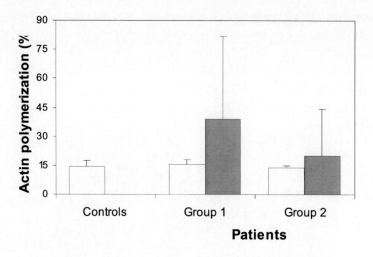

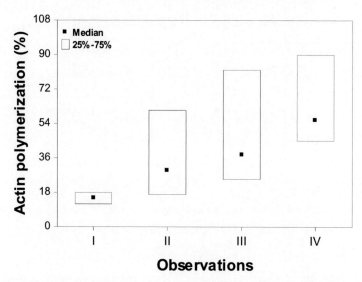

Fig. 5. Actin polymerization in peripheral blood neutrophils. (A) Percentages of neutrophils showing actin polarization in focal areas of the cytoplasm in healthy volunteers and in breast cancer patients before and during (striped columns) chemotherapy. Group 1 and group 2 include respectively patients treated with dose-dense chemotherapy and pegfilgrastim, and patients treated with standard chemotherapy. Data are shown as median values plus upper quartile. (B) Sequential evaluation of the percentages of neutrophils showing actin polarization in focal areas of the cytoplasm in patients treated with dose-dense chemotherapy and pegfilgrastim. Data are shown as median values plus interquartile range.

LAP Cytochemistry

We used LAP activity as an endogenous marker of the physiologic maturity of neutrophils (Katoh et al, 1992). In effect, the LAP activity of neutrophils in the bone marrow is low and that of neutrophils in the circulating pool as well as the marginal pool is high. A semiquantitative evaluation of neutrophil LAP activity was performed by a scoring method under the microscope.

Median basal level of LAP score in breast cancer patients was 115.5 (range 72-226). Chemotherapy increased the LAP score in both groups of patients, even though the rate of increase was significantly higher in pegfilgrastim-treated patients (more than 2-fold) (median 222, range 109-310), compared to patients treated with chemotherapy without pegfilgrastim support (median 161, range 106-196) (P=0.01) (Fig. 6A and 6B).

A)

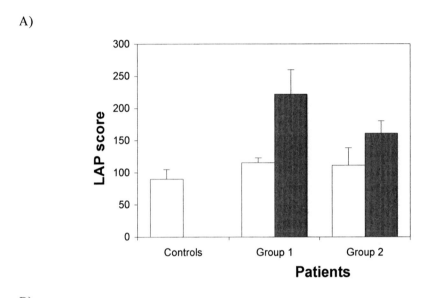

B)

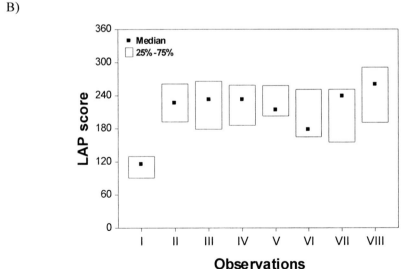

Figure 6. Neutrophil LAP activity. (A) LAP score in healthy volunteers and in breast cancer patients before and during (striped columns) chemotherapy. Group 1 and group 2 include respectively patients treated with dose-dense chemotherapy and pegfilgrastim, and patients treated with standard chemotherapy. Data are shown as median values plus upper quartile. (B) Sequential evaluation of the LAP score in patients treated with dose-dense chemotherapy and pegfilgrastim. Data are shown as median values plus interquartile range.

Correlation Between Parameters

In patients treated with dose-dense chemotherapy and pegfilgrastim there was a significant positive correlation between neutrophil count and both percentage of immature cells and LAP score (P=0.01, R=0.29 and P=0.006 and R=0.31 respectively), whereas neutrophil count was independent of the apoptotic rate. Moreover, rank correlation showed a tendential positive relationship between TUNEL positivity and actin polarization (P=0.09, R=0.27).

CONCLUSION

Neutropenia is a frequent and often dose limiting complication of cancer chemotherapy. Recombinant myeloid growth factors accelerate neutrophil recovery following chemotherapy resulting in decreased incidence of febrile neutropenia, possible reduction in hospitalization and antibiotic requirement, and improved ability to deliver planned treatment (Morstyn et al, 1988). G-CSF prophylaxis has proved to allow increased dose intensity of chemotherapy which may lead to improved survival in patients with chemosensitive tumors (ASCO, 2000). Increasing the dose of cytotoxic drugs through sequential administration has also been shown to improve survival in patients with primary breast cancer (Bonadonna et al, 1995; Citron et al, 2003; Henderson et al, 2003).

The main mechanism of action of G-CSF and GM-CSF consists in the enhancement of the production and maturation of granulocytes from common myeloid precursors in the bone marrow (Gabrilove, 1989). However, several reports suggest that cytokine-mediated neutrophilia is not simply due to enhanced hematopoiesis. Growth factors can also delay spontaneous granulocyte death and extend their life span. The explanation for their antiapoptotic effect is unclear: it may depend on the up- or downregulation of molecules associated with apoptosis, especially those of the mitochondrial intrinsic pathway (Maianski et al, 2002, 2004). Conversely, there is controversy as to whether G-CSF can inhibit also drug-induced apoptosis of granulocytes in vivo. Whereas both myeloid growth factors G-CSF and GM-CSF would be able to increase defective bactericidal activity of neutrophils from children with cancer treated by chemotherapy, only GM-CSF could correct their accelerated apoptosis (Lejeune et al., 1999).

Recently a covalent conjugate of recombinant G-CSF and polyethylene glycol (pegfilgrastim) has been introduced for the treatment of chemotherapy-induced neutropenia (Molineux et al, 1999). Its increased elimination half-life due to decreased serum clearance allows for administration in a single dose per chemotherapy cycle (Holmes et al, 2002). When administered 24 hours after chemotherapy, a single fixed dose of pegfilgrastim provides effective protection from neutropenic complications and is comparable to multiple filgrastim doses in terms of efficacy, safety and tolerability (Green et al, 2003). Data on biological effects of pegfilgrastim are scanty. We investigated some of them within two clinical trials aimed at evaluating the clinical usefulness of pegfilgrastim inserted into dose-dense regimens for breast cancer patients, as neoadjuvant and adjuvant setting.

Clinically, after stimulation with pegfilgrastim in all patients we observed stability of the absolute neutrophil count for the whole duration of treatment and no infectious events. As expected, chemotherapy induced apoptosis in peripheral blood neutrophils. However, in the group of pegfilgrastim-treated patients we noted a reduction of neutrophil apoptosis rate in comparison with that observed in control patients treated with standard chemotherapy courses without filgrastim support. Apparently, this effect on neutrophil survival was due not only to an increase of circulating immature cells; the concomitant finding of a significant increase of the leukocyte alkaline phosphatase activity, that can be considered as a sensitive marker of the last stages of terminal myeloid differentiation (Tsuruta et al, 1996; Dotti et al, 1999), rather suggests also an apoptosis delay of fully mature cells.

Very interestingly, in most pegfilgrastim stimulated neutrophils also persistent abnormalities of actin assembly, indicative of changes in cytoskeleton organization, were observed. These data confirm previous reports showing changes of the cytoskeleton structure and other morphological and biochemical alterations of neutrophils that were induced by in vivo administration of G-CSF (Carulli, 1997; Mattii et al, 2005; Carulli et al, 2006). The interaction of the growth factor with the specific receptor on the cell membrane could trigger a cascade of phenomena responsible for these modifications and consequently for an activation state of the cells. However, their influence on cell functional activity as well as, possibly, on cell survival remains to be elucidated. Alternatively, a direct relationship between drug-induced apoptosis and altered cytoskeletal functions could be hypothesized (White et al, 1993), since a tendential positive correlation was found between neutrophil AI and actin polymerization.

In conclusion, our results suggest that pegfilgrastim may improve the neutrophil count in patients with cancer exposed to CT by inhibiting the accelerated apoptosis and prolonging survival. This effect may, at least in part, be dependent on its influence on actin cytoskeleton organization.

REFERENCES

Adachi, S; Kubota, M; Lin, YW; Okuda, A; Matsubara, K; Wakazono, Y et al. (1994) In vivo administration of granulocyte colony-stimulating factor promotes neutrophil survival in vitro. *Eur J Haematol, 53,* 129-134.

Akgul, C; Moulding, DA; Edwards, SW. (2001) Molecular control of neutrophil apoptosis. *FEBS Letters, 487,* 318-322.

Altznauer, F; Martinelli, S; Yousefi, S; Thürig, C; Schmid, I; Conway, EM et al. (2004) Inflammation-associated cell cycle-independent block of apoptosis by survivin in terminally differentiated neutrophils. *J Exp Med, 199,* 1343-1354.

American Society of Clinical Oncology (ASCO) 2000 update of recommendation for the use of hematopoietic colony-stimulating factors: evidence based, clinical practice guidelines. (2000) *J Clin Oncol, 18,* 3558-3585.

Azzarà, A,; Carulli, G; Rizzuti-Gullaci, A; Capochiani, E; Petrini, M. (2001) Lenograstim and filgrastim effects on neutrophil motility in patients undergoing chemotherapy: evaluation by computer-assisted image analysis. *Am J Hematol, 66,* 306-307.

Baumann, R; Casaulta, C; Simon, D; Conus, S; Yousefi, S; Simon, HU. (2003) Macrophage migration inhibitory factor delays apoptosis in neutrophils by inhibiting the mitochondria-dependent death pathway. *FASEB J, 17*, 2221-2230.

Bonadonna, G; Valagussa, P; Moliterni, A; Zambetti, M; Brambilla, C. (1995) Adjuvant cyclophosphamide, methotrexate, and fluorouracil in node-positive breast cancer: the results of 20 years of follow-up. *N Engl J Med, 330*, 1253-1259.

Bruno, A; Conus, S; Schmid, I; Simon, HU. (2005) Apoptotic pathways are inhibited by leptin receptor activation in neutrophils. *J Immunol, 174*, 8090-8096.

Carulli G. (1997) Effects of recombinant human granulocyte colony-stimulating factor administration on neutrophil phenotype and functions. *Haematologica, 82*, 606-616.

Carulli, G; Mattii, L; Azzarà, A; Brizzi, S; Galimberti, S; Zucca, A et al. (2006) Actin polymerization in neutrophils from donors of peripheral blood stem cells: divergent effects of glycosylated and nonglycosylated recombinant human granulocyte colony-stimulating factor. *Am J Hematol, 81*, 318-323.

Chucklovin, A. (2000) Enhanced ex vivo apoptosis of peripheral granulocytes is a sufficient factor of neutropenia following myeloablative chemotherapy. *Leuk Res, 24*, 507-509.

Ciobotaro, P; Drucker, L; Neumann, A; Shapiro, H; Shapira, J; Radnay, J; Lishner, M. (2003) The effects of doxorubicin on apoptosis and adhesion molecules of normal peripheral blood leukocytes – an ex vivo study. *Anti-Cancer Drugs, 14*, 383-389.

Citron, ML; Berry, DA; Cirrincione, C; Hudis, C; Winer, EP; Gradishar, WJ et al. (2003) Randomized trial of dose-dense versus conventionally scheduled versus concurrent combination chemotherapy as postoperative adjuvant treatment of node-positive primary breast cancer: first report of Intergroup Trial C9741/Cancer and Leukemia Group B Trial 9741. *J Clin Oncol, 21*, 1431-1439.

Colott, F; Re, F; Polentarutti, N; Sozzani, S; Mantovani, A. (1992) Modulation of granulocyte survival and programmed cell death by cytokines and bacterial products. *Blood, 80*, 2012-2020.

Dibbert, B; Weber, M; Nikolaizik, WH; Vogt, P; Schöni, MH; Blaser, K; Simon, HU. (1999) Cytokine-mediated Bax deficiency and consequent delayed neutrophil apoptosis: a general mechanism to accumulate effector cells in inflammation. *Proc Natl Acad Sci USA, 96*, 13330-13335.

Dotti, G; Garattini, E; Borleri, G; Masuhara, K; Spinelli, O; Barbui, T; Rambaldi, A. (1999) Leukocyte alkaline phosphatase identifies terminally differentiated normal neutrophils and its lack in chronic myelogenous leukaemia is not dependent on p210 tyrosine kinase activity. *Br J Haematol, 105*, 163-172.

Gabrilove, JL. (1989) Introduction and overview of hematopoietic growth factors. *Semin Hematol, 26*, 1-4.

Goepel, F; Weinmann, P; Schymeinsky, J; Walzog, B. (2004) Identification of caspase-10 in human neutrophils and its role in spontaneous apoptosis. *J Leukoc Biol, 75*, 836-843.

Green, MD; Koelbl, H; Baselga, J; Galid, A; Guillem, V; Gascon,P et al. (2003) A randomized double-blind multicenter phase III study of fixed-dose single-administration pegfilgrastim versus daily filgrastim in patients receiving myelosuppressive chemotherapy. *Ann Oncol, 14*, 29-35.

Hall, A. (1998) Rho GTPases and the actin cytoskeleton. *Science, 279*, 509-514.

Hannun, AY. (1997) Apoptosis and the dilemma of cancer chemotherapy. *Blood*, *89*, 1845-1853.

Hasegawa, T; Suzuki, K; Sakamoto, C; Ohta, K; Nishiki, S; Hino, M et al. (2003) Expression of the inhibitor of apoptosis (IAP) family members in human neutrophils: up-regulation of cIAP2 by granulocyte colony-stimulating factor and overexpression of cIAP2 in chronic neutrophilic leukaemia. *Blood*, *101*, 1164-1171.

Hayhoe, FGJ and Quaglino, D. (1958) Cytochemical demonstration and measurement of leucocyte phosphatase activity in normal and pathological states by modified azo-dye coupling techniques. *Br J Haematol*, *4*, 375-389.

Henderson, IC; Berry, DA; Dimetri, DG; Cirrincione,CT; Goldstein, LJ; Martino, S et al. (2003) Improved outcomes from adding sequential Paclitaxel but not from escalating Doxorubicin dose in an adjuvant chemotherapy regimen for patients with node-positive primary breast cancer. *J Clin Oncol*, *21*, 976-983.

Holmes, FA; O'Shaughnessy, JA; Vukelja, S; Jones, SE; Shogan, J; Savin, M et al. (2002) Blinded randomized multicenter study to evaluate single administration pegfilgrastim once cycle versus daily filgrastim as an adjunct to chemotherapy in patients with high risk stage II or stage III/IV breast cancer. *J Clin Oncol*, *20*, 727-731.

Kamesaki, H. (1998) Mechanisms involved in chemotherapy-induced apoptosis and their implications in cancer chemotherapy. *Int J Hematol*, *68*, 29-43.

Katoh, M; Shirai, T; Shikoshi, K; Ishii, M; Saito, M; Kitagawa, S. (1992) Neutrophil kinetics shortly after initial administration of recombinant human granulocyte colony-stimulating factor: neutrophil alkaline phosphatase activity as an endogenous marker. *Eur J Haematol*, *49*, 19-24.

Kaufmann, HS and Earnshaw, CW. (2000) Induction of apoptosis by cancer chemotherapy. *Exp Cell Res*, *256*, 42-49.

Kerr, JFR; Winterford, CM; Harmon, BV. (1994) Apoptosis. Its significance in cancer and cancer therapy. *Cancer*, *73*, 2013-2026.

Kimhi, O; Drucker, L,; Neumann, A; Shapiro, H; Shapira, J; Yarkoni, S et al. (2004) Fluorouracil induces apoptosis and surface molecule modulation of peripheral blood leukocytes. *Clin Lab Haematol*, *26*, 327-333.

Leavey, PJ; Sellins, KS; Thurman, G; Elzi, D; Hiester, A; Silliman, CC et al. (1998) In vivo treatment with granulocyte colony-stimulating factor results in divergent effects on neutrophil functions measured in vitro. *Blood*, *92*, 4366-4374.

Lejeune, M; Ferster, A; Cantinieux, B; Sariban, E. (1998) Prolonged but reversible neutrophil dysfunctions differentially sensitive to granulocyte colony-stimulating factor in children with acute lymphoblastic leukaemia. *Br J Haematol*, *102*, 1284-1291.

Lejeune, M; Cantinieux, B; Harag, S; Ferster, A; Devalck, C; Sariban, E. (1999) Defective functional activity and accelerated apoptosis in neutrophils from children with cancer are differentially corrected by granulocyte and granulocyte-macrophage colony stimulating factors in vitro. *Br J Haematol*, *106*, 756-761.

Maianski, NA; Mul, FPJ; van Buul, JD; Roos, D; Kuijpers, TW. (2002) Granulocyte colony-stimulating factor inhibits the mitochondria-dependent activation of caspase-3 in neutrophils. *Blood*, *99*, 672-679.

Maianski, NA; Roos, D; Kuijpers, TW. (2004) Bid truncation, Bid/Bax targeting to the mitochondria, and caspase activation associated with neutrophil apoptosis are inhibited by granulocyte colony-stimulating factor. *J Immunol*, *172*, 7024-7030.

Mattii, L; Fazzi, R; Moscato, S; Segnani, G; Pacini, S; Galimberti, S et al. (2004) Carboxy-terminal fragment of osteogenic growth peptide regulates myeloid differentiation through RhoA. *J Cell Biochem*, *93*, 1231-1241.

Mattii, L; Azzarà, A; Fazzi, R; Carulli, G; Cimenti, M; Lecconi, N et al. (2005) Glycosylated or non-glycosylated G-CSF differently influence human granulocyte functions through RhoA. *Leuk Res*, *29*, 1285-1292.

Molineux, G; Kinstler, O; Briddell, B; Hartley, C; McElroy, P; Kerzic, P et al. (1999) A new form of Filgrastim with sustained duration in vivo and enhanced ability to mobilize PBPC in both mice and humans. *Exp Hematol*, *27*, 1724-1734.

Molloy, EJ; O'Neill, A; Grantham, JJ; Sheridan-Pereira, M; Fitzpatrick, JM; Webb, DW; Watson, RWG. (2005) Granulocyte colony-stimulating factor and granulocyte-macrophage colony-stimulating factor have differential effects on neonatal and adult neutrophil survival and function. *Ped Res*, *57*, 806-812.

Morstyn, G; Campbell, L; Souza, LM; Alton, NK; Keech, J; Green, M et al (1988) Effect of granulocyte colony-stimulating factor on neutropenia induced by cytotoxic chemotherapy. *Lancet*, *1*, 667-672.

Moulding, DA; Akgul, C; Derouet, M; White, MRH; Edwards, SW. (2001) BCL-2 family expression in human neutrophils during delayed and accelerated apoptosis. *J Leukoc Biol*, *70*, 783-792.

Negoescu, A ; Lorimier, P ; Labat-Moleur, F ; Drouet, C ; Robert, C ; Guillermet, C et al. (1996) In situ apoptotic cell labeling by the TUNEL method : improvement and evaluation on cell preparations. *J Histochem Cytochem*, *44*, 959-968.

Schmitt, AC and Lowe, WS. (1999) Apoptosis and therapy. *J Pathol*, *187*, 127-137.

Tsangaris, GT; Moschovi, M; Mikraki, V; Vrachnou, E; Tzortzatou-Stathopoulou, F. (1996) Study of apoptosis in peripheral blood of patients with acute lymphoblastic leukemia during induction therapy. *Anticancer Res*, *16*, 3133-3140.

Tsuruta, T; Tani, K; Shimane, M; Ozawa, K; Takahashi, S; Tsuchimoto, D et al. (1996) Effects of myeloid cell growth factors on alkaline phosphatase, myeloperoxidase, defensin and granulocyte colony-stimulating factor receptor mRNA expression in haemopoietic cells of normal individuals and myeloid disorders. *Br J Haematol*, *92*, 9-22.

Villunger, A; O'Reilly, LA; Holler, N, Adams, J; Strasser, A. (2000) Fas ligand, Bcl-2, granulocyte colony-stimulating factor, and p38 mitogen-activated protein kinase: regulators of distinct cell death and survival pathways in granulocytes. *J Exp Med*, *192*, 647-658.

Vose, JM; Armitage, JO. (1995) Clinical applications of hematopoietic growth factors. *J Clin Oncol*, *13*, 1023-1035.

Watson, RWG; O'Neill, A; Brannigen, AE; Coffey, R; Marshall, JC; Brady, HR; Fitzpatrick, JM. (1999) Regulation of Fas antibody induced neutrophil apoptosis is both caspase and mitochondrial dependent. *FEBS Letters*, *453*, 67-71.

Weinmann, P; Scharffetter-Kochanek, K; Forlow, SB; Peters, T; Walzog, B. (2003) A role for apoptosis in the control of neutrophil homeostasis in the circulation: insights from CD18-deficient mice. *Blood, 101,* 739-746.

Wesche, DE; Lomas-Neira, JL; Perl, M; Chung, CS; Ayala, A. (2005) Leukocyte apoptosis and its significance in sepsis and shock. *J Leukoc Biol, 78,* 325-337.

Whyte, MKB; Meagher, LC; MacDermot, J; Haslett, C. (1993) Impairment of function in aging neutrophils is associated with apoptosis. *J Immunol, 150,* 5124-5134.

Wulf, E; Deboben, A; Bautz, FA; Faulstich, H; Wieland, T. (1979) Fluorescent phallotoxin, a tool for the visualization of cellular actin. *Proc Natl Acad Sci USA, 76,* 4498-4502.

In: Cell Apoptosis Research Trends
Editor: Charles V. Zhang, pp. 263-276

ISBN: 1-60021-424-X
© 2007 Nova Science Publishers, Inc

Chapter X

Rapamycin Controls Multiple Signalling Pathways Involved in Cancer Cell Survival

Maria Fiammetta Romano, Simona Romano, Maria Mallardo, Rita Bisogni and Salvatore Venuta

Department of Biochemistry and Medical Biotechnology,
University Federico II, Naples, Italy.
Department of Clinical and Experimental Medicine, University Magna Graecia,
Catanzaro, Italy.

ABSTRACT

Suppression of apoptosis by survival signals is considered a hallmark of malignant transformation and resistance to anti-cancer therapy. The phosphoinositide-3 kinase (PI3k)/Akt pathway and NF-κB transcription factors are potent mediators of tumour cell survival. The carbocyclic lactone-lactam antibiotic rapamycin, a widely used immunosuppressant, inhibits the oncogenic transformation of human cells induced by PI3k or Akt by blocking the downstream mTOR kinase. However, inhibition of the PI3k/Akt/mTOR cascade may not be the only mechanism whereby rapamycin exerts anticancer effects. We previously demonstrated that rapamycin inhibits NF-κB by acting on FKBP51, a large immunophilin whose isomerase activity is essential for the functioning of the IKK kinase complex. This suggested that rapamycin may be effective also against neoplasias that express the tumour suppressor PTEN, which, by reducing cellular levels of phosphatidyl-inositol triphosphate, antagonizes the action of PI3k. To address this issue, we over-expressed PTEN in a human melanoma cell line characterized by high phospho-Akt and phospho-mTOR levels, and examined the effect of rapamycin on the apoptotic response to the NF-κB inducer doxorubicin versus cisplatin, which does not activate NF-κB. Rapamycin increased both cisplatin- and doxorubicin-induced apoptosis. Transient transfection of PTEN remarkably decreased phospho-mTOR levels and increased sensitivity to cisplatin's cytotoxic effect. Under these conditions, rapamycin failed to enhance cisplatin-induced apoptosis. This finding supports the notion that inhibition of a survival pathway increases the efficacy of cytotoxic drugs, and suggests that the pro-apoptotic effect of the rapamycin-cisplatin association requires activated mTOR. Rapamycin retained the capacity to enhance doxorubicin-induced apoptosis in cells over-expressing PTEN, which confirms our earlier observation that inhibition of the PI3k/Akt/mTOR pathway is not involved in the effect exerted by the rapamycin-doxorubicin association. These findings indicate that constitutive activation of mTOR is sufficient but not necessary for rapamycin's anti-cancer effect. Finally, we show that a decrease in FKBP51 expression levels, obtained with the small interfering RNA technique in the leukemic cell line Jurkat, increased doxorubicin-induced apoptosis, suggesting that this rapamycin ligand is involved in resistance to chemotherapy-induced apoptosis.

In conclusion, rapamycin affects more than one signalling survival pathway and more than one target. Our data may impact on the synthesis of rapamycin derivatives. Thus far, rapamycin derivatives used in clinical trials have been tested for their mTOR-inhibiting effect. Our study opens the door to a novel class of anti-cancer drugs that specifically target immunophilins.

INTRODUCTION

Apoptosis is the predominant mechanism by which cancer cells die when subjected to chemotherapy [1,2]. The resistance of tumour cells to anticancer agents can result from the development of survival signals [3]. Signalling pathways responsible for cell survival can be constitutively activated in cancer cells due to mutation or loss of tumour-suppressor genes [4] and may even be induced by chemotherapy itself [5]. Understanding how tumour cells evade apoptotic events may provide a new paradigm for cancer therapy [6]. The phosphoinositide-3 kinase (PI3k)/Akt pathway and NF-κB transcription factors are considered potent mediators of cell survival in cancer [7-9].

The PI3k/Akt pathway plays an important role in regulating glucose metabolism in normal cells [10], it is inactive in resting cells and often deregulated in cancer cells [4,7]. Extracellular survival signals, delivered as soluble factors or through cell attachment, can inhibit apoptosis by activating this pathway [10-12]. Upon growth factor binding, transmembrane receptor tyrosine kinases undergo auto- or trans-phosphorylation, which creates binding sites for PI3k in their cytoplasmic domains. This, in turn, enables recruitment of active PI3k to the inner surface of the plasma membrane [10-12], where PI3k causes membrane phosphoinositides to generate phosphatidylinositol 3,4-bisphosphate and phosphatidylinositol 3,4,5-trisphosphate (PIP3), and activates the serine/threonine kinase Akt [10-12]. Termination of the PIP3 signal occurs through the action of PTEN, the inositol 3-phosphatase and tensin homologue deleted on chromosome 10 in many tumours [4]. Essentially, PTEN dephosphorylates phosphoinositides at the D3 position.

Akt controls a host of signalling molecules, including tuberin, the product of the TSC2 gene [13]. Phosphorylation of tuberin inactivates the tuberous sclerosis complex formed by hamartin (produced by the TSC1 gene) and tuberin, thereby activating the mammalian target of rapamycin (mTOR) [13]. mTOR is a serine-threonine kinase, whose effector molecules are the ribosomal protein S6 kinase [14] and the eukaryotic initiation factor 4E binding protein 1

(4E-BPI) [15], which regulate ribosome biogenesis and the translation of proteins involved in cell cycle progression and proliferation [14-16]. In many human cancers, the PI3k/Akt pathway is targeted by genomic aberrations including mutations, amplifications and rearrangements [4,7,17]. Whatever the mechanism, activation of this pathway results in disturbance of the control of cell growth and survival and defective apoptosis, which contributes to a competitive growth advantage, metastatic competence and therapy resistance [7].

Another important factor involved in oncogenesis and chemoresistance is NF-κB [5,9,18]. The NF-κB transcription complex belongs to the Rel family, which is constituted by five mammalian Rel/NF-κB proteins that exert transcriptional activity: RelA (p65), c-Rel, RelB, NF-κB1 (p50/p105) and NF-κB2 (p52/100) [18,19]. RelA, c-Rel and RelB are synthesized as mature proteins, whereas p50 and p52 are first synthesized as large precursors (p105 and p100) that are processed by the proteosome [18,19]. The activity of NF-κB is controlled by shuttling from the cytoplasm to the nucleus in response to cell stimulation. NF-κB dimers containing RelA or c-Rel are retained in the cytoplasm through interaction with inhibitors of NF-κB (IκBs) [15-17]. In response to a variety of stimuli, IκBs are phosphorylated by the activated IκB kinase (IKK) complex, a ~900-kDa multiprotein complex responsible for signal-dependent phosphorylation of IκB [18-20]. This complex contains two catalytic subunits, IKKα and IKKβ, and an essential regulatory subunit NEMO or IKKγ. IκB phosphorylation is followed by rapid ubiquitin-dependent degradation by the 26S proteosome [18-20]. This allows NF-κB dimers to translocate to the nucleus, where they stimulate the expression of target genes. Once in the nucleus, NF-κB dimers are further modified, mostly through phosphorylation of Rel proteins, to optimize their transcriptional activity [6]. NF-κB's oncogenic activity is closely linked to its anti-apoptotic function [9,18,19,21]. Activation of NF-κB has proved to be a pivotal mechanism of tumour chemoresistance [5,9,18]. This is supported by the observation that inhibition of NF-κB sensitizes cancer cell lines to chemotherapy [9,21]. NF-κB induces resistance to cancer therapies by modulating the regulation of various antiapoptotic genes [18], including the inhibitors of apoptosis (IAPs) [22], the caspase 8 inhibitory protein (cFLIP) [23], A1/Bfl1 [24], and TNF-receptor-associated factor (TRAF) 1 and 2 [25].

Rapamycin, the product of *Streptomyces hygroscopicus*, targets the PI3k/Akt [26,27] and NF-κB signaling pathways [28,29]. Although conventionally used as an immunosuppressant agent, rapamycin is an effective anticancer agent that both decreases cell proliferation and increases apopotosis [14,30]. Rapamycin is associated with a much lower risk of cancer recurrence versus other immunosuppressant agents because it controls the growth of primary and metastatic tumours and angiogenesis [30]. Its anti-cancer activity is classically ascribed to binding to FK506 binding protein (FKBP)12 [31]. This results in a complex that, in turn, binds to mTOR and inhibits its kinase activity [26].

Rapamycin inhibits the oncogenic transformation of human cells induced by PI3k or Akt [27]. Consequently, its major therapeutic indication is neoplasias that lack the tumour suppressor gene PTEN [17]. However, we recently reported data suggesting that rapamycin can exert anticancer effects also by inhibiting NF-κB [28,29]. We demonstrated that rapamycin reduces the phosphorylating activity of IKK on its IκB substrate [28] by blocking FKBP51, which is an important cofactor of the IKK-α subunit [32]. Because of its effect on

NF-κB, rapamycin sensitizes melanoma [28] and leukemic cells [29] to the action of anthracyclins. Thus, it is conceivable that rapamycin may be effective also in PTEN-positive tumours when associated with NF-κB-inducing chemotherapeutic drugs. To address this issue, we transfected PTEN in a human melanoma cell line characterized by high levels of phospho-Akt and -mTOR, and investigated the pro-apoptotic effect of rapamycin in association with cytotoxic agents that either induce or do not induce NF-κB. We also attempted to determine the role of FKBP51 as a factor of resistance to chemotherapy by down-modulating the protein level using the small interfering (si) RNA technique [33].

EFFECT OF RAPAMYCIN ON CANCER CELL APOPTOSIS

Rapamycin Enhances Apoptosis of Cancer Cells Expressing Activated MTor

A loss of PTEN protein and function has been implicated in the early stages of melanomagenesis [34]. Therefore, the PI3k/Akt survival pathway is likely to be constitutively activated in this tumour. In accordance with this concept, we found the active form of both Akt and mTOR in our human melanoma cell line [28], as demonstrated by the decrease of phospho-protein levels after incubation with the PI3k inhibitor wortmannin (Fig.1).

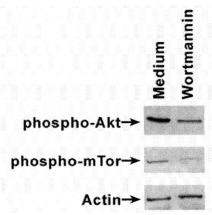

Figure 1. *Akt and mTOR are constitutively activated in melanoma cells.* Western blot assay of phospho-Akt (Ser 473) and phospho-mTOR (Ser2448) expression levels of lysates from melanoma cells incubated with and without wortmannin (1 μM), for 3 hrs. Whole cell lysates were prepared by homogenization in modified RIPA buffer (150 mM sodium-chloride, 50 mM Tris-HCl, pH 7.4, 1 mM ethylenediamine tetraacetic acid, 1 mM phenylmethylsulfonyl fluoride, 1% Triton X-100, 1% sodium deoxycholic acid, 0.1% sodium dodecylsulfate, 5 μg/ml of aprotinin, 5 μg/ml of leupeptin). Cell debris was removed by centrifugation. The cell lysate was boiled for 5 min in 1x SDS sample buffer (50 mM Tris-HCl pH 6.8, 12.5% glycerol, 1% sodium dodecylsulfate, 0.01% bromophenol blue) containing 5% beta-mercaptoethanol, run on 10% SDS polyacrylamide gel electrophoresis, transferred onto a membrane filter (Cellulosenitrate, Schleider and Schuell, Keene, NH) and incubated with the primary antibody. Anti-phospho-Akt (Ser473) and anti-phospho-mTOR (Ser2448) were rabbit polyclonal antibodies (Cell Signaling Technology, Beverly, MA). After a second incubation with peroxidase-conjugated goat anti-rabbit IgG (Santa Cruz Biotechnology, Santa Cruz, CA) the blots were developed with the ECL system (Amersham Pharmacia Biotech, Piscataway, NJ). Actin was used as the control for loading.

Survival pathways can be activated also by anti-cancer agents [35]. As shown in Figure 2, the anthracyclin doxorubicin, but not the alkylating cisplatin, induced NF-κB transcription factors in melanoma cells. Treatment of tumour cells with rapamycin did not stimulate apoptosis per se, but strikingly enhanced apoptosis induced by both chemotherapeutics (Figs 3 and 4).

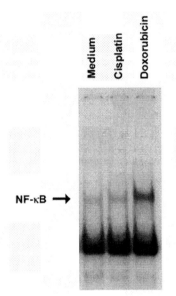

Figure 2. *Effect of cisplatin and doxorubicin on NF-κB activation.* Electrophoretic mobility shift assay of nuclear extracts from melanoma cells cultured with or without cisplatin (25 μM) or doxorubicin (3 μM) for 5 hrs. The band indicated by the arrow corresponds to NF-κB complexes. Nuclear extracts were prepared by cell pellet homogenization in two volumes of 10 mM Hepes, pH 7.9, 10 mM KCl, 1.5 mM $MgCl_2$, 1 mM EDTA, 0.5 mM DTT, 0.5 mM PMSF and 10% glycerol v/v. Nuclei were centrifuged at 1,000 g for 5 min, washed and resuspended in two volumes of the above-specified solution. KCl was added to reach 0.39 M KCl. Nuclei were extracted at 4°C for 1 h and centrifuged at 10,000 g for 30 min. The supernatant was clarified by centrifugation and stored at –80°C. The NF-κB consensus 5'-CAACGGCAGGGGAATCTCCCTCTCCTT-3' oligonucleotide was end-labeled with [γ-^{32}P] ATP (Amersham Pharmacia Biotech) using a polynucleotide kinase (Roche, Basel Switzerland). End-labeled DNA fragments were incubated at room temperature for 15 min with 5 μg of nuclear protein, in the presence of 1 μg poly(dI-dC), in 20 μl of a buffer consisting of 10 mM Tris-HCl, pH 7.5, 50 mM NaCl, 1 mM EDTA, 1 mM DTT and 5% glycerol v/v. Protein-DNA complexes were separated from free probe on a 6% polyacrylamide w/v gel run in 0.25X Tris borate buffer at 200 mV for 3 hrs at room temperature. The gels were dried and exposed to X-ray film (Kodak AR).

Rapamycin Enhances Apoptosis of Cancer Cells Reconstituted with PTEN

To investigate the effect of rapamycin on chemotherapy-induced apoptosis in the context of PTEN over-expression, we transiently co-transfected PTEN and green fluorescent protein (GFP) in melanoma cells, and measured apoptosis induced by doxorubicin and cisplatin in the absence and in the presence of rapamycin. As shown in Figure 5, phospho-mTOR levels, determined in GFP-gated cells by flow cytometry, were clearly lower in PTEN-transfected cells than in both control cells and cells transfected with mutated PTEN.

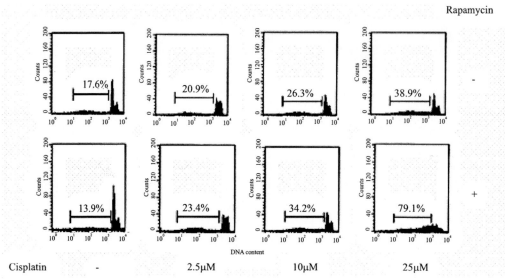

Figure 3. *Rapamycin enhances cisplatin-induced apoptosis.* Measurement of apoptosis of melanoma cells cultured with cisplatin at different doses, in the absence or the presence of 100 ng/ml rapamycin. Apoptosis, using the propidium iodide incorporation assay, was evaluated in permeabilized cells by flow cytometry. The cells were harvested after 24 hrs of culture, washed in PBS and resuspended in 500 μl of a solution containing 0.1% sodium citrate, 0.1 % Triton X-100 and 50μg/mL propidium iodide (Sigma Aldrich, Italy). After incubation at 4°C for 30 min in the dark, cell nuclei were examined with flow cytometry. DNA content was recorded on a logarithmic scale. The percentage of the elements in the hypodiploid region was calculated.

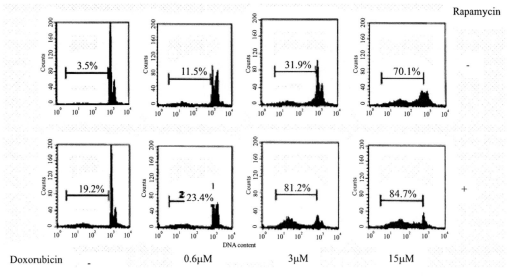

Figure 4. *Rapamycin enhances doxorubicin-induced apoptosis.* Measurement of apoptosis of melanoma cells cultured with different doses of doxorubicin, in the absence or the presence of 100 ng/ml rapamycin. Apoptosis, using the propidium iodide incorporation assay, was evaluated in permeabilized cells by flow cytometry. Cells were harvested after 24 hrs of culture, washed in PBS and resuspended in 500 μl of a solution containing 0.1% sodium citrate, 0.1% Triton X-100 and 50 μg/mL propidium iodide (Sigma Aldrich, Italy). After incubation at 4°C for 30 min in the dark, cell nuclei were examined by flow cytometry. DNA content was recorded on a logarithmic scale. The percentage of the elements in the hypodiploid region was calculated.

This demonstrates that the protein encoded by the plasmid was functional. Subsequently, we incubated the cells with cisplatin or doxorubicin, in the absence and in the presence of rapamycin, and measured apoptosis using the TdT-mediated dUTP nick end-labelling (TUNEL) assay [36].

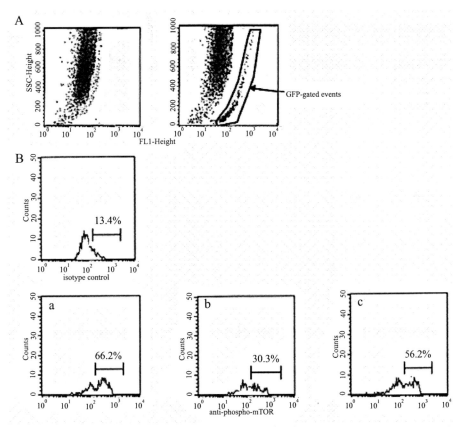

Figure 5. *PTEN-transfection decreases phospho-mTOR levels.* **A** Flow cytometric dot plots of melanoma cells non-encoding or encoding GFP. A gate was placed on GFP$^+$ cells. **B** Flow cytometric histograms of phospho-mTOR expression in GFP-gated cells. (**a**), cells transfected with GFP; (**b**), cells co-transfected with GFP+ PTEN; (**c**), cells co-transfected with GFP+mutated PTEN. Staining of GFP-gated cells with isotype control Ig is also shown. Melanoma cells in the logarithmic growth phase were resuspended in serum-free RPMI 1640 and transfected with 5 μg of plasmid encoding GFP plus 15 μg of plasmid encoding PTEN or mutated PTEN, by electroporation at 250 mV and 960 μF using the Gene Pulser (Bio-Rad Laboratories, Hercules, CA, USA). The cells were transferred to six-well plates in 10% FCS RPMI 1640 supplemented with antibiotic and glutamine at 37°C in a 5% CO_2 humidified atmosphere. The cDNA encoding for PTEN or mutated PTEN [48] was kindly provided by Dr. David Stokoe (Cancer Research Institute, UCSF, CA). Three days later, the cells were collected, fixed, permeabilized and an indirect immunofluorescence with anti-phospho-mTOR antibody (Cell Signaling Technology, Beverly, MA) was performed using a PE-conjugated secondary antibody.

GFP-positive cells were gated (Fig. 5A), and we evaluated the percentage of cells incorporating dUTP (apoptotic cells) using flow cytometry. As shown in Figure 6, rapamycin significantly enhanced cisplatin-induced apoptosis in both control cells (p=0.006) and cells transfected with mutated PTEN (p=0.03), but not in cells transfected with PTEN (p=0.2). Moreover, cisplatin-induced apoptosis was greater (p=0.04) in PTEN-transfected cells than in control cells. There was no difference in cisplatin-induced apoptosis between control cells and cells transfected with mutated PTEN. This result indicated that inhibition of mTOR increases the efficacy of cisplatin, and that the pro-apoptotic effect of the rapamycin-cisplatin association depended on the presence of activated mTOR. Unlike cisplatin-treated cells, doxorubicin cultures were sensitive to rapamycin pro-apoptotic effect also under conditions of PTEN over-expression. Indeed, rapamycin significantly (p=0.02) enhanced anthracyclin-induced apoptosis in PTEN-transfected cells (Fig. 7). These findings support our previous studies [26,27] that the cooperative effect between rapamycin and NF-κB-inducing drugs occurs irrespective of mTOR inhibition.

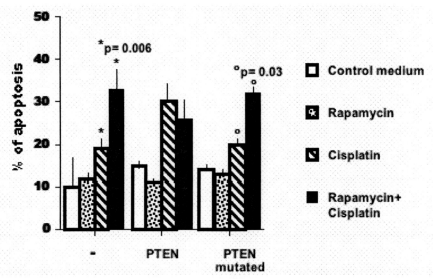

Figure 6. *Effect of rapamycin on cisplatin-induced apoptosis in PTEN-transfected cells.* Graphic representation of apoptosis percentages determined in flow cytometry; error bars indicate standard deviations. Melanoma cells in the logarithmic growth phase were resuspended in serum-free RPMI 1640 and transfected (see legend to Fig. 5). Three days later, rapamycin (100 ng/ml) and or cisplatin (25 μM) was added to the cultures and after incubation for a further 9 hrs, TUNEL was performed with the In Situ Cell Death Detection Kit TMR red (Roche, Basel, Switzerland), according to the manufacturer's instructions. Briefly, cells were fixed with 2% paraformaldehyde in PBS v/v, for 15 min at room temperature, washed and permeabilized with 0.1% TRITON x-100 in PBS, for 2 min in ice. After a second wash, the cells were incubated with TMR red labelling solution for 1 hr at 37°C and examined by flow cytometry. The percentage of GFP-gated cells incorporating dUTP was calculated.

The Rapamycin-Binding Protein FKBP51 is Responsible for Chemoresistance

FKBPs are the first cellular target of rapamycin and FKBP51 controls chemotherapy-induced NF-κB activation [28,29]. We recently reported that rapamycin did not increase doxorubicin-induced apoptosis in the leukemic cell line Jurkat, which hyper-expresses the p65 subunit of NF-κB [29]. This illustrated the importance of inactivation of this transcription factor in rapamycin regulation of apoptosis. To verify whether FKBP51 was involved in resistance to doxorubicin, we down-modulated FKBP51 expression levels using the siRNA technique in Jurkat cells, and investigated doxorubicin-induced apoptosis by propidium iodide incorporation by flow cytometry. Treatment of the cells with FKBP51 siRNA efficiently reduced the expression levels of this protein (Fig. 8A) and enhanced doxorubicin-induced apoptosis by more than 60% (Fig. 8B).

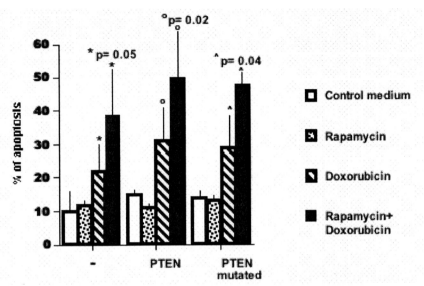

Figure 7. *Effect of rapamycin on doxorubicin-induced apoptosis in PTEN-transfected cells.* Graphic representation of apoptosis percentages determined by flow cytometry; error bars indicate standard deviations. Melanoma cells in the logarithmic growth phase were resuspended in serum-free RPMI 1640 and transfected (see legend to Fig.5). Three days later, rapamycin (100 ng/ml) and or doxorubicin (3 μM) was added to the cultures and after incubation for a further 9 hrs, TUNEL was performed with the In Situ Cell Death Detection Kit TMR red (Roche, Basel, Switzerland), according to the manufacturer's instructions. Briefly, cells were fixed with 2% paraformaldehyde in PBS v/v, for 15 min at room temperature, washed and permeabilized with 0.1% TRITON x-100 in PBS, for 2 min in ice. After a second wash, the cells were incubated with TMR red labelling solution for 1 hr at 37°C and examined with flow cytometry. The percentage of GFP-gated cells incorporating dUTP was calculated.

CONCLUSION

Modern anti-cancer therapy is based on the use of innovative agents able to antagonize signalling pathways that are deregulated in cancer. A number of rapamycin analogs have been developed over recent years, the most notable being the cell cycle inhibitor 779 (CCI-779), RAD-001 and AP23573 [37]. Rapamycin and its analogs are the most promising candidates for the treatment of tumours that lack the suppressor function of PTEN [8,17]. However, our data suggest that constitutive activation of mTOR is sufficient but not necessary for the anti-cancer effect of rapamycin [28,29]. Several lines of evidence support the view that rapamycin inhibits NF-κB [38-40] and, as a consequence, it inhibits the expression of anti-apoptotic genes that are under NF-κB transcriptional control [28,29] and that are involved in the resistance of cancer cells to chemotherapy.

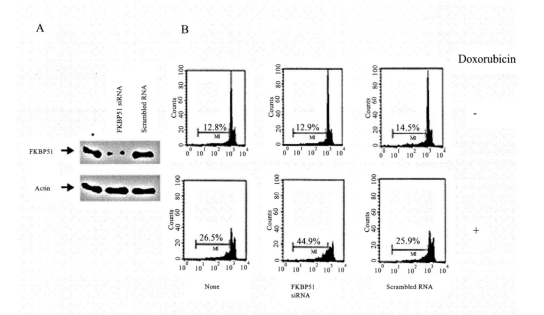

Figure 8. *Downmodulation of FKBP51 increases doxorubicin-induced apoptosis.* **A** Western blot assay of FKBP51 expression levels in cell lysates obtained from non-transfected Jurkat cells and from Jurkat cells transfected with specific siRNA or the scrambled oligonucleotide as control. Cells were incubated for 24 hrs in six-well plates in medium without antibiotics before transfection of the oligonucleotide 5'-ACCUAAUGCUGAGCUUAUAdTdT-3' corresponding to the sense strand of the target sequence 5'-AAACCUAAUGCUGAGCUUAUA-3' of human FKBP51 (Dharmacon Research Inc., Boulder, CO) or of a scrambled duplex as control (Dharmacon Research Inc.). The siRNA or the scrambled oligo was transfected at the final concentration of 50 nM using Metafectene (Biontex, Munich, Germany) according to the manufacturer's recommendations. Three days later, whole cell lysates were prepared by homogenization in modified RIPA buffer and assayed in Western blot to determine FKBP51 expression levels. Anti-FKBP51 was a goat polyclonal antibody (Sigma Aldrich, Italy). Actin was used as the control for loading. **B** Flow cytometric histograms of DNA content, bar indicates hypodiploid region (apoptosis). Cells transfected as described in **A** were cultured with rapamycin (100 ng/ml) and/or doxorubicin (3 µM). Twenty-four hrs later, cells were harvested, washed in PBS and resuspended in 500 µl of a solution containing 0.1% sodium citrate, 0.1% Triton X-100 and 50 µg/mL propidium iodide (Sigma Aldrich, Italy). After incubation at 4°C for 30 min in the dark, cell nuclei were analyzed in flow cytometry. DNA content was recorded on a logarithmic scale, and he percentage of the elements in the hypodiploid region was calculated.

In the present study, we transfected PTEN in a human melanoma cell line to abate the expression of activated mTOR, and investigated the apoptosis-enhancing effect which rapamycin exerts in association with two cytotoxic drugs: doxorubicin, which induces NF-κB, and cisplatin that does not. We found that rapamycin retains the ability to enhance apoptosis in PTEN-transfected cells cultured with doxorubicin but not in the same cells cultured with cisplatin. Consequently, it appears that the cooperative effect between rapamycin and cisplatin requires the presence of activated mTOR. This is in agreement with reports that inhibition of mTOR enhances the cytotoxic effect of cisplatin [41]. Indeed, resistance to cisplatin treatment was attributed to overexpression of elongation factor alpha and of genes involved in ribosomal biogenesis that leads to the production of repair and/or survival proteins [42]. Differently, the cytotoxicity of the NF-κB-inducer doxorubicin was enhanced by rapamycin even in the context of PTEN reconstitution. Our previous finding that the large immunophilin FKBP51 is involved in the control of chemotherapy-induced NF-κB activation [28,29] provided a mechanism for NF-κB down-modulation by rapamycin. In agreement with these findings, we now show that FKBP51 down-modulation increases doxorubicin-induced apoptosis, an observation that implicates FKBP51 in resistance to anthracyclin compounds.

As mentioned above, FKBPs are the first target of rapamycin and, in fact, binding to FKBP12 is crucial for mTOR inhibition. FKBPs are immunophilins, which are abundant, cytosolic proteins endowed with inherent peptidyl-prolyl cis-trans isomerase activity. Rapamycin very specifically binds to FKBPs and inhibits their isomerase activity, which is important for several biological functions of the cell, namely, response to transforming growth factor (TGF)-β [43], control of intracellular calcium release [44], sensitivity to glucocorticoids [45], and IKK kinase complex function [32]. FKBP12 interacts with the cytoplasmic domain of TGF-β type I receptor [43] and acts as a molecular guardian of this receptor to prevent it from leaky signalling under sub-optimal ligand concentrations [43]. Moreover, FKBP12 interacts stoichiometrically with multiple intracellular calcium release channels thereby controlling cellular calcium influx [44]. FKBP51 and FKBP52 regulate the affinity of glucocorticoid receptors that form complexes with Hsp-90 and Hsp-binding co-chaperones [45]. Finally, FKBP51 is a co-factor of IKKα [32]. Given the biological relevance of this class of proteins, it is not surprising that rapamycin exerts effects independent of PI3k/Akt/mTOR inhibition.

The identification of increased FKBP12 expression in high-grade childhood astrocytoma [46] and of FKBP51 in idiopathic myelofibrosis, which is responsible for growth factor independence of bone marrow progenitor proliferation [47], suggest that FKBPs are involved in cancerogenesis. In line with these findings, our studies implicate FKBP51 in resistance to apoptosis, thus opening the door to the development of a new class of anti-cancer drugs that specifically target immunophilins.

Our research provides the first demonstration that FKBPs are involved in the control of apoptosis. FKBP rotamase activity is essential for several biological processes in mammalian cells. Therefore, it cannot be excluded that, besides affecting NF-κB, FKBPs act on other regulatory steps of the apoptotic machinery. There is a need for studies designed to elucidate the role of immunophilins in apoptosis regulation.

REFERENCES

[1]   Brown, JM; Attardi LD. (2005). The role of apoptosis in cancer development and treatment response. *Nature Reviews Cancer, 5,* 231-237.

[2]   Fridman, JS; Lowe, SW. *(2003).* Control of apoptosis by p53. *Oncogene, 22,* 9030-9040.

[3]   Green, DR; Evan, GI. (2002). A matter of life and death. *Cancer Cell, 1,* 19-30.

[4]   Wu, X; Senechal, K; Neshat, MS; Whang, YE; Sawyers, CL. (1998). The PTEN/MMAC1 tumor suppressor phosphatase functions as a negative regulator of the phosphoinositide 3-kinase/Akt pathway. *Proc Natl Acad Sci U S A, 95,* 15587-91.

[5] Cusack, JC; Liu, R; Baldwin, AS. (1999). NF-kappa B and chemoresistance: potentiation of cancer drugs via inhibition of NF-kappa B. *Drug Resist Update, 2,* 271-73.

[6] Petak, I; Houghton, JA, and Kopper, L. (2006). Molecular targeting of cell death signal transduction pathways in cancer. *Current Signal Transduction Therapy, 1,* 113-131.

[7] Hennessy, BT; Smith, DL; Ram, PT; Lu, Y; and Mills GB. (2005). Exploiting the PI3K/AKT pathway for cancer drug discovery. *Nature Reviews Drug Discovery, 4,* 988-1004.

[8] Bjornsti, M-A; Houghton, PJ. (2004). The Tor pathway: a target for cancer therapy. *Nature Cancer Reviews, 4,* 335-48.

[9] Nakanishi, C; Toi, M. (2005). Nuclear factor-kappaB inhibitors as sensitizers to anticancer drugs. *Nat Rev Cancer, 5,* 297-309.

[10] Roques, M; Vidal, H. (1999). A phosphatidylinositol 3-Kinase/p70 ribosomal S6 protein kinase pathway is required for the regulation by insulin of the p85alpha regulatory subunit of phosphatidylinositol 3-kinase gene expression in human muscle cells. *J Biol Chem, 274,* 34005-10.

[11] Burgering, BM; Coffer, PJ. (1995). Protein kinase B (c-Akt) in phosphatidylinositol-3-OH kinase signal transduction. *Nature, 376,* 599-602.

[12] Varticovski, L; Harrison-Findik, D; Keeler, ML; Susa, M. (1994). Role of PI 3-kinase in mitogenesis. *Biochim Biophys Acta, 1226,* 1-11. Review.

[13] Tee, AR; Fingar, DC; Manning, BD; Kwiatkowski, DJ; Cantley, LC; Blenis, J. (2002). Tuberous sclerosis complex-1 and -2 gene products function together to inhibit mammalian target of rapamycin (mTOR)-mediated downstream signaling. *Proc Natl Acad Sci USA, 99,* 13571-6.

[14] Fingar, DC; Salama, S; Tsou, C; Harlow, E; Blenis, J. (2002). Mammalian cell size is controlled by mTOR and its downstream targets S6K1 and 4EBP1/eIF4E. *Genes Dev, 16,* 1472-1487.

[15] Gingras, AC; Raught, B; Gygi, SP; Niedzwiecka, A; Miron, M; Burley, SK; Polakiewicz, RD; Wyslouch-Cieszynska, A; Aebersold, R; Sonenberg, N. (2001). Hierarchical phosphorylation of the translation inhibitor 4E-BP1. *Genes Dev, 15,* 2852-64.

[16] Gingras, AC; Raught, B; Sonenberg, N. (2001). Regulation of translation initiation by FRAP/mTOR. *Genes De 15,* 807-26.

[17] Neshat, MS; Mellinghoff, IK; Tran, C; Stiles, B; Thomas, G; Petersen, R; Frost, P; Gibbons, JJ; Wu, H; Sawyers, CL. (2001). Enhanced sensitivity of PTEN-deficient tumors to inhibition of FRAP/mTOR. *Proc Natl Acad Sci USA, 98,* 10314-9.

[18] Karin, M; Cao, X; Greten, FR; Li, Z-W. (2002). NF-kappaB in cancer: from innocent bystander to major culprit. *Nature Reviews Cancer, 2,* 301-310.

[19] Ghosh, S; May, MJ; and Kopp, EB. (1998). NF-κB and Rel proteins: evolutionarily conserved mediators of immune response. *Annu Rev Immunol, 16,* 225-260.

[20] Baldwin, AS Jr. (1996). The NF-□B and I□B proteins: new discoveries and insights. *Annu Rev Immunol, 14,* 649-81.

[21] Wang, CY; Cusack, JC Jr; Liu, R; Baldwin, AS Jr. (1999). Control of inducible chemoresistance: enhanced anti-tumor therapy through increased apoptosis by inhibition of NF-kappaB. *Nat Med, 5,* 412-7.

[22] Roy, N; Deveraux, QL; Takahashi, R; Salvesen, GS; and Reed, JC. (1997). The c-IAP-1 and c-IAP-2 proteins are direct inhibitors of specific caspases. *The EMBO Journal,* 16, 6914–6925.

[23] Thome, M; Schneider, P; Hofmann, K; Fickenscher, H; Meinl, E; Neipel, F; Mattmann, C; Burns, K; Bodmer, JL; Schroter. M; Scaffidi, C; Krammer, PH; Peter ME; Tschopp, J. (1997). Viral FLICE-inhibitory proteins (FLIPs) prevent apoptosis induced by death receptors. *Nature* 386, 517–520.

[24] Wang, C-Y; Guttridge, DC; Mayo, MW; and Baldwin, AS Jr. (1999). NF-κB Induces Expression of the Bcl-2 Homologue A1/Bfl-1 To Preferentially Suppress Chemotherapy-Induced Apoptosis. *Mol Cell Biol.* 19, 5923–5929.

[25] Rothe, M., Wong, S.C., Henzel, W.J., Goeddel, D.V. (1994). A novel family of putative signal transducers associated with the cytoplasmic domain fo the 75 kDa tumor necrosis factor receptor. *Cell.* 78, 681-692.

[26] Brown, EJ; Albers, MW; Shin, TB; Ichikawa, K; Keith, CT; Lane, WS; Schreiber, SL. (1994). A mammalian protein targeted by G1-arresting rapamycin-receptor complex. *Nature, 369,* 756-758.

[27] Aoki, M; Blazek, E; Vogt, PK. (2001). A role of the kinase mTOR in cellular transformation induced by the oncoproteins P3k and Akt. *Proc Natl Acad Sci USA, 98,* 136-41.

[28] Romano, MF; Avellino, R; Petrella, A; Bisogni, R; Romano, S; Venuta, S. (2004). Rapamycin inhibits doxorubicin-induced NF-□B/Rel nuclear activity and enhances apoptosis in melanoma. *Eur J Cancer, 40,* 2829-36.

[29] Avellino, R; Romano, S; Parasole, R; Bisogni, R; Lamberti, A; Poggi, V; Venuta, S; and Romano, MF. (2005). Rapamycin stimulates apoptosis of childhood acute lymphoblastic leukemia cells. *Blood, 106,* 1400-6.

[30] Guba, M; von Breitenbuch, P; Steinbauer, M; Koehl, G; Flegel, S; Hornung, M; Bruns, CJ; Zuelke, C; Farkas, S; Anthuber, M; Jauch, KW; Geissler, EK. (2002). Rapamycin inhibits primary and metastatic tumor growth by antiangiogenesis: involvement of vascular endothelial growth factor. *Nat Med, 8,* 128-135.

[31] Siekierka JJ, Hung SH, Poe M, Lin CS, Sigal NH. (1989). A cytosolic binding protein for the immunosuppressant FK506 has peptidyl-prolyl isomerase activity but is distinct from cyclophilin. *Nature, 341,* 755-757.

[32] Bouwmeester T, Bauch A, Ruffner H, Angrand PO, Bergamini G, Croughton K, Cruciat C, Eberhard D, Gagneur J, Ghidelli S, Hopf C, Huhse B, Mangano R, Michon AM, Schirle M, Schlegl J, Schwab M, Stein MA, Bauer A, Casari G, Drewes G, Gavin AC, Jackson DB, Joberty G, Neubauer G, Rick J, Kuster B, Superti-Furga G. (2004). A physical and functional map of the human TNF-alpha/NF-kappaB signal transduction pathway. *Nat Cell Biol, 6,* 97-105.

[33] Chi J-T, Chang HI, Wang NN, Chang DS, Dunphy N, and O. Brown P. (2003). Genomewide view of gene silencing by small interfering RNAs. *Proc Natl Acad Sci USA , 100,* 6343–6346.

[34] Wu H, Goel V, Haluska FG. (2003). PTEN signaling pathways in melanoma. *Oncogene, 22,* 3113-22. Review.

[35] Laurent G, Jaffrezou J-P. (2001). Signaling pathways activated by daunorubicin. *Blood, 98,* 913-924.

[36] Li, X; Traganos, F; Melamed, MR, et al. (1995). Single-step procedure for labeling DNA strand breaks with fluorescein- or BODIPY-conjugated deoxynucleotides: detection of apoptosis and bromodeoxyuridine incorporation. *Cytometry,* 20:172-80.

[37] Mita MM, Mita A, Rowinsky EK. (2003). The molecular target of rapamycin (mTOR) as a therapeutic target against cancer. *Cancer Biol Ther, 2,* S169-77. Review.

[38] Lai JH, Tan TH. (1994). CD28 signaling causes a sustained down-regulation of I kappa B alpha which can be prevented by the immunosuppressant rapamycin. *J. Biol. Chem, 269,* 30077-30080.

[39] Conejo, R., Valverde, A.M., Benito, M., Lorenzo, M. (2001). Insulin produces myogenesis in C2C12 myoblasts by induction of NF-kappaB and downregulation of AP-1 activities. *J. Cell. Physiol, 186,* 82-94.

[40] Tunon MJ, Sanchez-Campos S, Gutierrez B, Culebras JM, Gonzalez-Gallego J. (2003). Effects of FK506 and rapamycin on generation of reactive oxygen species, nitric oxide production and nuclear factor kappa B activation in rat hepatocytes. *Biochem Pharmacol, 66,* 439-45.

[41] Wu C, Wangpaichitr M, Feun L, Kuo MT, Robles C, Lampidis T, Savaraj N. (2005). Overcoming cisplatin resistance by mTOR inhibitor in lung cancer. *Mol Cancer. 20,* 4-25.

[42] Johnsson A, Zeelenberg I, Min Y, Hilinski J, Berry C, Howell SB, Los G. (2000). Identification of genes differentially expressed in association with acquired cisplatin resistance. *Br J Cancer, 83,* 1047-54.

[43] Wang T, Li BY, Danielson PD, Shah PC, Rockwell S, Lechleider RJ, Martin J, Manganaro T, Donahoe PK.☐ (1996). The Immunophilin FKBP12 Functions as a Common Inhibitor of the TGF Family Type I Receptors. *Cell. 86,* 435- 444 .

[44] Samso M, Shen X, Allen PD. (2006). Structural characterization of the RyR1-FKBP12 interaction. *J Mol Biol, 356,* 917-27.

[45] Sinars CR, Cheung-Flynn J, Rimerman RA, Scammell JG, Smith DF, Clardy J. (2003). Structure of the large FK506-binding protein FKBP51, an Hsp90-binding protein and a component of steroid receptor complexes. *Proc Natl Acad Sci USA, 100,* 868-73.

[46] Khatua S, Peterson KM, Brown KM, et al. (2003). Overexpression of the EGFR/FKBP12/HIF-2alpha pathway identified in childhood astrocytomas by angiogenesis gene profiling. Cancer Res, 63, 1865-1870.

[47] Giraudier S, Chagraoui H, Komura E, et al. (2002). Overexpression of FKBP51 in idiopathic myelofibrosis regulates the growth factor independence of megakaryocyte progenitors. *Blood, 100,* 2932-2940.

[48] Xu Z, Stokoe D, Kane LP, Weiss A. (2002). The inducible expression of the tumor suppressor gene PTEN promotes apoptosis and decreases cell size by inhibiting the PI3K/Akt pathway in Jurkat T cells. *Cell Growth Differ,* 285-96.

Index

B

C

D

E

H

I

J

O

P

T

U

V

W